T0039130

METHODS IN MOLECULAR BIOLOGY

Series Editor
John M. Walker
School of Life Sciences
University of Hertfordshire
Hatfield, Hertfordshire, AL10 9AB, UK

For further volumes:
http://www.springer.com/series/7651

Dengue

Methods and Protocols

Edited by

Radhakrishnan Padmanabhan

Department of Microbiology and Immunology, Georgetown University, Washington, DC, USA

Subhash G. Vasudevan

Emerging Infectious Diseases Program, Duke-NUS Graduate Medical School, Singapore, Singapore

Humana Press

Editors
Radhakrishnan Padmanabhan
Department of Microbiology and Immunology
Georgetown University
Washington, DC, USA

Subhash G. Vasudevan
Emerging Infectious Diseases Program
Duke-NUS Graduate Medical School
Singapore, Singapore

ISSN 1064-3745   ISSN 1940-6029 (electronic)
ISBN 978-1-4939-0347-4   ISBN 978-1-4939-0348-1 (eBook)
DOI 10.1007/978-1-4939-0348-1
Springer New York Heidelberg Dordrecht London

Library of Congress Control Number: 2014933052

Printed on acid-free paper

Humana Press is a brand of Springer
Springer is part of Springer Science+Business Media (www.springer.com)

Preface

Infection by flaviviruses such as dengue virus serotypes (DENV 1–4), Japanese encephalitis virus (JEV), tick-borne encephalitis virus (TBE), yellow fever virus (YFV), and West Nile virus (WNV) impacts millions of lives and causes tens of thousands of mortalities each year. Recent studies on global dengue burden indicated that there are at least 100 million human symptomatic infections annually. This original estimate has recently been revised in 2013 to about three times higher than the dengue burden estimate of the World Health Organization. The urban-breeding *Aedes aegypti* mosquito has spread the DENV to more than 100 countries around the world and ~50 % of the world's population is now estimated to be at risk. Dengue is a global public health emergency especially since there is no preventative vaccine or antiviral treatment for dengue disease. Usually, infection with any one of the four DENV serotypes leads to mild self-limiting dengue fever (DF) with lifelong immunity to that specific serotype. Epidemiological evidence suggests that 90 % of the severe and potentially fatal dengue diseases, dengue hemorrhagic fever (DHF), or dengue shock syndrome (DSS) occur during secondary heterotypic infections where the protective antibodies from a previous infection become pathogenic through the Antibody Dependent Enhancement (ADE) phenomenon. The co-circulation of multiple serotypes in dengue epidemic countries increases the risk of severe dengue diseases due to ADE. Dengue has also reappeared in the United States of America: the combination of a low immunity in the population, increased mosquito vector activity, and the continuous introduction of virus from the endemic countries forms the right ingredient for explosive epidemics.

This edition of methods and protocols for dengue research is aimed at providing the increasing number of dengue researchers a one-stop protocol book contributed by some of the leading laboratories working on dengue. Chapters on dengue virus isolation from clinical samples, quantification of human antibodies against the virus, and assays to quantify the virus particles are included. The widely used mouse model to study dengue pathogenesis, vaccine, and antiviral efficacies is also described. New technologies to study the conformation of *cis*-acting elements in dengue viral RNA genome that contribute to its function in translation and replication by novel computational and experimental methods are described in this book for the first time. The dynamic dengue RNA molecule from its initial biogenesis to its final most stable conformation through multiple intermediate folding pathways is analyzed by the predictive Massively Parallel Genetic Algorithm (MPGAfold) with frequencies of occurrence of each stage. *S*elective 2′-*H*ydroxyl *A*cylation analyzed by *P*rimer *E*xtension (SHAPE) analyzes the conformation of RNA experimentally. High-throughput SHAPE combines a novel chemical probing technology with reverse transcription, capillary electrophoresis, and secondary structure prediction software to determine RNA structure at a single nucleotide resolution. Next Generation Sequencing methodologies described here utilize high-throughput and massively parallel sequencing to track the viral genomes constantly changing under selective pressure imposed by environment. Cutting-edge cryo-electron microscopy technology reveals how the viruses also change their surface morphologies when they are subjected to environmental conditions under which the viruses grow and replicate their genomes. Moreover, the three-dimensional structures of the viral proteins are important for their function. One of the modern methods to achieve this objective, Small Angle X-ray Scattering (SAXS), is described here. Reverse genetic systems for different dengue virus serotypes to study viral replication using different reporter systems and virus-like particles to study viral entry, replication, and assembly are also described. The viral RNA codes for a number of enzymes that are important for its replication. Methods are described to measure quantitatively the various enzyme activities that are useful to screen for antivirals. Genome-wide screening methods and discovery of human and insect host cell proteins that are involved in virus life cycle are also included. The book contains 24 chapters, and we sincerely hope that the protocols contributed by the authors will form a valuable resource for dengue researchers.

We would like to thank Professor John Walker, the Chief Editor of the series, for his guidance. We also thank the help and guidance of the editors at Springer, especially David Casey, Tamara Cabrero, Patrick Marton, Anne Meagher, and Paul Wehn, as well as Priya Ranganathan, Project Manager at SPi Global, India, for their hard work in bringing these chapters to the final stage.

Washington, DC, USA   *Radhakrishnan Padmanabhan*
Singapore, Singapore   *Subhash G. Vasudevan*

Contents

PART I   MODERN QUANTITATIVE VIROLOGICAL METHODS

PART II   REVERSE GENETIC SYSTEMS TO STUDY VIRUS REPLICATION
          AND EVOLUTION

Contributors

SOFIA L. ALCARAZ-ESTRADA • *Division de Medicina Genomica, Centro Medico Nacional "20 de Noviembre"-ISSSTE, Mexico D.F., Mexico*

RUKLANTHI DE ALWIS • *Department of Microbiology and Immunology, University of North Carolina School of Medicine, Chapel Hill, NC, USA; Southeast Regional Center of Excellence for Biodefense and Emerging Infectious Diseases Research, University of North Carolina School of Medicine, Chapel Hill, NC, USA*

ROSA DEL ANGEL • *Departamento de Infectómica y Patogénesis Molecular, CINVESTAV-IPN, México, D.F., Mexico*

PAULINE POH KIM AW • *Genome Institute of Singapore, Singapore, Singapore*

NICHOLAS J. BARROWS • *Program in Cell and Molecular Biology, Department of Molecular Genetics and Microbiology, Center for RNA Biology, Duke University Medical Center, Durham, NC, USA*

SIWAPORN BOONYASUPPAYAKORN • *Faculty of Medicine, Department of Microbiology and Immunology, Chulalongkorn University, Bangkok, Thailand*

SHELTON S. BRADRICK • *Department of Molecular Genetics and Microbiology, Center for RNA Biology, Duke University Medical Center, Durham, NC, USA*

KUAN RONG CHAN • *Emerging Infectious Diseases Program, Duke-NUS Graduate Medical School, Singapore, Singapore*

KYUNG H. CHOI • *Department of Biochemistry and Molecular Biology, University of Texas Medical Branch, Galveston, TX, USA*

MILLY M. CHOY • *Emerging Infectious Diseases Program, Duke-NUS Graduate Medical School, Singapore, Singapore*

JOHN E. CONNOLLY • *Singapore Immunology Network (SIgN), A*Star, Singapore, Singapore*

ANDREW D. DAVIDSON • *Faculty of Medical and Veterinary Sciences, School of Cellular and Molecular Medicine, University of Bristol, Bristol, UK*

KIMBERLY A. DOWD • *Viral Pathogenesis Section, Laboratory of Viral Diseases, National Institute of Allergy and Infectious Diseases, National Institutes of Health, Bethesda, MD, USA*

JOHANNA E. FRASER • *Nuclear Signalling Laboratory, Department of Biochemistry and Molecular Biology, Monash University, Clayton, Australia*

MARIANO A. GARCIA-BLANCO • *Department of Molecular Genetics and Microbiology, Center for RNA Biology, Duke University Medical Center, Durham, NC, USA; Department of Medicine, Duke University Medical Center, Durham, NC, USA*

STUART F.J. LE GRICE • *HIV Drug Resistance Program, RT Biochemistry Section, Frederick National Laboratory for Cancer Research, Frederick, MD, USA*

DUANE J. GUBLER • *Emerging Infectious Diseases Program, Duke-NUS Graduate Medical School, Singapore, Singapore*

J. GUNARATNE • *Mass Spectrometry and Systems Biology Laboratory, Institute of Molecular and Cell Biology, Singapore, Singapore*

MARTIN LLOYD HIBBERD • *Genome Institute of Singapore, Singapore, Singapore*

LONG TRUONG HOANG • *Genome Institute of Singapore, Singapore, Singapore*

YIN HOE YAU • *Division of Chemical Biology and Biotechnology, School of Biological Sciences, Nanyang Technological University, Singapore, Singapore*

SHARON F. JAMISON • *Department of Molecular Genetics and Microbiology, Center for RNA Biology, Duke University Medical Center, Durham, NC, USA*

DAVID A. JANS • *Nuclear Signalling Laboratory, Department of Biochemistry and Molecular Biology, Monash University, Clayton, Australia*

WOJCIECH K. KASPRZAK • *Basic Science Program, Leidos Biomedical Research, Inc., Frederick, MD, USA; Basic Research Laboratory, National Cancer Institute, Frederick, MD, USA*

KITTI CHAN WING KI • *Emerging Infectious Diseases Program, Duke-NUS Graduate Medical School, Singapore, Singapore*

SO YOUNG KIM • *Department of Molecular Genetics and Microbiology, Center for RNA Biology, Duke University Medical Center, Durham, NC, USA*

MANOJ KRISHNAN • *Emerging Infectious Diseases Program, Duke-NUS Graduate Medical School, Singapore, Singapore*

HUIGUO LAI • *Department of Microbiology and Immunology, Georgetown University, Washington, DC, USA*

CHIN CHIN LEE • *Emerging Infectious Diseases Program, Duke-NUS Graduate Medical School, Singapore, Singapore*

SHEE MEI LOK • *Emerging Infectious Diseases Program, Duke-NUS Graduate Medical School, Singapore, Singapore*

MARK MANZANO • *Department of Microbiology–Immunology, Feinberg School of Medicine, Northwestern University, Chicago, IL, USA*

MARC MORAIS • *Department of Biochemistry and Molecular Biology, University of Texas Medical Branch, Galveston, TX, USA*

NICOLE J. MORELAND • *School of Biological Sciences, University of Auckland, Auckland, New Zealand*

SWATI MUKHERJEE • *Viral Pathogenesis Section, Laboratory of Viral Diseases, National Institute of Allergy and Infectious Diseases, National Institutes of Health, Bethesda, MD, USA*

NIRANJAN NAGARAJAN • *Genome Institute of Singapore, Singapore, Singapore*

KAITING NG • *Singapore Immunology Network (SIgN), A*Star, Singapore, Singapore*

ENG EONG OOI • *Emerging Infectious Diseases Program, Duke-NUS Graduate Medical School, Singapore, Singapore*

JANAK PADIA • *Prime Time Life Sciences, LLC, Germantown, MD, USA*

RADHAKRISHNAN PADMANABHAN • *Department of Microbiology and Immunology, Georgetown University, Washington, DC, USA*

JAMES PEARSON • *Department of Molecular Genetics and Microbiology, Center for RNA Biology, Duke University Medical Center, Durham, NC, USA*

THEODORE C. PIERSON • *Viral Pathogenesis Section, Laboratory of Viral Diseases, National Institute of Allergy and Infectious Diseases, National Institutes of Health, Bethesda, MD, USA*

EMILY PLUMMER • *Division of Vaccine Discovery, La Jolla Institute for Allergy and Immunology, La Jolla, CA, USA*

STEPHEN M. RAWLINSON • *Nuclear Signalling Laboratory, Department of Biochemistry and Molecular Biology, Monash University, Clayton, Australia*

PAOLA FLOREZ DE SESSIONS • *Genome Institute of Singapore, Singapore, Singapore*

OCTOBER M. SESSIONS • *Emerging Infectious Diseases Program, Duke-NUS Graduate Medical School, Singapore, Singapore*

BRUCE A. SHAPIRO • *Basic Research Laboratory, National Cancer Institute, Frederick, MD, USA*

SUSANA GEIFMAN SHOCHAT • *Division of Chemical Biology and Biotechnology, School of Biological Sciences, Nanyang Technological University, Singapore, Singapore*

SUJAN SHRESTA • *Division of Vaccine Discovery, La Jolla Institute for Allergy and Immunology, La Jolla, CA, USA*

ARAVINDA M. DE SILVA • *Department of Microbiology and Immunology, University of North Carolina School of Medicine, Chapel Hill, NC, USA; The Southeast Regional Center of Excellence for Biodefense and Emerging Infectious Diseases Research, University of North Carolina School of Medicine, Chapel Hill, NC, USA*

CAROLINE LE SOMMER • *Department of Molecular Genetics and Microbiology, Center for RNA Biology, Duke University Medical Center, Durham, NC, USA*

JOANNA SZTUBA-SOLINSKA • *HIV Drug Resistance Program, RT Biochemistry Section, Frederick National Laboratory for Cancer Research, Frederick, MD, USA*

JOANNE L. TAN • *Emerging Infectious Diseases Program, Duke-NUS Graduate Medical School, Singapore, Singapore*

HWEE CHENG TAN • *Emerging Infectious Diseases Program, Duke-NUS Graduate Medical School, Singapore, Singapore*

KAH HIN TAN • *Emerging Infectious Diseases Program, Duke-NUS Graduate Medical School, Singapore, Singapore*

MOON Y. F. TAY • *Emerging Infectious Diseases Program, Duke-NUS Graduate Medical School, Singapore, Singapore*

TADAHISA TERAMOTO • *Department of Microbiology and Immunology, Georgetown University, Washington, DC, USA*

SUBHASH G. VASUDEVAN • *Emerging Infectious Diseases Program, Duke-NUS Graduate Medical School, Singapore, Singapore*

KYLIE M. WAGSTAFF • *Nuclear Signalling Laboratory, Department of Biochemistry and Molecular Biology, Monash University, Clayton, Australia*

CHUNXIAO WANG • *Nuclear Signalling Laboratory, Department of Biochemistry and Molecular Biology, Monash University, Clayton, Australia*

ALEX M. WARD • *Emerging Infectious Diseases Program, Duke-NUS Graduate Medical School, Singapore, Singapore*

SATORU WATANABE • *Emerging Infectious Diseases Program, Duke-NUS Graduate Medical School, Singapore, Singapore*

ANDREAS WILM • *Genome Institute of Singapore, Singapore, Singapore*

SUMMER ZHANG • *Emerging Infectious Diseases Program, Duke-NUS Graduate Medical School, Singapore, Singapore*

Part I

Modern Quantitative Virological Methods

Chapter 1

Dengue Virus Growth, Purification, and Fluorescent Labeling

Summer Zhang, Kuan Rong Chan, Hwee Cheng Tan, and Eng Eong Ooi

Abstract

The early events of the dengue virus life cycle involve virus binding, internalization, trafficking, and fusion. Fluorescently labeled viruses can be used to visualize these early processes. As dengue virus has 180 identical copies of the envelope protein attached to the membrane surface and is surrounded by a lipid membrane, amine-reactive (Alexa Fluor) or lipophilic (DiD) dyes can be used for virus labeling. These dyes are highly photostable and are ideal for studies involving cellular uptake and endosomal transport. To improve virus labeling efficiency and minimize the nonspecific labeling of nonviral proteins, virus concentration and purification precede fluorescent labeling of dengue viruses. Besides using these viruses for single-particle tracking, DiD-labeled viruses can also be used to distinguish serotype-specific from cross-neutralizing antibodies. Here the details of virus concentration, purification, virus labeling, applications, and hints of troubleshooting are described.

Key words Dengue virus, Concentration, Purification, Fluorescence labeling, Alexa Fluor, DiD

1 Introduction

Virus concentration and purification techniques have been routinely used to [1] improve virus titer to achieve a higher multiplicity of infection for cells and [2] separate virus from the host cell milieu in which it has been grown for subsequent studies. Purification is especially vital in studies looking at host response to infection where the remnants from host cell milieu may act as antigens that trigger immune responses independent of the virus. It is also essential in electron microscopy and fluorescent labeling experiments.

Concentration of virus particles is required as a starting point for virus purification. It can be achieved by several methods such as ultracentrifugation, molecular sieve filtration, or with precipitation via the addition of polyethylene glycol-6000 (PEG). All these procedures are carried out in the cold. However, ultrafiltration requires a constant stirring to prevent clogging of the filter.

Radhakrishnan Padmanabhan and Subhash G. Vasudevan (eds.), *Dengue: Methods and Protocols*, Methods in Molecular Biology, vol. 1138, DOI 10.1007/978-1-4939-0348-1_1, © Springer Science+Business Media, LLC 2014

It also results in substantial loss of viruses as they get trapped in the filter itself, while PEG concentration, on the other hand, requires more steps and is hence technically more demanding. Ultracentrifugation is the simplest approach although the resuspension of the virus pellet may require a little technical finesse to minimize loss of viable virions.

Following concentration, the virus can then be purified. A density medium is typically used during purification to separate virus particles from other contaminants by rate-zonal or isopycnic centrifugation. Sucrose is suitable for most rate-zonal centrifugation procedures and is commonly used for purification of many viruses, including enveloped viruses (influenza virus and Japanese encephalitis virus) and non-enveloped virus (adenovirus) [3–6].

In this chapter, we will describe how dengue virus can be cultured using mosquito (C6/36) or mammalian (Vero) cell lines, concentrated by centrifugation and purified through a sucrose cushion. We will further describe how the purified dengue virus can be labeled with an amine-reactive (Alexa Fluor) or lipophilic (1,1′-dioctadecyl-3,3,3′,3′-tetramethylindodicarbocyanine, 4-chlorobenzenesulfonate Salt, DiD) dye to provide valuable tools to answer important questions on virus cell entry, receptor/co-receptor usage and trafficking by fluorescent imaging. Fluorescently labeled dengue virus can also be used for quantitative analyses by flow cytometry or sorting of live infected cells by fluorescence-activated cell sorting.

2 Materials

2.1 Virus Culture

1. Either C6/36 (derived from *Aedes albopictus*) or Vero (derived from African green monkey kidney) cell lines can be used to produce dengue virus to high titer.

2. Maintenance medium (cell culture medium with reduced FBS).

3. Dengue virus.

2.2 Virus Purification

1. Centrifuges (low and high speed, ultracentrifuge) and respective centrifugation tubes.

2. 0.2 μm filter unit.

3. Sterile HNE buffer, pH 7.4 (5 mM Hepes, 150 mM NaCl, 0.1 mM EDTA in deionized water). Filter sterilize with 0.2 μm filter unit.

4. Sterile 30 % sucrose solution prepared in HNE buffer. Filter sterilize with 0.2 μm syringe filter.

2.3 Fluorescent Labeling

2.3.1 Amine-Reactive Dye Labeling (Alexa Fluor)

1. Amine-reactive dye (Alexa Fluor succinimidyl ester, Molecular Probes, Invitrogen).

2. Reaction buffer, 0.2 M sodium bicarbonate buffer, pH 8.3. Dissolve 0.84 g of sodium bicarbonate powder in 50 mL of

deionized water. Adjust pH, if necessary, to 8.3 and sterilize solution with 0.2 μm syringe filter.

3. Stop reagent, 1.5 M hydroxylamine, pH 8.5. Dissolve 1.05 g of hydroxylamine hydrochloride in 5 mL of deionized water. Adjust pH to 8.5 with 5 M sodium hydroxide and top up to 10 mL with deionized water (*see* **Note 1**). Sterilize solution with 0.2 μm syringe filter.

4. HNE buffer, pH 7.4 (5 mM Hepes, 150 mM NaCl, 0.1 mM EDTA). Filter sterilize and degas with 0.2 μm filter unit.

5. Purified dengue virus prepared in HNE buffer.

6. Sephadex G-25 or G-50 size exclusion column (e.g., prepacked PD-10 disposable columns from GE Healthcare).

7. Vero cell line, anti-dengue envelope protein antibody and appropriate secondary antibody for immunofluorescence assay (IFA).

8. 1× PBS buffer, pH 7.

9. 3 % paraformaldehyde solution prepared in 1× PBS

10. Permeabilization solution (0.1 % saponin, 5 % bovine serum albumin in 1× PBS).

11. Mounting solution (Mowiol 4-88 containing 2.5 % DABCO).

2.3.2 Lipophilic Dye Labeling (DiD)

1. DiD dye (1,1′-dioctadecyl-3,3,3′,3′-tetramethylindodicarbocyanine, 4-chlorobenzenesulfonate salt, Molecular Probes, Invitrogen).

2. HNE buffer, pH 7.4 (5 mM Hepes, 150 mM NaCl, 0.1 mM EDTA). Filter sterilize and degas with 0.2 μm filter unit.

3. Purified dengue virus prepared in HNE buffer.

4. Sephadex G-25 or G-50 size exclusion column (e.g., prepacked PD-10 disposable columns from GE Healthcare).

2.3.3 Assessing Antibody Titer for Complete Neutralization in Monocytes

1. THP-1 cell line (ATCC).

2. Maintenance medium (cell culture medium with reduced FBS).

3. Antibody or convalescent serum sample.

4. Dengue virus.

2.3.4 Distinguishing Serotype-Specific from Cross-Reactive Antibodies in Monocytes

1. DiD-labeled dengue virus.

2. Maintenance medium.

3. Antibody or sera.

4. 1× PBS.

5. 12 % paraformaldehyde solution prepared in 1× PBS.

6. Cytospin centrifuge, funnels, and positively charged slides.

7. Coverslips.

8. Permeabilization solution (0.1 % saponin, 5 % bovine serum albumin in 1× PBS).

9. Primary antibodies against LAMP-1 and secondary antibodies.

10. Mounting solution (Mowiol 4-88 containing 2.5 % DABCO).

3   Methods

3.1   Virus Culture

1. Grow Vero or C6/36 cells to a monolayer of ≥80 % confluency in growth medium (sufficient to produce 500 mL to 1 L of virus culture).

2. Infect the cells with dengue virus at multiplicity of infection (moi) 0.1 in maintenance medium (appropriate volume to just cover the monolayer of cells to allow good adsorption). Rock the flasks every 15 min for 1 h and add fresh maintenance medium for culture. Harvest culture supernatant when ≥75 % of the cells shows cytopathic effect or syncytial formation. This usually takes about 5–7 days depending on the serotype/strain of virus used.

3. Clarify the pooled harvest by low-speed centrifugation at $1,200 \times g$ for 20 min at 4 °C to remove cellular debris.

4. Filter the clarified culture supernatant through a 0.2 μm filter to remove smaller aggregates.

3.2   Virus Purification

1. Concentrate the resulting virus particles in the supernatant by centrifugation at $30,000 \times g$ for 2 h at 4 °C (Beckman high-speed JA25.50 rotor). Repeat this step, if necessary, to concentrate the entire volume of virus culture.

2. Discard the supernatant and gently resuspend the pellet in HNE buffer, pH 7.4. Carefully layer the virus suspension over 2.5 mL (10–20 % of total volume of the tube) of 20–30 % sucrose solution prepared in the HNE buffer. Subject the virus to ultracentrifugation in Beckman SW41Ti rotor at $80,000 \times g$ overnight at 4 °C.

3. Discard the supernatant and gently resuspend the virus pellet in HNE buffer (*see* **Note 2**). Remove any insoluble aggregates by brief centrifugation and store the clarified virus in aliquots at –80 °C. Determine the virus titer in plaque forming units per milliliter (pfu/mL) by plaque assay.

3.3   Fluorescent Labeling

3.3.1   Amine-Reactive Dye Labeling

A simple method of Alexa Fluor labeling dengue virus has been previously described [9]. User can also refer to a video protocol for a better understanding on how the procedure is done [8].

Alexa Fluor Labeling

1. The Alexa Fluor succinimidyl esters are supplied as lyophilized powder. Reconstitute the dye to 1 mM solution in 0.2 M sodium bicarbonate reaction buffer (*see* **Note 3**). Minimize exposure to light from this step onwards.

2. Dilute 3×10^8 pfu of purified dengue virus to 1 mL in sodium bicarbonate buffer. This can be scaled up or down according to needs.

3. Add 100 μL of 1 mM Alexa Fluor dye to the diluted dengue virus while stirring gently. Incubate the mixture for 1 h at room temperature for the labeling to take place. Invert the tube 4–6 times to mix every 15 min followed by a brief centrifugation.

4. Add 100 μL 1.5 M hydroxylamine stop reagent to the labeled virus, while stirring gently, to neutralize any unbound Alexa Fluor succinimidyl esters. Incubate the mixture for an additional 1 h at room temperature. Invert the tube 4–6 times to mix every 15 min followed by a brief centrifugation.

5. Apply the labeled virus mixture to a size exclusion column to separate the labeled virus from the unbound free dye (*see* **Note 4**). Aliquot and store at –80 °C. Determine the titer of labeled virus by standard plaque assay.

Determine Labeling Efficiency by Confocal Microscopy

1. Seed 5×10^4 Vero cells per well on glass coverslips in 24-well plate the day prior to experiment. Infect cell monolayer with Alexa Fluor labeled virus at an approximate moi of 1 for 10 min at 37 °C.

2. Remove the inoculums and wash cells twice with 1× PBS.

3. Fix cells with 3 % paraformaldehyde for 30 min at room temperature.

4. Wash cells thrice with 1× PBS and permeabilize for 30 min at room temperature with permeabilization solution.

5. Pick up the coverslips from wells using a sharp-tip forceps, drain excess permeabilization solution on paper towel, and invert onto 25 μL of anti-dengue envelope protein antibody. Incubate for 1 h at room temperature in a humid chamber, protected from light.

6. Wash coverslips thrice in 1× PBS, drain excess buffer, and invert onto 25 μL of secondary antibody. Incubate for 45 min at room temperature in a humid chamber in the dark.

7. Wash coverslips thrice in 1× PBS, rinse once in deionized water, and mount onto glass slide with approximately 8 μL of Mowiol mounting solution. Allow mounting solution to set overnight at 4 °C and view with confocal microscope.

8. The degree of labeling can then be estimated from the colocalization of the labeled virus with anti-dengue envelope protein staining by the overlap coefficient or Pearson's correlation (Fig. 1a, b). Proper labeling should have a labeling efficiency

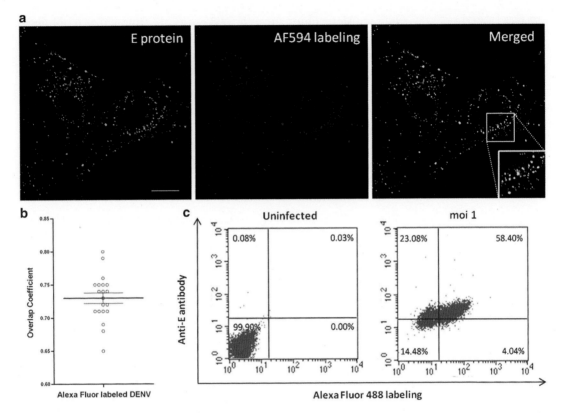

Fig. 1 Estimating the degree of labeling. (**a**) Vero cells grown on coverslips the day prior were infected with AF594-labeled dengue virus at moi of 1 for 10 min at 37 °C. The cells were subsequently fixed and labeled with anti-E antibody and examined for co-localization of E protein (*green*) and AF594 labeling (*red*). Fluorescent signals were visualized under ×63 magnification using Zeiss LSM710 confocal microscope. Scale bar is 10 μm. *Yellow* indicates areas of co-localization, as shown in the *inset* (Image adapted from Zhang et al.) [8]. (**b**) Efficiency of labeling can be determined by analyzing the overlap coefficient between E protein staining and Alexa Fluor label in individual cells. (**c**) Raji B cells transfected with DC-SIGN were infected with moi 1 of AF488-labeled dengue virus at 37 °C for 10 min, fixed and stained for E protein, and analyzed using FACS Calibur. The degree of labeling can be determined by the ratio of E protein^{+ve}AF488^{+ve}/E protein^{+ve} (Figure adapted from Zhang et al.) [9]

ranging between 65 and 80 %. Alternatively, the degree of labeling can also be determined by flow cytometry using other dengue virus permissive cells (e.g., Raji B cells transfected with DC-SIGN) (Fig. 1c).

3.3.2 Lipophilic Dye Labeling (DiD)

DiD labeling of viruses have been described in several other reports [1, 4, 5]. For a more detailed description on DiD labeling of dengue virus, we recommend reading the publication by Ayala-Nunez et al. [1]. Here, we provide the simplified protocol of DiD labeling dengue virus which have worked in our laboratory.

1. Dissolve DiD dye in DMSO to 1 mg/mL concentration. This can be aliquoted and stored at –20 °C.

2. Warm HNE and DiD dye at 37 °C before start of the experiment.

3. Dilute 2×10^8 pfu of purified virus to 1.35 mL in HNE buffer and add 150 μL of 1 mg/mL of DiD dye (*see* **Note 5**).

4. Incubate at 37 °C for 30 min. Gently mix the dye-virus mixture once after 15 min.

5. Apply the labeled virus mixture to a size exclusion column to separate the labeled virus from the unbound free dye (*see* **Note 4**). The labeled virus can be stored at 4 °C for up to 2 days, but for more consistent results, we recommend that the labeled virus be used on the same day of labeling (*see* **Note 6**).

3.4 Potential Applications

DiD dye is quenched when they are at a close proximity to each other, but its fluorescence increases exponentially when they are diluted out. The unique properties of DiD dye has been cleverly employed to visualize dengue virus entry and fusion in endosomal compartments of BS-C-1 cells by single-virus tracking [1]. Here we describe a platform where we further adapted the DiD-labeled dengue virus to investigate how antibodies neutralize the virus in host immune cells. Previously, we have shown that cross-reactive but not serotype-specific antibodies require high antibody concentration to co-ligate FcγRIIB to inhibit uptake and infection in monocytes [2]. These differences can hence be potentially used to distinguish serotype-specific from cross-neutralizing antibodies [7]. The method involves (1) determining the highest antibody or sera dilution that confers complete neutralization in monocytes and (2) detecting the presence or absence of co-localization of neutralized antibody-virus immune complexes with late endosome or lysosome compartments under confocal immunofluorescence.

3.4.1 Assessing Antibody Titer for Complete Neutralization in Monocytes

1. Incubate 2×10^5 pfu/100 μL of dengue virus with 10 μL twofold serially diluted antibodies or dengue immune sera for 1 h at 37 °C in a 96-well plate (*see* **Note 7**).

2. Dilute THP-1 cells in maintenance media to 2×10^5 cells/mL. Add 100 μL of cells to the antibody-opsonized dengue virus prepared in **step 1**.

3. Incubate for 72 h postinfection at 37 °C.

4. Aspirate 150 μL of supernatant from individual wells to a separate 96-well plate and store the plate at −80 °C.

5. Perform standard plaque assay to determine the highest dilution of antibodies or sera that results in complete neutralization (with no visible plaques).

3.4.2 Distinguishing Serotype-Specific from Cross-Reactive Antibodies in Monocytes

1. Incubate 2×10^5 pfu/100 μL of DiD-labeled dengue virus (described in Subheading 3.3.2) with 10 μL of the highest dilution of antibodies/sera that results in complete neutralization for 1 h at 37 °C in a 96-well plate. For dengue virus-only infection, 10 μL of maintenance media is added instead of antibodies/sera.

2. Dilute THP-1 cells in maintenance media to 2×10^5 cells/mL.

3. Prechill dengue virus or antibody-opsonized dengue virus and THP-1 cells after **steps 1** and **2,** respectively, at 4 °C for at least 5 min.

4. Add 100 μL of THP-1 cells to the dengue virus or antibody-opsonized dengue. Synchronize the mixture at 4 °C for 20 min.

5. Incubate for 30 min at 37 °C.

6. Add 50 μL/well of 12 % paraformaldehyde and incubate for 30 min at 4 °C.

7. Transfer mixture to cytospin funnel. Cytospin cells onto positively charged slides by centrifugation at 800 rpm for 3 min.

8. Fix cells with 3 % paraformaldehyde for 30 min at room temperature.

9. Wash cells thrice with 1× PBS and permeabilize for 30 min at room temperature with permeabilization solution.

10. Drain excess permeabilization solution with paper towels and incubate the cells with 25 μL of antihuman LAMP-1 antibodies for 1 h at room temperature in a humid chamber.

11. Wash slides thrice in 1× PBS, drain excess buffer, and add 25 μL of secondary antibody (e.g., anti-mouse Alexa 488 conjugated and antihuman Cy3 conjugated). Incubate for 45 min at room temperature in a humid chamber in the dark.

12. Wash slides thrice in 1× PBS, rinse once in deionized water, and mount a round coverslip with approximately 8 μL of Mowiol mounting solution. Allow mounting solution to set overnight at 4 °C and view with a confocal microscope.

13. At levels that permit virus neutralization, serotype-specific antibodies and dengue virus will co-localize with the LAMP-1 compartments within 30 min after infection. In contrast, when only cross-reactive antibodies are present, neutralization occurs only with the formation of viral aggregates that co-ligate FcγRIIB to inhibit phagocytosis in monocytes. Hence, a strong antibody signal will be detected on the cell surface with significantly reduced DiD signal that is only evident upon uptake of dengue immune complexes (Fig. 2). Virus only serves as a baseline control to distinguish these contrasting phenotypes mediated by serotype-specific or cross-reactive antibodies.

4 Notes

1. Adjusting the pH of the hydroxylamine solution can be tricky. Freshly prepared hydroxylamine solution has a low pH of around 3. Sodium hydroxide solution can be added in large

droplets to quickly raise the pH to ~7. Thereafter, addition of small amounts of sodium hydroxide solution would lead to a big increment in the pH. Therefore, small droplets should be added at a time and allowed for the pH to stabilize between drops, to fine-tune the pH to 8.5.

2. The virus pellet would be very compact and difficult to resuspend after ultracentrifugation. Gently flow HNE buffer over

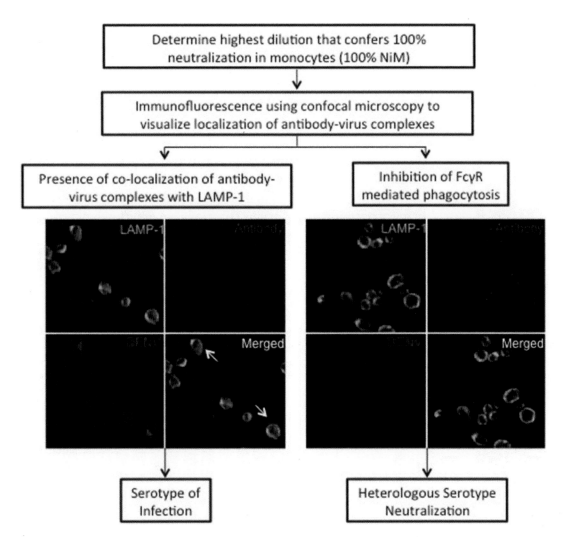

Fig. 2 Overview of procedure used to distinguish serotype-specific from cross-neutralizing antibodies. THP-1 cells were used to determine the highest dilution of antibodies or sera that confers complete neutralization. This neutralizing titer was then used to opsonize DiD-labeled dengue viruses. Serotype-specific antibodies are able to neutralize dengue virus in the presence of uptake, with observed strong co-localization of antibody-virus complexes with LAMP-1. In contrast, cross-reactive antibodies (heterologous serotype neutralization) neutralize at levels that form viral aggregates and inhibit phagocytosis. LAMP-1 is indicated in *green*, human antibodies in *red*, and DiD-labeled viruses in *blue* (Figure adapted from Wu et al.) [7] (Color figure online)

the pellet using a Pasteur pipette to dissolve the pellet slowly. Care has to be taken not to dislodge the pellet in the process as the virus would not be dissolved sufficiently and aggregated virus has to be removed, thus resulting in lower yield. In the event that partial or whole pellet gets dislodged in the resuspension process, try to maximize recovery of virus by homogenizing the pellet fragment(s).

3. Although Alexa Fluor succinimidyl esters are used here, this protocol should also work for any amine-reactive dyes with similar labeling chemistry. Molecular Probes recommends that Alexa Fluor succinimidyl esters be reconstituted in high-quality, anhydrous dimethyl sulfoxide (DMSO) or dimethylformamide (DMF). However, in our hands, DMSO drastically reduces the titer of dengue virus postlabeling; hence, 0.2 M sodium bicarbonate buffer is used in place of DMSO [9]. Alexa Fluor succinimidyl esters dissolved in aqueous solutions should be used immediately as they hydrolyze to nonreactive free acids. This labeling method should also work for other enveloped viruses, with optimization.

4. Different columns would have different bed volumes and flow rate. Thus, a preliminary experiment should be carried out to determine the optimal fractions where the bulk of virus loaded would elute using the column of your choice (Fig. 3). If using prepacked PD-10 column, equilibrate the column

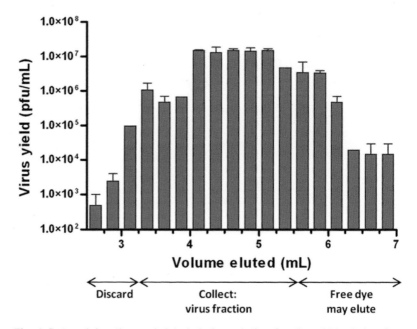

Fig. 3 Determining the peak labeled virus elution fraction. 250 μL fractions are collected and titered by standard plaque assay

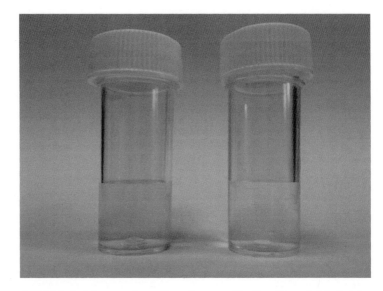

Fig. 4 DiD-labeled dengue virus after column purification. Properly labeled dengue virus would have a *slight blue tinge* (*left*) compared to no DiD control (*right*)

with 25 mL of HNE buffer (this can be done during incubation time at the above steps). Apply 2 mL of the labeled virus to the column and start collecting the flow through once the labeled virus enters the matrix. Fill the column with HNE buffer after all of the labeled virus has entered the matrix and as required. Discard the first 3.25 mL and collect the next 2 mL of labeled virus. Stop collection if free dye starts eluting.

5. DiD dye is not water soluble and has to be reconstituted in DMSO. To prevent viruses from being inactivated by DMSO present in the DiD dye, we recommend diluting the viruses with HNE buffer first before adding the DiD dye.

6. The labeled viruses obtained will have a slight tinge of blue (Fig. 4). If this assay is done the first time, we recommend titering the labeled virus fractions by standard plaque assay as well as measuring the efficiency of labeling as described in Ayala-Nunez et al. [1]. Briefly, fluorescence spectrometry (650–750 nm) can be used to assess the incorporation of DiD dye into the virus particles in the presence or absence of the detergent octaethyleneglycol monododecyl ether (C12E8). Addition of the detergent would increase the florescence significantly.

7. An aliquot of the dengue virus should be kept for back-titration to ensure that the viruses added are indeed at multiplicity of infection 10.

Acknowledgements

This work was supported by Singapore National Research Foundation under its Clinician-Scientist Award administered by the National Medical Research Council to EEO.

References

1. Ayala-Nunez NV, Wilschut J, Smit JM (2011) Monitoring virus entry into living cells using DiD-labeled dengue virus particles. Methods 55:137–143

2. Chan KR, Zhang SL, Tan HC, Chan YK, Chow A, Lim AP, Vasudevan SG, Hanson BJ, Ooi EE (2011) Ligation of Fc gamma receptor IIB inhibits antibody-dependent enhancement of dengue virus infection. Proc Natl Acad Sci U S A 108:12479–12484

3. Killington RA, Stokes A, Hierholzer JC (1996) Virus purification. In: Mahy BWJ, Kangro HO (eds) Virology Methods Manual. Academic Press Limited, London

4. Lakadamyali M, Rust MJ, Babcock HP, Zhuang X (2003) Visualizing infection of individual influenza viruses. Proc Natl Acad Sci U S A 100:9280–9285

5. Melikyan GB, Barnard RJ, Abrahamyan LG, Mothes W, Young JA (2005) Imaging individual retroviral fusion events: from hemifusion to pore formation and growth. Proc Natl Acad Sci U S A 102:8728–8733

6. Srivastava AK, Aira Y, Mori C, Kobayashi Y, Igarashi A (1987) Antigenicity of Japanese encephalitis virus envelope glycoprotein V3 (E) and its cyanogen bromide cleaved fragments examined by monoclonal antibodies and Western blotting. Arch Virol 96:97–107

7. Wu RS, Chan KR, Tan HC, Chow A, Allen JC Jr, Ooi EE (2012) Neutralization of dengue virus in the presence of Fc receptor-mediated phagocytosis distinguishes serotype-specific from cross-neutralizing antibodies. Antiviral Res 96(3):340–343

8. Zhang S, Tan HC, Ooi EE (2011) Visualizing dengue virus through Alexa Fluor labeling. J Vis Exp (53):e3168

9. Zhang SL, Tan HC, Hanson BJ, Ooi EE (2010) A simple method for Alexa Fluor dye labelling of dengue virus. J Virol Methods 167:172–177

Chapter 2

Isolation and Titration of Dengue Viruses by the Mosquito Inoculation Technique

Milly M. Choy and Duane J. Gubler

Abstract

Mosquito inoculation is a highly sensitive technique for isolation and titration of dengue virus (DENV) from sera, human tissues, wild animals, or mosquitoes. It has been under utilized since it was described 40 years ago because most dengue laboratories do not have access to an insectary to rear mosquitoes. This technique requires good eye-hand coordination while doing manipulation under a stereoscopic microscope, and extensive practice is needed to become proficient at inoculating mosquitoes. Following inoculation, mosquitoes are held for 10 days to allow dengue virus to replicate and disseminate to tissues throughout the mosquitoes. They are then harvested and examined for the presence of viral antigens in head tissue by either immunofluorescence assay (IFA) or PCR (polymerase chain reaction). The mosquito infectious dose 50 (MID_{50}) is calculated using the method of Reed and Muench to quantitate the virus. This method can be used for other arboviruses as well as for dengue.

Key words Dengue, Mosquito inoculation, 50 % end-point titration, Virus isolation, Virus titer, Arboviruses

1 Introduction

Dengue is a reemerging disease of global significance. The environmental changes associated with rapid population growth, unplanned urbanization, increased international air travel, and lack of effective mosquito control are considered to be the principal drivers for the geographic spread of both the viruses and their *Aedes* mosquito vectors [1, 2]. The public health impact of dengue is enormous, given that dengue virus (DENV) infection causes significantly more human disease than any other arboviruses. It is estimated that over 400 million infections, 100 million cases of symptomatic dengue infection, and over two million cases of dengue hemorrhagic fever/dengue shock syndrome occur worldwide annually [2, 3]. Other arboviruses such as Japanese encephalitis virus, West Nile virus, Zika virus, Ross River virus, and chikungunya virus have also emerged to cause major epidemics in recent years.

Radhakrishnan Padmanabhan and Subhash G. Vasudevan (eds.), *Dengue: Methods and Protocols*, Methods in Molecular Biology, vol. 1138, DOI 10.1007/978-1-4939-0348-1_2, © Springer Science+Business Media, LLC 2014

Table 1

Comparative dengue virus isolation rates using mosquito inoculation and C6/36 *Aedes albopictus* cell cultures, showing the influence of virus strains on isolation rates

Method of isolation	Aruba, 1985 DENV-1	%	Mozambique, 1985 DENV-3	%
Mosquito cells	13/29	45	2/28	7
Mosquito inoculation	14/29	48	5/28	18

DJ Gubler, 1985, unpublished data. *Note*: mosquito head squashes and cell cultures were screened for DENV by IFA [21]

The chi square values of the Aruba and Mozambique data are 0.5 and 2.67 respectively, both non-significant at 1 degree of freedom

DENV has been among the most difficult viruses to isolate and propagate because the viruses do not infect and replicate well in most laboratory animals. Studies on dengue in the 1940s to 1960s used suckling mice and various tissue culture systems for isolation and assay of DENV [4–8]. Although the use of mammalian and insect cell culture systems improved sensitivity [4, 6, 8], which is influenced by both the cell culture system and the strain of DENV, many unpassaged DENV do not produce cytopathic effects (CPE) when grown in these cells. The lack of a sensitive isolation and assay system that could be used for unpassaged wild-type viruses prevented rapid advancement in dengue research.

The development of the mosquito inoculation technique in the early 1970s provided a highly sensitive method for the isolation, propagation, and quantitation of DENV [9]. This method proved to be 10–1,000 times more sensitive in detecting dengue viruses than the commonly used plaque assay, depending on the serotype or strain of virus [9]. The influence of the virus strain is illustrated by the isolation rates of DENV from human sera in two 1985 epidemics in Aruba (DENV-1) ($\chi^2 = 0.5$, *ns*) and Mozambique (DENV-3) ($\chi^2 = 2.67$, *ns*) (Table 1). Both epidemics were explosive with severe disease and fatalities [10, 11], but isolation rates were very different. The isolation rate of DENV-1 was nearly identical by mosquito inoculation (48 %) and C6/36 *Aedes albopictus* cells (45 %), but for DENV-3 ($\chi2 = 2.67$, *ns*), the isolation rate was much higher by mosquito inoculation (18 %) than by C6/36 cells (7 %). The higher sensitivity of mosquito inoculation is consistent regardless of the virus serotypes ($\chi^2 = 11.1$, $p < 0.001$) (Table 2). Of 193 patient sera screened by the two methods during the 1986 epidemic in Puerto Rico (DENV-1, DENV-2, and DENV-4), 106 (61.3 %) dengue cases were detected by mosquito inoculation compared to 93 (53.8 %) by C6/36 cell cultures. As DENV can be frequently recovered from serum in mosquitoes when the same serum tested negative in both mammalian and insect cell culture, the use of mosquitoes is especially important when attempting to propagate viruses with low replication efficiency from mildly symptomatic illness and to isolate viruses from sera, tissues of naturally infected humans, wild animals, or field-caught mosquitoes.

Table 2
Comparison of mosquito inoculation and C6/36 *Aedes albopictus* cell cultures for isolation of DENV-1, DENV-2, and DENV-4 from human sera during an epidemic in Puerto Rico, 1986[a]

		Mosquito inoculation		
		Positive	Negative	Totals
Mosquito cell culture	Positive	93	0	93 (53.8 %)
	Negative	13	67	80 (46.2 %)
	Totals	106 (61.7 %)	67 (38.7 %)	173 (100 %)

[a]DJ Gubler, 1986, unpublished data. *Note*: mosquito head squashes and cell cultures were screened for DENV by IFA [21]

The quantitation of virus in sera, human tissues, animals, mosquitoes, and cell culture is calculated using the method of Reed and Muench [12] and expressed as the dose that infects 50 % of the mosquitoes inoculated (MID_{50}). Using this method, many studies in the South Pacific islands, Indonesia, Thailand, Sri Lanka, and Puerto Rico showed a high rate of isolation of DENV from primary clinical samples. Equally important was the demonstration of considerable variation in viremia levels of different DENV strains and serotypes in patients showing a correlation with disease severity [13–19]. This technique has also been used as an assay for other arboviruses.

An accurate measure of infectious virus is critical to fully understand virus-host relationships, disease severity, viral fitness, and pathogenesis, as well as for development of effective diagnostic tests, vaccines, and therapeutics. The mosquito inoculation technique is arguably the most sensitive isolation and quantitative assay for infectious DENV in sera, autopsy tissues, mosquitoes, and in vitro systems. In this chapter, we describe the mosquito inoculation technique used to determine the MID_{50} titer of DENV. Although the primary vector for dengue is *Aedes aegypti*, other species from the subgenus *Stegomyia* such as *Aedes albopictus* and *Aedes polynesiensis* as well as non-blood-feeding *Toxorhynchites* species can be used for this assay. Mosquitoes are reared in the laboratory under BSL-2 conditions as described previously [20].

The mosquito inoculation technique itself is technically easy, but requires good eye-hand coordination while doing the manipulation under a stereoscopic microscope. Extensive practice is required to ensure good survival of the inoculated mosquitoes. Once a person has become proficient at inoculating mosquitoes, the method becomes less labor intensive. A critical step of this proficiency is preparation of glass capillary needles, an often-overlooked step for ensuring good survival of mosquitoes after inoculation. Following inoculation, mosquitoes are placed in cartons inside security cages, provided a 10 % sucrose solution in a locked incubator at 28 °C and held for 10 days to allow dengue virus to replicate, disseminate, and infect tissues throughout the mosquito. Surviving mosquitoes are then harvested and examined for the presence of viral antigens in head tissue by either immunofluorescence assay (IFA) [21] or polymerase chain reaction. IFA is usually used for DENV detection, as it is the most rapid and economical way to detect positive mosquitoes. Virus is quantitated by calculating the MID_{50} using the method of Reed and Muench [12].

2   Materials

2.1 Preparation of Glass Capillary Needles for Mosquito Inoculation

1. 6-inch glass capillaries (1/0.75 OD/ID mm) (World Precision Instruments).
2. PC-10 puller (Narishge Co. Ltd)*.
3. Alcohol lamp.
4. Jeweler's forceps.
5. Stereoscopic microscope.

2.2 Mosquito Inoculation

1. Viral suspension.
2. Leibovitz's L-15 medium.
3. 3–5-day-old mosquitoes.
4. Glove box.
5. Inoculation apparatus.
   (a) 30-ml glass syringe with Luer lock tip (Sigma-Aldrich)*.
   (b) Plastic tubing.
   (c) Three-way stopcock (Becton Dickinson)*.
   (d) Brass steel needle holder with compression fittings (custom made, Sky Engineering Pte Ltd)*.
6. Glass capillary needles with markings.
7. Stereoscopic microscope.
8. Jeweler's forceps.
9. Pint holding cartons with lid (Science Supplies)*.
10. BugDorm Rearing Cage (Bioquip Product, Inc.)*.
11. Insect growth chamber (Darwin Chambers Co.)*.

2.3 Dengue Virus Detection in Mosquitoes by Indirect Immunofluorescence on Mosquito Head Squashes

1. 12-well (6 mm) teflon-coated microscope slide.
2. Razor blade.
3. Pellet pestle.
4. Flavivirus cross-reactive monoclonal antibody 4G2, HB112 (ATCC)*.

5. Sheep anti-mouse IgG FITC (Meridian Life Science)*.

6. 1× PBS.

7. Slide holder.

8. Glycerol mounting fluid.

9. Coverslips.

10. Fluorescence microscope.

   *Mention of commercial products in this chapter is for information only; it does not imply recommendation or endorsement by the authors.

3 Methods

All procedures can be carried out in an air-conditioned BSL-2 laboratory unless otherwise specified.

3.1 Preparation of Glass Capillary Needles

1. Use the single-pull mode of a PC-10 puller to pull a glass capillary vertically. Set the heater level to approximately 55 °C to obtain a cone-shape tip (*see* **Note 1**).

2. Bend the glass capillary needle slightly over an alcohol lamp. Mark off the untapered portion of the needle under a stereoscopic microscope at 1.0 mm intervals with a thin marker and a ruler (Fig. 1).

3. Break off the tip of the needle at the appropriate diameter (approximately 0.5 μm), with a jeweler's forceps under a stereoscopic microscope (*see* **Note 2**).

3.2 Mosquito Inoculation of Dengue Viruses

3.2.1 Dengue Virus Isolation

1. Aspirate 50 3–5-day-old mosquitoes and hold them in tubes on ice for at least 30 min to anesthetize them. Separate the mosquitoes into males or females to be used for experiments into a 50-ml falcon tube (*see* **Note 3**).

2. Ensure all mosquitoes have been properly immobilized by holding them on ice throughout experimentation.

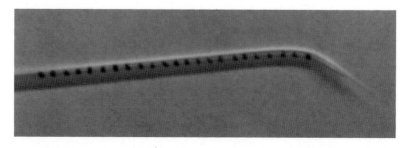

Fig. 1 Tip of glass needle used for inoculation. Each division is marked 1.0 mm apart

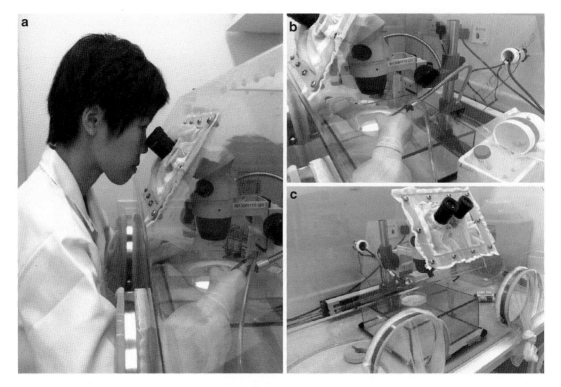

Fig. 2 Inoculation apparatus in use (**a**, **b**). Inoculation setup within a glove box (**c**)

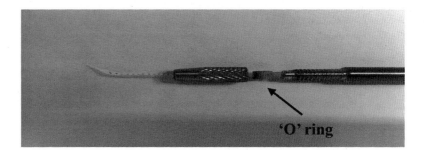

'O' ring

Fig. 3 Needle holder holding "O" ring to obtain airtight seal

3. Set up the inoculation apparatus within a glove box (Fig. 2, *see* **Note 4**) and place a glass capillary needle in the needle holder (Fig. 3, *see* **Note 5**).

4. Load the needle with inoculum (from **Step 1**) by immersing the tip in the viral sample and withdrawing the plunger of the syringe (*see* **Note 6**).

5. Place 3–5 mosquitoes on a petri dish held on a tall stand under a stereoscopic microscope and position the mosquito on the dorsal aspect of its thorax.

6. Impale the mosquito on the underside of the neck (Fig. 4, *see* **Note 7**).

Fig. 4 Ventral aspect of head and neck of male mosquito with "X" showing preferred site of inoculation (Adapted from [9])

7. Bring the marked portion of the glass capillary needle into view under the microscope and observe the fluid meniscus when the plunger of the syringe is depressed until the desired amount has been inoculated (*see* **Note 8**).

8. Place mosquito within a holding carton after inoculation by using the wall of the carton or a jeweler's forceps to dislodge the mosquito. Place the carton on ice to ensure that the mosquitoes are immobilized and contained. Block the exit of mosquitoes in the carton using cotton.

9. Inoculate 10–20 mosquitoes to form a virus seed pool.

10. Secure the holding carton with tape and store the cartons within a larger BugDorm Rearing Cage at the end of the experiment.

11. Dispose any unwanted or unused mosquito into 70 % ethanol in a 50-ml falcon tube. Place all experimental waste that came into contact with virus in a solution of Virkon for decontamination.

12. Incubate the mosquitoes at 28 °C and 80 % humidity for 10 days in an insect growth chamber. Allow mosquitoes access to 10 % sucrose and water.

3.2.2 Dengue Virus Titration

1. Perform tenfold serial dilutions (10^{-1} to 10^{-6}) of viral sample with Leibovitz's L-15 medium. Place all viral samples on ice.

2. Perform mosquito inoculation as described (Subheading 3.2.1).

3. Inoculate enough mosquitoes so that at least 5 mosquitoes are available for testing each dilution of the viral sample.

4. Incubate the mosquitoes at 28 °C and 80 % humidity for 10 days in an insect growth chamber. Allow mosquitoes access to 10 % sucrose and water.

3.3   Dengue Virus Detection in Mosquitoes by Indirect Immunofluorescence on Mosquito Head Squashes

1. After incubation for 10 days, harvest surviving mosquitoes and store at −80 °C.

2. Label and number the wells of a 12-well (6 mm) teflon-coated microscope slide. Include a positive and a negative control on the slide.

3. Remove mosquitoes from the freezer and place them in a petri dish on an ice tray.

4. Remove the head of the mosquitoes using a razor blade or scalpel. Place one mosquito head per well (*see* **Note 9**).

5. Squash the mosquito head using a pellet pestle (*see* **Note 10**). Allow to dry before picking off excess chitin with jeweler's forceps.

6. Using a slide holder, submerge the slides in cold 80 % acetone. Allow cells to fix for 10 min at room temperature. Rinse once in PBS.

7. Let the slides air-dry completely. Place inside biosafety cabinet for faster drying.

8. Pipette 25 µl of flavivirus cross-reactive monoclonal antibody 4G2 (HB112) (diluted 1:10) onto each well (*see* **Note 11**).

9. Dampen paper towels to line a plastic incubation container.

10. Place the slides in the incubation container. Gently place the lid on the container and incubate at 37 °C for 30 min.

11. Remove and rinse slides in staining dish with 1× PBS. Incubate for 5 min using a magnetic stirrer.

12. Let slides air-dry.

13. Pipette 25 µl of anti-mouse IgG FITC onto each well.

14. Place the slides in the incubation container. Gently place the lid on the container and incubate at 37 °C for 30 min.

15. Remove and rinse slides twice in staining dish with 1× PBS. Incubate for 5 min using a magnetic stirrer.

16. Let slides air-dry completely.

17. Place a drop of mounting fluid onto the slide and gently lay down the coverslip.

18. View slides using a fluorescent microscope (*see* **Note 12**).

19. Record the number of infected mosquitoes for each dilution.

20. Calculate virus titers using the method of Reed and Muench and express the titers in terms of the dose that infects 50 % of the mosquitoes inoculated (MID_{50} per ml).

4 Notes

1. The needles can be pulled using a simple alcohol lamp if a needle puller is not available.

2. The tip of the needle is to be made as small as possible and yet large enough to minimize plugging with debris contained in the inocula. A small tip is advantageous both for ease in piercing the exoskeleton of the mosquito and in controlling the amount of inoculum. It is also advantageous to make the tapered portion of the needle as short as possible for better control during inoculation.

3. Male mosquitoes are used for titration experiments as they offer a significant advantage in safety since they cannot transmit the infection should they escape in the laboratory. Although male mosquitoes are as sensitive as females to dengue infection, they are less able to tolerate larger inocula and are less resistant to toxic effects of certain undiluted human sera. Female mosquitoes are only used when dealing with large inocula of up to 1 µl or inoculation of undiluted human sera.

4. Escape of infected mosquito within the insectary is prevented by handling them in a glove box with glove ports to allow access to the containment area. The needle holder is attached by a plastic tubing to a 30-ml syringe held by the base of a tall stand. The tubing is attached to the syringe by a three-way stopcock, which allows the plunger of the syringe to be manipulated when necessary, without changing the pressure in the plastic tubing.

5. Airtight seal of the needle to the brass needle holder is made via the use of a compression fitting. The holders are not available commercially but can be made by most machine shops.

6. Start from the highest to the lowest dilution so that the same needle can be used. Change a needle after each viral sample.

7. It is also possible to inoculate female mosquitoes through the membranous area anterior to the mesepisternum and below the spiracle without compromising the survival rate of the mosquitoes after inoculation (Fig. 5).

8. A distance of 1 mm on the markings of the glass needle is normally recommended for inoculation (0.4 µl). After inoculation, it is necessary to check that all inoculum has entered the insect and transfixation has not occurred.

9. Test at least 5 or more mosquitoes for dengue infection at each dilution of a viral sample. It is always a good practice to store the remainder of the labeled mosquito in a 1.5-ml Eppendorf tube at −80 °C for future testing and possible virus isolation.

Fig. 5 Lateral aspect of thorax of female mosquito with showing the preferred site of inoculation (Adapted from [9])

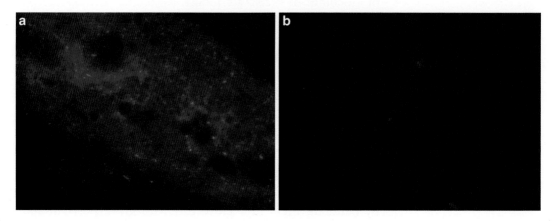

Fig. 6 Example of a positive head squash (**a**) and a negative head squash (**b**)

10. Alternatively, one can coat a coverslip with Rain-X. Allow the coverslip to dry. Place the coverslip over the 12-well microscope slide with mosquito heads on them. Use the end of a pencil to squash the heads.

11. Primary antibodies specific for each dengue serotypes can also be used depending on the experiment.

12. Care should be taken in interpreting whether or not a head squash is positive. The preparation typically has a lot of tissue/chitin that autofluoresces. A distinctive positive squash should contain some doughnut-shaped cells (Fig. 6).

References

1. Kyle JL, Harris E (2008) Global spread and persistence of dengue. Annu Rev Microbiol 62:71–92

2. Gubler DJ (2011) Dengue, Urbanization and Globalization: The Unholy Trinity of the 21(st) Century. Trop Med Health 39:3–11

3. Bhatt S, Gething PW, Brady OJ, Messina JP, Farlow AW, Moyes CL, Drake JM, Brownstein JS, Hoen AG, Sankoh O, Myers MF, George DB, Jaenisch T, Wint GR, Simmons CP, Scott TW, Farrar JJ, Hay SI (2013) The global distribution and burden of dengue. Nature 496:504–507

4. Igarashi A (1978) Isolation of a Singh's Aedes albopictus cell clone sensitive to Dengue and Chikungunya viruses. J Gen Virol 40:531–544

5. Halstead SB, Sukhavachana P, Nisalak A (1964) In Vitro Recovery of Dengue Viruses from Naturally Infected Human Beings and Arthropods. Nature 202:931–932

6. Sukhavachana P, Nisalak A, Halstead SB (1966) Tissue culture techniques for the study

of dengue viruses. Bull World Health Organ 35:65–66

7. Hammon WM, Rudnick A, Sather GE (1960) Viruses associated with epidemic hemorrhagic fevers of the Philippines and Thailand. Science 131:1102–1103

8. Yuill TM, Sukhavachana P, Nisalak A, Russell PK (1968) Dengue-virus recovery by direct and delayed plaques in LLC-MK2 cells. Am J Trop Med Hyg 17:441–448

9. Rosen L, Gubler DJ (1974) The use of mosquitoes to detect and propagate dengue viruses. Am J Trop Med Hyg 23:1153–1160

10. Gubler DJ, Sather GE, Kuno G, Cabral JR (1986) Dengue 3 virus transmission in Africa. Am J Trop Med Hyg 35:1280–1284

11. Pinheiro FP (1989) Dengue in the Americas. 1980-1987. Epidemiol Bull 10:1–8

12. Reed LJ, Muench HA (1938) A simple method of estimating fifty percent endpoints. Am J Hyg 27:493–497

13. Gubler DJ, Reed D, Rosen L, Hitchcock JR (1978) Epidemiologic, clinical, and virologic observations on dengue in the Kingdom of Tonga. Am J Trop Med Hyg 27:581–589

14. Gubler DJ, Rosen L (1977) Quantitative aspects of replication of dengue viruses in Aedes albopictus (Diptera: Culicidae) after oral and parenteral infection. J Med Entomol 13:469–472

15. Gubler DJ, Suharyono W, Lubis I, Eram S, Gunarso S (1981) Epidemic dengue 3 in central Java, associated with low viremia in man. Am J Trop Med Hyg 30:1094–1099

16. Gubler DJ, Suharyono W, Lubis I, Eram S, Sulianti SJ (1979) Epidemic dengue hemorrhagic fever in rural Indonesia. I. Virological and epidemiological studies. Am J Trop Med Hyg 28:701–710

17. Gubler DJ, Suharyono W, Tan R, Abidin M, Sie A (1981) Viraemia in patients with naturally acquired dengue infection. Bull World Health Organ 59:623–630

18. Kuberski T, Rosen L, Reed D, Mataika J (1977) Clinical and laboratory observations on patients with primary and secondary dengue type 1 infections with hemorrhagic manifestations in Fiji. Am J Trop Med Hyg 26:775–783

19. Vaughn DW, Green S, Kalayanarooj S, Innis BL, Nimmannitya S, Suntayakorn S, Endy TP, Raengsakulrach B, Rothman AL, Ennis FA, Nisalak A (2000) Dengue viremia titer, antibody response pattern, and virus serotype correlate with disease severity. J Infect Dis 181:2–9

20. Gerberg EJ (1970) Manual for mosquito rearing and experimental techniques. AMA Bulletin 5:1–109

21. Kuberski TT, Rosen L (1977) A simple technique for the detection of dengue antigen in mosquitoes by immunofluorescence. Am J Trop Med Hyg 26:533–537

Chapter 3

Measuring Antibody Neutralization of Dengue Virus (DENV) Using a Flow Cytometry-Based Technique

Ruklanthi de Alwis and Aravinda M. de Silva

Abstract

Dengue virus (DENV) is an emerging virus that threatens over two-third of the world's population. The specific diagnosis of dengue infection by serology is based on assays that detect DENV-specific antibodies including neutralizing antibodies (Abs). Neutralizing Abs are an important, if not the main, mechanism of protection from natural dengue virus (DENV) infection as well. The current gold-standard assay for measuring neutralizing Ab responses against DENV is the plaque reduction neutralization assay (PRNT). However, this assay is slow and laborious and utilizes physiologically irrelevant cell lines. Here, we describe a relatively high-throughput, flow cytometry-based neutralization assay for DENV that has been optimized for use with a human monocytic suspension cell line, U937 + DC-SIGN, or the more commonly used adherent monkey kidney cells, Vero-81.

Key words Dengue virus, Serotypes, Neutralization assay, Flow cytometry, Human monocytic cells, U937 cells (with DC-SIGN), Vero cells

1 Introduction

Dengue is a rapidly emerging global health problem that causes over 390 million infections per year in the tropical and subtropical regions of the world. The dengue virus (DENV) complex has four serotypes, and people are commonly exposed to repeat infections with different serotypes. The presence of four serotypes complicates the pathogenesis of dengue and poses unique problems for vaccine developers [1, 2]. The neutralizing antibody response is an important mechanism of protection against DENV infections and therefore an area of intense study [1]. We need cost-effective, accurate, high-throughput DENV neutralization assays for diagnostic use and for evaluating vaccine-induced responses.

Traditionally, plaque reduction (PRNT) or foci reduction (FRNT) neutralization assays have been widely used to measure the neutralization potency of monoclonal and polyclonal antibodies (i.e., in both natural and vaccine settings). However, PRNT/FRNT

Radhakrishnan Padmanabhan and Subhash G. Vasudevan (eds.), *Dengue: Methods and Protocols*, Methods in Molecular Biology, vol. 1138, DOI 10.1007/978-1-4939-0348-1_3, © Springer Science+Business Media, LLC 2014

assays are time consuming and cannot be easily applied to high-throughput screening of monoclonal antibodies or serum panels. Furthermore, plaque/foci counting is subjective and often inaccurate. Additionally, PRNT/FRNT assays are dependent on the plaque or foci-forming properties of viruses. Different DENV strains form plaque/foci with different size and efficiency. More importantly, PRNT or FRNT assays utilize nonhuman, non-Fc receptor-bearing cells. During a natural infection in humans, DENV is believed to mainly replicate in immune cells of myeloid origin including Fc receptor-bearing monocytes and macrophages. Recent studies comparing neutralization data obtained using Fc receptor versus non-Fc receptor-bearing cells have shown that the lack of Fc receptors made the assay significantly less accurate in predicting the DENV serotype of infection from DENV-immune human sera [3–5].

The present chapter describes a high-throughput DENV neutralization assay. This assay has been previously tested with all four DENV serotypes as well as multiple strains within each serotype [6–9]. Furthermore, when this flow cytometry-based neutralization technique using U937 + DC-SIGN cells was evaluated using a coded panel of 26 human immune sera, it successfully predicted previous DENV infection history with greater than 92 % accuracy (Table 1). The outlined protocol is optimized for the use of the traditional monkey kidney cell line, Vero, and the Fc receptor-bearing monocytic-like human cell line, U937, expressing dendritic cell-specific intracellular adhesion molecule-3-grabbing non-integrin (DC-SIGN). U937 cell line stably transfected with DC-SIGN was kindly provided by Mark Heise (at University of North Carolina, Chapel Hill), and we are in the process of depositing these cells with the American Type Culture Collection (ATCC). DC-SIGN is an important attachment factor for DENV infection and greatly improves the efficiency with which DENV infects U937 cells [10, 11]. The present protocol includes description of cell maintenance, growth of virus stocks, titration of virus for use in the neutralization assay, neutralization assay, staining of infected cells, and data acquisition and analysis. The 96-well format and the data acquisition through flow cytometry make the present assay relatively high throughput and ideal for neutralization screening of large samples of monoclonal antibodies and polyclonal sera.

2   Materials

1. U937 cells stably expressing dendritic cell-specific intracellular adhesion molecule-3-grabbing non-integrin (DC-SIGN). These U937 + DC-SIGN have been deposited into the American Type Culture Collection (ATCC). Vero-81 cells can also be obtained from ATCC.

Table 1
The U937 + DC-SIGN flow cytometry-based neutralization assay accurately predicted the DENV-immune status of 24 human sera from a panel of 26 sera that had prior been tested with the traditional PRNT assay

#	Infection type (based on the PRNT assay)	50 % Neutralization titer from the U937 + DC-SIGN flow cytometry-based assay				Infection type
		DENV1	DENV2	DENV3	DENV4	
1	JE	<20	<20	<20	<20	Dengue naïve
2		<20	<20	<20	<20	Dengue naïve
3	Naïve	<20	<20	<20	<20	Dengue naïve
4		<20	66	52	<20	Poly
5		<20	<20	<20	<20	Dengue naïve
6		<20	<20	25	<20	Dengue naïve
7	DEN-1	443	34	28	46	DEN-1
8		>1,280	45	140	<20	DEN-1
9		111	45	77	<20	Poly
10	DEN-2	20	>1,280	60	32	DEN-2
11		<20	423	61	<20	DEN-2
12		<20	>1,280	34	<20	DEN-2
13	DEN-3	<20	75	>1,280	<20	DEN-3
14		<20	172	>1,280	<20	DEN-3
15		<20	30	188	<20	DEN-3
16	DEN-4	<20	32	60	>1,280	DEN-4
17		<20	153	367	>1,280	DEN-4
18		<20	26	91	685	DEN-4
19	Poly	929	356	277	40	Poly
20		279	927	1,054	226	Poly
21		596	411	96	57	Poly
22		45	>1,280	>1,280	>1,280	Poly
23		261	1,208	740	225	Poly
24		104	826	265	296	Poly
25		254	>1,280	>1,280	157	Poly
26		28	485	176	43	Poly

2. *Growth medium* for U937 + DC-SIGN cells is made of Roswell Park Memorial Institute (RPMI) medium 1640 supplemented with 10 % of heat-inactivated fetal bovine serum (FBS), 2 mM L-glutamine, 1 % MEM nonessential amino acid (NEAA) solution, antibiotics (20 U/ml penicillin, 20 µg/ml streptomycin, 0.05 µg/ml amphotericin B), and 0.055 mM 2-mercaptoethanol (*see* **Note 1** *for Vero growth media*).

3. 2 and 24 h *infection media* for U937 + DC-SIGN cells consist of RPMI medium 1640 supplemented with 2 % FBS, 2 mM L-glutamine, 1 % MEM nonessential amino acid solution, antibiotics (20 U/ml penicillin, 20 µg/ml streptomycin, 0.05 µg/ml amphotericin B), and 0.055 mM 2-mercaptoethanol. In addition, add 20 mM HEPES for the 2 h infection media (*see* **Note 2** *for Vero infection media*).

4. Titered infectious DENV *virus stocks* from all four serotypes (*see* **Note 3** *for details on preparation of infectious DENV virus stocks* and **Notes 4** and **5** *for details on titration of DENV virus stocks on U937+DC-SIGN and Vero cells, respectively*). The WHO reference strain viruses, DENV1 (West Pac 74), DENV2 (S-16803), DENV3 (CH54389), and DENV4 (TVP-360), have been most frequently used in this assay. The assay has also been used with many other strains of DENV [7, 9, 12].

5. Polyclonal serum heat inactivated at 55 °C for 30 min.

6. Monoclonal antibody filter-sterilized with a 0.22 µm filter.

7. 4 % paraformaldehyde (PFA): Dilute 16 % paraformaldehyde 1:4 using 1× PBS, pH 7.4. Store at –20 °C.

8. FACS buffer: Measure out 200 mL of 1× PBS, pH 7.4 into a sterile storage bottle. Measure out 0.2 g of sodium azide and 5.0 mL of 35 % bovine serum albumin (BSA) into the storage bottle. Add 1× PBS (pH 7.4) up to 1 L, filter sterilize with a 0.45 µm filter, and store at 4 °C.

9. 5 % saponin: Dissolve 2.5 g of saponin in 50 mL of 1× PBS, pH 7.4 Filter-sterilize with a 0.45 µm filter and store at 4 °C.

10. Permeabilization (PERM) buffer: Add 12.5 mL of 35 % BSA and 0.5 mL of filter-sterilized 5 % saponin to 370 mL of FACS buffer. Store at 4 °C.

11. Blocking buffer: 1 % w/v normal mouse serum in PERM buffer.

12. The 2H2-secreting HB-114 hybridoma (D3-2H2-9-21) can be purchased from ATCC (*see* **Note 6**), and the hybridoma cultured and 2H2 antibody purified using previously described protocols [13]. Or purified 2H2 can be purchased from Millipore (Cat# MAB8705).

13. Oregon Green 488 conjugation to 2H2: The 2H2 monoclonal antibody was fluorescently labeled with Oregon Green 488 using an Oregon Green 488 protein labeling kit (Cat# O-10241, Invitrogen) as per manufacturer's instructions.

3   Methods

3.1 Passaging/ Subculturing Cells

1. Mix *U937+DC-SIGN* cells evenly by pipetting the cell-containing medium, and remove an aliquot for cell counting. Mix cells 1:20 with 0.4 % Trypan Blue, load on a hemacytometer,

and estimate the live cell count. Cells in log phase should have a cell viability of >95 %.

2. Split the *U937+ DC-SIGN* cells into a T-75 or T-175 flask at a density between 5×10^4 and 2×10^5 cells/mL. The cell culture will be ready for splitting or experimental use when it reaches a cell density of 8.0×10^5–1.0×10^6 cells/mL (*see* **Note 7**), i.e., approximately 5 and 3 days post seeding for an initial density split of 0.5×10^5 and 2×10^5 cells/mL, respectively (*see* **Note 8** *for plating Vero cells for experiment*).

3.2 Neutralization Assay with U937 + DC-SIGN Cells

1. Set up plate layout as appropriate to your experimental question. In Fig. 1 we provide an example of a typical layout for measuring neutralizing antibody in a serum sample. Conduct calculations for dilution of antibody, virus, and cells. Each plate should contain a couple of wells of the following controls: (1) no-antibody treated wells (i.e., virus + cells only); (2) no-virus, no-antibody treated wells (i.e., cells only); and a DENV naïve serum control (*see* **Notes 9–11**).

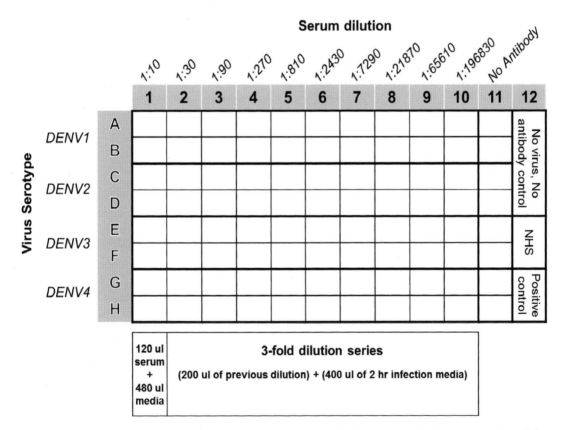

Fig. 1 Example 96-well plate layout for a neutralization assay with human polyclonal serum against all four serotypes of DENV. Serum dilution calculations shown above are for a threefold dilution series. Both NHS and positive control used at a final dilution of 1:10. NHS: DENV naive human serum

2. As calculated in the above step, serially dilute filter-sterilized antibodies or heat-inactivated polyclonal serum in 0.65 mL dilution tubes containing 2 h infection medium. Then, transfer 40 μL of diluted monoclonal antibody or human serum into the appropriate wells of a 96-well U-bottom plate.

3. Dilute DENV stock in 2 h infection medium so that 40 μL contains enough virus to infect between 10 and 15 % of the cells in a single well of a 96-well plate. Then, using a multi-channel pipette, transfer 40 μL of the appropriate virus into the appropriate wells of the 96-well plate from **step 2**, and mix by pipetting. Remember to change pipette tips between each transfer. For no-antibody treated control wells, add 40 μL of 2 h infection medium instead of diluted antibody, and for cells-only wells, add 80 μL of 2 h infection medium. Once plate is set up, incubate it at 37 °C for 1 h (*see* **Note 12** *for instructions on conducting the neutralization assay using Vero cells*).

4. Pellet *U937+DC-SIGN* cells at $500 \times g$ for 5 min, remove supernatant, and gently resuspend cell pellet in fresh 2 h infection medium at a density of 2.5×10^6 cells/mL. At the end of the 1 h incubation in **step 3**, add 20 μL of resuspended cells into each experimental well (i.e., 5×10^4 cells/well), and incubate for 2 h at 37 °C incubator containing 5 % CO_2.

5. At the end of 2 h, remove plate from the 37 °C incubator, spin at $500 \times g$, carefully aspirate supernatant, and resuspend in 200 μL/well of fresh 2 h infection medium.

6. Repeat **step 5**, and add 200 μL/well of fresh 24 h infection medium. Place plate at 37 °C incubator containing 5 % CO_2 for 22 h.

3.3 Fixing and Staining Cells

1. At 24 h postinfection, pellet U937 + DC-SIGN cells at $500 \times g$ for 5 min, aspirate the supernatant, and resuspend cells in 150 μL/well of sterile 1× PBS, pH 7.4.

2. Pellet cells at $500 \times g$ for 5 min and aspirate supernatant. Then, resuspend in 50 μL/well of 4 % PFA and fix cells at room temperature for 10 min.

3. After fixing, pellet the cells at $750 \times g$ for 5 min, aspirate supernatant and resuspend with 150 μL/well of PERM buffer.

4. Repeat **step 3**.

5. After two washes with PERM buffer (i.e., **steps 3** and **4**), cells are pelleted at $750 \times g$ for 5 min and resuspend in 40 μL/well of blocking buffer. Cells are blocked for 30 min at room temperature or overnight at 4 °C.

6. Add 20 μL/well of 2H2-conjugated Oregon Green 488 at a final concentration determined by prior titration (i.e., the final concentration should account for the 40 μL/well of blocking buffer

already in each well). In addition, to assess DC-SIGN expression, add 1 μL/well of anti-CD209 conjugated to phycoerythrin (PE) to two wells containing uninfected cells. Avoid exposure to light by wrapping plates with aluminum foil. Incubate plates at 37 °C for 1 h on a plate shaker set for gentle shaking.

7. After staining with antibody, pellet cells at $750 \times g$ for 5 min, aspirate supernatant, and resuspend in 150 μL/well of PERM buffer.

8. Repeat **step 7**.

9. Pellet cells at $750 \times g$ for 5 min, aspirate supernatant and resuspend in a sufficient volume of FACS buffer to obtain a cell concentration between 200 and 450 cells/μL. Generally, 125 μL/well of FACS buffer will place the final cell concentration within the specified range (*see* **Note 13**).

3.4 Data Acquisition and Analysis

1. Percentage of infected cells can be measured using flow cytometry capable of exciting at both 533 and 488 nm, such as an LSR with a high-throughput sampler (HTS) (BD Biosciences) or Guava easyCyte HT flow cytometer (Millipore).

2. Single cells (M1) are differentiated from duplets, clumps, or debris by gating on a dot plot of forward scatter versus side scatter. To assess DC-SIGN expression in the cells, the single cells are then gated on CD209-PE. Additionally, to estimate the percentage of infected cells (M2), the single cells (M1) are also separately gated on DENV-Oregon Green 488 nm fluorescence versus forward scatter. Set the cell count to 1×10^4 cells/well of single-gated cells. Initial gates and flow cytometry parameters are set using an uninfected, cells-only sample well.

3. Once data acquisition is complete, the data is opened on a flow cytometry data analysis software such as FlowJo, Summit, or guavaSoft. Go through every acquired sample and make readjustments of gates as deem necessary, and then export the percentage infection values (M2) to a Microsoft excel sheet.

4. First calculate percent infection values (I_{s-0}), by subtracting any background percent infection value obtained for cells-only samples (I_0), from percent infection values of all experimental samples (I_s), i.e., $I_{s-0} = I_s - I_0$ (*see* **Notes 14–16**).

5. For each DENV serotype, average the percent infection values obtained from wells treated with cells and virus only (DV_{Av}). Then, calculate the percent neutralization of each sample (N_x) from their respective percent infection sample values (S) by applying the following equation: $N_x = \left\{ 100 - \left[100 \left(\dfrac{s}{DV_{Av}} \right) \right] \right\}$

6. Using GraphPad Prism software, the percent neutralization values are then tabulated and plotted against Log(1/Dilution)

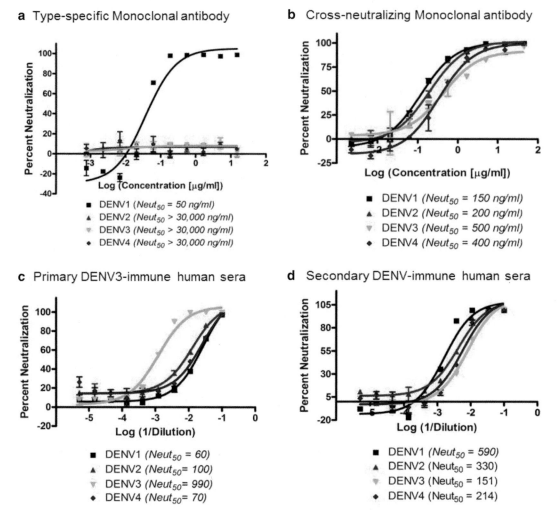

a Type-specific Monoclonal antibody

- DENV1 *(Neut$_{50}$ = 50 ng/ml)*
- DENV2 *(Neut$_{50}$ > 30,000 ng/ml)*
- DENV3 *(Neut$_{50}$ > 30,000 ng/ml)*
- DENV4 *(Neut$_{50}$ > 30,000 ng/ml)*

b Cross-neutralizing Monoclonal antibody

- DENV1 *(Neut$_{50}$ = 150 ng/ml)*
- DENV2 *(Neut$_{50}$ = 200 ng/ml)*
- DENV3 *(Neut$_{50}$ = 500 ng/ml)*
- DENV4 *(Neut$_{50}$ = 400 ng/ml)*

c Primary DENV3-immune human sera

- DENV1 *(Neut$_{50}$ = 60)*
- DENV2 *(Neut$_{50}$= 100)*
- DENV3 *(Neut$_{50}$= 990)*
- DENV4 *(Neut$_{50}$= 70)*

d Secondary DENV-immune human sera

- DENV1 *(Neut$_{50}$ = 590)*
- DENV2 *(Neut$_{50}$ = 330)*
- DENV3 *(Neut$_{50}$ = 151)*
- DENV4 *(Neut$_{50}$ = 214)*

Fig. 2 Neutralization curves of monoclonal antibodies and human sera against all 4 DENV serotypes generated using the flow cytometry-based neutralization assay with U937 + DC-SIGN cells; (**a**) type-specific monoclonal antibody, (**b**) cross-neutralizing monoclonal antibody, (**c**) primary DENV3-immune human serum, and (**d**) secondary DENV-immune human serum

for polyclonal serum or Log[Concentration (μg/mL)] for monoclonal antibodies. As shown in Fig. 2, the plotted data is then fitted to a sigmoidal dose–response curve and 50 % neutralization values (Neut$_{50}$) are obtained from the graph.

4   Notes

1. For Vero cell growth media, Dulbecco's modified eagle medium (DMEM)/F12, supplemented with 1 % (v/v) of 7.5 % sodium bicarbonate, and similar amounts of FBS, L-glutamine, and antibiotics as described for U937 + DC-SIGN growth medium.

Table 2
Dilution calculations for virus titration in U937 + DC-SIGN and Vero cells

U937 + DC-SIGN cells				Vero cells[a]			
Virus (µl)/well	Virus (µl)	Media (µl)	Total volume (µl)[b]	Virus (µl)/well	Virus (µl)	Media (µl)	Total volume (µl)[b]
0.25	1	159	160	5	20	180	200
0.5	2	158	160	6	24	176	200
0.75	3	157	160	8	32	168	200
1.0	4	156	160	10	40	160	200
2.0	8	152	160	15	60	140	200
4.0	16	144	160	20	80	120	200
6.0	24	136	160	30	120	80	200
8.0	32	128	160	35	140	60	200
10.0	40	120	160	40	160	40	200
12.0	48	112	160	50	200	0	200

[a]It is usually observed that a significantly greater amount of DENV of the same viral stock is required to obtain a 7–15 % infection rate in Vero cells, as compared to U937 + DC-SIGN cells. Therefore, as shown in table, larger volumes of virus are used for virus titration in Vero cells than in U937 + DC-SIGN cells
[b]Total volume calculated for 4 wells worth. Virus titration in U937 + DC-SIGN and Vero cells receive 40 and 50 µL/well, respectively

2. For *Vero* cells, 2 h and 24 h infection medium recipe is similar to U937 + DC-SIGN formula, except substitute DMEM/F12 medium for RPMI and 1 % (v/v) of 7.5 % sodium bicarbonate for 0.055 mM 2-mercaptoethanol.

3. *Preparation of infectious DENV virus stocks.* Remove the supernatant from a T-175 flask with an 85 % confluent monolayer of C6/36 cells. Estimating 10^8 cells per T-175 flask, infect cell monolayer with a DENV serotype at a multiplicity of infection (MOI) of 0.01 in 4 mL of C6/36 infection media. Incubate flask at 28 °C for 2 h, with rocking every 15 min. At the end of 2 h, add the remaining 21 mL of fresh infection media to the T-175 flask and incubate at 28 °C. Virus supernatant is harvested 6–7 days postinfection (dpi), clarified at $1,200 \times g$ for 30 min at 4 °C, and stored in aliquots at –80 °C.

4. *Titration of infectious DENV virus on U937 + DC-SIGN cells.* Thaw an aliquot (>250 µL) of DENV virus to be titrated on ice. As shown in Table 2, using the appropriate 2 h infection medium and sterile 0.65 mL dilution tubes, dilute virus sufficient for four wells. Using the plate layout in Fig. 3, transfer 40 µL/well of diluted virus to the plate. Then, transfer 40 µL

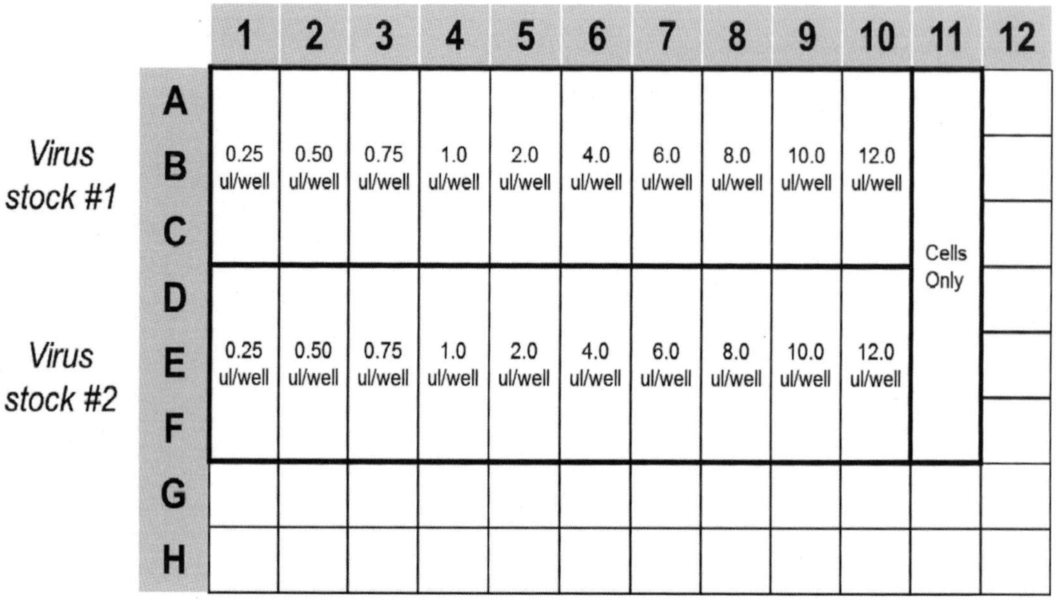

	1	2	3	4	5	6	7	8	9	10	11	12
A												
B	0.25 ul/well	0.50 ul/well	0.75 ul/well	1.0 ul/well	2.0 ul/well	4.0 ul/well	6.0 ul/well	8.0 ul/well	10.0 ul/well	12.0 ul/well		
C											Cells Only	
D												
E	0.25 ul/well	0.50 ul/well	0.75 ul/well	1.0 ul/well	2.0 ul/well	4.0 ul/well	6.0 ul/well	8.0 ul/well	10.0 ul/well	12.0 ul/well		
F												
G												
H												

Virus stock #1 (rows A–C) and Virus stock #2 (rows E–F)

Fig. 3 Example 96-well plate layout for DENV virus titration in U937 + DC-SIGN cells

of 2 h infection medium to each well and mix by pipetting up and down. Incubate 96-well plate at 37 °C in an incubator containing 5 % CO_2 for 1 h. Pellet cells at $500 \times g$ for 5 min, remove supernatant, and gently resuspend cell pellet in fresh 2 h infection medium up to a density of 2.5×10^6 cells/mL. Add 20 µL of resuspended cells into each experimental well (i.e., 5×10^4 cells/ well), and incubate for 2 h in a 37 °C incubator containing 5 % CO_2. Remove plate from the 37 °C incubator, spin at $500 \times g$ for 5 min, carefully aspirate supernatant and resuspend in 200 µL/ well of fresh 2 h infection medium. Aspirate, add 200 µL/well of fresh 24 h infection medium, and place plate in a 37 °C incubator containing 5 % CO_2 for 22 h. For fixing and staining of cells and acquisition and analysis of data, follow instructions on Subheadings 3.3 and 3.4. From the virus titration in Table 2, the volume of virus per well that leads to a percent infection of between 7 and 15 % of total cells will be used in the neutralization assay (*see* **Notes 7–9**).

5. *Titration of infectious DENV virus on Vero cells.* Thaw an aliquot (>950 µL) of DENV virus to be titrated on ice. As shown in Table 2, using the Vero cells 2 h infection medium and sterile dilution tubes, dilute virus sufficient for four wells. Using the volumes (µL/well) shown in Table 1, set up plate layout similar to the one shown in Fig. 3. Transfer 50 µl of diluted virus to the appropriate wells in a U-bottom 96-well plate, and add 50 µL of 2 h Vero infection medium to each well followed by mixing.

Incubate plate in a 37 °C incubator containing 5 % CO_2 for 1 h. Gently aspirate medium from the Vero cell-coated 96-well plate. Then transfer diluted virus from the U-bottom 96-well plate to the Vero cell-coated 96-well plate, while keeping the plate layout the same between the two. Incubate infection for 2 h in a 37 °C incubator containing 5 % CO_2. At 2 h postinfection, aspirate the medium from each well and wash cells with 200 μL of fresh 2 h infection medium. Aspirate the medium and add 200 μL/well of fresh 24 h infection medium. Place plate in a 37 °C incubator containing 5 % CO_2 for 22 h. For fixing and staining of cells and acquisition and analysis of data, follow instructions on Subheadings 3.3 and 3.4. Take caution not to scrape the bottom of the well while aspirating media off of the Vero cell-coated wells. Scraping will lead to loss of cells, large variations in percentage infection across duplicate wells, and very low cell densities, making data acquisition slow during flow cytometry.

6. DENV has two surface exposed glycoproteins on the viral membrane, E and prM. The prM protein is highly conserved among all four DENV serotypes and elicits many cross-reactive antibodies during DENV infections. 2H2 is a mouse monoclonal antibody that binds relatively well to the prM of all four DENV serotypes and therefore used to detect DENV-infected cells regardless of the serotype of infection [14, 15].

7. Do not let the density of U937 + DC-SIGN exceed 10.0×10^5 cells/ml. Cell death rapidly increases and DC-SIGN expression and susceptibility to DENV infection decreases at cell densities over 12.0×10^5 cells/ml.

8. For neutralization experiment with Vero cells, trypsinize cells, count cell density, dilute cells up to 6.25×10^4 cells/mL, and add 200 μL (i.e., 1.25×10^4 Vero cells/well) of diluted cells into each well in a flat-bottom 96-well plate. Each well will reach a cell density of 5.0×10^4 cells in 48 h and be ready for use in the neutralization assay.

9. Example plate layout in Fig. 1 is for a neutralization assay of a human polyclonal serum against all four serotypes of DENV. Greater than eight dilutions (as shown in Fig. 1) are recommended to generate an accurate neutralization curve. Since percentage cell infection data from this assay is analyzed using GraphPad Prism, more than eight dilutions/concentrations are required for the software to accurately fit a sigmoidal dose-response curve from the data.

10. During estimation of antibody dilutions, remember to calculate for 2× of final concentration of serum. To account for dilution and pipetting loss, during all calculations estimate the total number of wells needed as 0.5× the total number of experimental wells.

11. Set up dilution of monoclonal antibodies similarly. An example dilution series of a prior untested monoclonal antibody being tested by the present neutralization assay would consist of 10 threefold dilutions starting at 45 μg/mL.

12. *Neutralization assay using Vero cells:* Dilute a prior titrated virus stock to 50 μL/well using 2 h infection medium. Following a plate layout similar to the one shown in Fig. 1, transfer 50 μL of diluted virus to each appropriate well in a U-bottom 96-well plate. Transfer 50 μL of diluted monoclonal antibody or human serum into the appropriate wells in the plate, mix by pipetting and incubate plate at 37 °C in an incubator containing 5 % CO_2 for 1 h. After the 1 h incubation, aspirate medium from Vero cell-coated wells, transfer the antibody-virus mixture to the aspirated wells, and incubate for 2 h at 37 °C in an incubator containing 5 % CO_2. At 2 h postinfection, aspirate medium from each well and wash cells with 200 μL of fresh 2 h infection media. Aspirate again and add 200 μL/well of fresh 24 h infection medium. Place plate in a 37 °C incubator containing 5 % CO_2 for 22 h. At 24 h postinfection, wash each Vero cell-containing well once with 150 μL of sterile 1× PBS, pH 7.4. Trypsinize cells with 50 μL/well of 0.05 % trypsin for 10 min in a 37 °C incubator containing 5 % CO_2. Stop trypsinization with 100 μL/well of 2 % infection medium, and tap plate gently in a horizontal direction to dislodge cells. Confirm under a light microscope that cells are in suspension, before transferring cells into a U-bottom 96-well plate. Then fix and stain trypsinized Vero cells by following instructions on Subheading 3.3 (**steps 2–9**).

13. If a Guava easyCyte HT flow cytometer (Millipore) instrument is used for data acquisition, it is important to keep the cell concentration of the final cell resuspension under 500 cells/μL. Higher cell densities tend to result in clogging of the capillary flow cell in the Guava easyCyte HT flow cytometer.

14. When the cell infection rates exceed 30 %, the percentage of cells infected is observed to no longer be proportional to the volume of virus added. This observed deviation from linearity may be due to cells either being limiting or differentially susceptible to DENV infection [9].

15. Unfortunately, as compared to the traditional PRNT or FRNT assays, this flow-based neutralization technique is limited by the relatively higher titer of virus required for the assay. Generally, virus stocks greater than 1.0×10^5 ffu/ml are required to obtain percent infections of greater than 7 %, while keeping the volume of virus per well to less than 40 μL.

16. Even virus stocks stored at −80 °C will reduce in infectious titer over time. Therefore, it is recommended that virus stocks be re-tittered once every year before use in this neutralization assay.

Acknowledgments

We thank Anne Broadwater for excellent technical assistance. These studies were supported by a Pediatric Dengue Vaccine Initiative Targeted Research Grant and by National Institutes of Health Grant U54 AI057157 (PI: F. Sparling) from the Southeastern Regional Center of Excellence for Emerging Infections and Biodefense.

References

1. Murphy BR, Whitehead SS (2011) Immune response to dengue virus and prospects for a vaccine. Annu Rev Immunol 29:587–619. doi:10.1146/annurev-immunol-031210-101315

2. Simmons CP, Farrar JJ, Nguyen v V, Wills B (2012) Dengue. N Engl J Med 366(15):1423–1432. doi:10.1056/NEJMra1110265

3. Moi ML, Lim CK, Kotaki A, Takasaki T, Kurane I (2010) Discrepancy in dengue virus neutralizing antibody titers between plaque reduction neutralizing tests with Fcgamma receptor (FcgammaR)-negative and FcgammaR-expressing BHK-21 cells. Clin Vaccine Immunol 17(3):402–407. doi:10.1128/CVI.00396-09

4. Moi ML, Lim CK, Chua KB, Takasaki T, Kurane I (2012) Dengue virus infection-enhancing activity in serum samples with neutralizing activity as determined by using FcgammaR-expressing cells. PLoS Negl Trop Dis 6(2):e1536. doi:10.1371/journal.pntd.0001536

5. Wu RS, Chan KR, Tan HC, Chow A, Allen JC Jr, Ooi EE (2012) Neutralization of dengue virus in the presence of Fc receptor-mediated phagocytosis distinguishes serotype-specific from cross-neutralizing antibodies. Antiviral Res 96(3):340–343. doi:10.1016/j.antiviral.2012.09.018

6. Wahala WM, Kraus AA, Haymore LB, Accavitti-Loper MA, de Silva AM (2009) Dengue virus neutralization by human immune sera: role of envelope protein domain III-reactive antibody. Virology 392(1):103–113. doi:10.1016/j.virol.2009.06.037

7. Wahala WM, Donaldson EF, de Alwis R, Accavitti-Loper MA, Baric RS, de Silva AM (2010) Natural strain variation and antibody neutralization of dengue serotype 3 viruses. PLoS Pathog 6(3):e1000821. doi:10.1371/journal.ppat.1000821

8. Diamond MS, Harris E, Ennis FA (2012) Defeating dengue: a challenge for a vaccine. Nat Med 18(11):1622–1623. doi:10.1038/nm.2997

9. Kraus AA, Messer W, Haymore LB, de Silva AM (2007) Comparison of plaque- and flow cytometry-based methods for measuring dengue virus neutralization. J Clin Microbiol 45(11):3777–3780. doi:10.1128/JCM.00827-07

10. Navarro-Sanchez E, Altmeyer R, Amara A, Schwartz O, Fieschi F, Virelizier JL, Arenzana-Seisdedos F, Despres P (2003) Dendritic-cell-specific ICAM3-grabbing non-integrin is essential for the productive infection of human dendritic cells by mosquito-cell-derived dengue viruses. EMBO Rep 4(7):723–728. doi:10.1038/sj.embor.embor866

11. Sakuntabhai A, Turbpaiboon C, Casademont I, Chuansumrit A, Lowhnoo T, Kajaste-Rudnitski A, Kalayanarooj SM, Tangnararatchakit K, Tangthawornchaikul N, Vasanawathana S, Chaiyaratana W, Yenchitsomanus PT, Suriyaphol P, Avirutnan P, Chokephaibulkit K, Matsuda F, Yoksan S, Jacob Y, Lathrop GM, Malasit P, Despres P, Julier C (2005) A variant in the CD209 promoter is associated with severity of dengue disease. Nat Genet 37(5):507–513. doi:10.1038/ng1550

12. Zompi S, Santich BH, Beatty PR, Harris E (2011) Protection from secondary dengue virus infection in a mouse model reveals the role of serotype cross-reactive B and T cells. J Immunol 188(1):404–416. doi:jimmunol.1102124, [pii] 10.4049/jimmunol.1102124

13. Yokoyama WM, Christensen M, Santos GD, Miller D (2006) Production of monoclonal antibodies. Curr Protoc Immunol Chapter 2:Unit 2 5. doi:10.1002/0471142735.im0205s74

14. Henchal EA, Gentry MK, McCown JM, Brandt WE (1982) Dengue virus-specific and flavivirus group determinants identified with monoclonal antibodies by indirect immunofluorescence. Am J Trop Med Hyg 31(4):830–836

15. Kaufman BM, Summers PL, Dubois DR, Cohen WH, Gentry MK, Timchak RL, Burke DS, Eckels KH (1989) Monoclonal antibodies for dengue virus prM glycoprotein protect mice against lethal dengue infection. Am J Trop Med Hyg 41(5):576–580

Chapter 4

Dengue Virus Purification and Sample Preparation for Cryo-Electron Microscopy

Joanne L. Tan and Shee Mei Lok

Abstract

Cryo-electron microscopy (cryo-EM) is a valuable tool used to study the structures of icosahedral viruses without having to resort to crystallization. During the last few decades, significant progress has been made where virus structures previously resolved only to low resolution have now breached the sub-nanometer threshold. Critical to such excellent results are the acquisition of highly purified virus samples and well-frozen samples in vitreous ice. With the virus particles locked in their native conformations, cryo-EM together with single-particle analysis can then be deployed to study the structures of the viruses in their fully hydrated states.

Key words Cryo-EM, Dengue virus, Virus purification, Potassium tartrate gradient centrifugation, Vitrobot

1 Introduction

Cryo-electron microscopy is utilized to study virus structures. The shape of the virus and the arrangement of its capsomeres can be visualized at low resolution (20–40 Å), while α-helices and β-sheets become apparent at sub-nanometer resolutions (<10 Å). C-α backbones as well as some side chains also start to become more evident when viruses are resolved to near-atomic resolution (<4.5 Å) [1]. The quality of the images is, however, dependent on the purity and quality of an adequately concentrated sample.

Structures of the immature and mature forms of dengue virus have been studied extensively. The immature dengue virus contains three structural proteins: capsid (C), precursor membrane (prM), and envelope (E) proteins [2]. The capsid protein packages the viral genomic RNA to form the nucleocapsid, while the prM and E proteins form heterodimers that organize into 60 trimeric spikes [3, 4]. These spikes project outwards from the lipid envelope of the virion, and in these trimeric structures, the pr portion of the prM proteins cap the E protein fusion loops. During the maturation

Radhakrishnan Padmanabhan and Subhash G. Vasudevan (eds.), *Dengue: Methods and Protocols*, Methods in Molecular Biology, vol. 1138, DOI 10.1007/978-1-4939-0348-1_4, © Springer Science+Business Media, LLC 2014

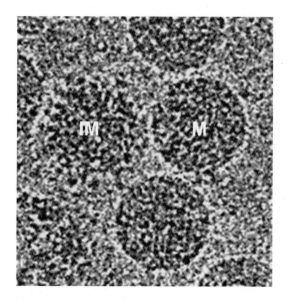

Fig. 1 Cryo-electron micrograph of a dengue 1 virus preparation showing an immature (IM) and a mature (M) virion particle

process, the immature virus particle passes through the trans-Golgi network (TGN) from the endoplasmic reticulum. The acidic environment in the TGN initiates the rearrangement of the spiky prM/E heterodimers into 90 E dimers that cover the virus surface in a herringbone formation [5]. At this stage, a host protease, furin, cleaves the pr peptide from the M protein [6]. However, the pr peptide still remains bound to the fusion loop of the E protein until the virus is released into the extracellular environment, forming the infectious mature dengue virus particle [7]. The function of the pr molecule is thought to prevent the newly synthesized virus particles from fusing back into the cell while moving through the acidic TGN compartments.

The preparations of mature and immature virus vary slightly. Growth of immature virus involves the addition of ammonium chloride (NH_4Cl) to the growth media during cell culture. This increases the pH of the TGN, preventing the rearrangement of the trimers and cleavage of viral prM molecule by furin protease. The immature virus particle thus remains spiky, while the mature virion appears smooth (Fig. 1).

Potassium tartrate gradients have been successfully used to purify dengue for cryo-EM as well as atomic force microscopy [8, 9]. This gradient, of low cost and toxicity, serves to separate the virus from cytoplasmic debris.

To freeze the virus on cryo-EM grids, a thin layer of freshly purified sample is applied to the grid and then plunge-frozen into liquid ethane. This procedure minimizes the formation of ice crystals that may potentially damage the structure of the particles and allows the retention of the sample close to its original conformational state [10, 11]. Furthermore, since electron beams penetrate thin ice layers more easily, better contrast cryo-EM images can be obtained [1].

This chapter describes in detail the large-scale preparation of dengue viruses, its purification, and also the procedure to freeze samples on cryo-EM grids.

2 Materials

2.1 Virus Culture

1. RPMI 1640 supplemented with 10 % fetal bovine serum (FBS), 25 mM HEPES, and 2.05 mM L-glutamine, pH 7.0.

2. Minimal essential medium (MEM) without HEPES, pH 7.0, supplemented with 10 % (v/v) FBS.

3. T175 tissue culture flasks.

4. Two- and ten-cell stacks (Thermo Scientific).

2.2 Virus Purification and Concentration

1. NTE buffer (1,000 mL): 120 mM NaCl, 12 mM Tris–Cl, 1 mM EDTA, pH 8.0.

2. 40 % (w/v) PEG 8000 (500 mL): 200 g PEG 8000, topped up with NTE buffer.

3. 24 % (w/v) sucrose (20 mL): 4.8 g sucrose topped up with NTE buffer.

4. 40 % (w/v) potassium tartrate-30 % glycerol (100 mL): 40 g potassium sodium tartrate, 30 mL glycerol, topped up with NTE buffer.

5. Light torch.

6. 18 G needle.

7. 5 mL syringe.

8. Gradient maker (Hoefer).

9. 100 kDa MWCO centrifugal filter (Millipore).

10. Centrifuges and rotors: benchtop microfuge and centrifuge, low speed centrifuge with a ≥1 L handling capacity, ultra-centrifuge.

2.3 Virus Quantitation

1. BSA standards: 0.25, 0.5, 1, and 2 mg/mL concentrations.

2. SDS-PAGE gel (4–15 %) (BioRad).

3. Bio-Safe Coomassie Blue stain (BioRad).

2.4 Cryo-EM Sample Preparation and Grid Freezing

1. Plasma glow discharger (Quorum Technologies).

2. Vitrobot Mark IV (FEI).

3. Vitrobot cup (FEI).

   4. Ethane cup (FEI).

   5. Metal spindle (thermal conductor) (FEI).

   6. Vitrobot tweezers (FEI).

   7. Long tweezers.

   8. Liquid cryogens: nitrogen and ethane gas cylinders with attached flow regulators.

   9. Lacey carbon grids (Ted Pella Inc).

   10. Cryo-transfer dewar (SPI Supplies).

   11. Filter paper (Ted Pella Inc).

   12. Kimwipes.

   13. Hairdryer.

   14. Yellow tips.

3 Methods

The entire process of virus purification spans several days. To prevent the degradation of the virus, samples should be frozen on cryo-EM grids within 72 h after purification. The following methods have been adapted from previously published procedures [5, 7].

3.1 Virus Culture

Two methods are listed below for the large-scale production of immature and mature virus particles. Incubations are carried out at 29 °C with 5 % carbon dioxide. All centrifugation steps are carried out at 4 °C.

   1. For mature dengue virus production, C6/36 *Aedes albopictus* cells were cultured in RPMI media supplemented with 10 % (v/v) FBS in two- and ten-cell stacks to around 90 % confluence (*see* **Note 1**). The total volume of media should be 1.5 L. For immature dengue virus production, C6/36 cells were cultured in MEM (without HEPES) with 10 % (v/v) FBS to approximately 90 % confluence [7] (*see* **Note 2**).

   2. Ensure that the media in the cell stacks are level (*see* **Note 3**). Incubate the stacks for 2 days or until the cells have reached 70–90 % confluence.

   3. Defrost the virus stock (*see* **Note 4**) and dilute it to 300 mL with either RPMI or MEM medium supplemented with 2 % (v/v) FBS. Discard the media from the cell stacks before infection. Aliquot proportional amounts of the virus stock into the cell stacks and incubate them for 1.5 h. Rock the stacks from side to side every 15 min.

   4. The inoculum is removed from both cell stacks and replaced with a total of 1.5 L of fresh media containing 2 % (v/v) FBS.

For immature virus production, replace the medium with 1.5 L of MEM containing 2 % (v/v) FBS and 20 mM filter-sterilized NH_4Cl (*see* **Note 5**).

5. Harvest at 48–60 h or 4 days post-infection for immature and mature viruses, respectively. Centrifuge at $9,000 \times g$ for 45 min.

3.2 Virus Purification and Concentration

1. The supernatant is transferred to a large flask and swirled very gently with an adequate amount of 40 % (w/v) PEG 8000 to achieve a final concentration of 8 % (w/v) PEG (*see* **Note 6**). Leave the flask at 4 °C overnight.

2. Centrifuge the suspension at $14,000 \times g$ for 1 h.

3. The pellet is gently resuspended in 10 mL NTE buffer using a pipette. Rinse the centrifuge bottle with a further 5 mL of NTE buffer after the initial 10 mL has been moved to another tube (*see* **Note 7**).

4. Purify the virus suspension through a sucrose cushion by transferring it into an ultracentrifuge tube. Gently add 5 mL of 24 % (w/v) sucrose to the bottom of the suspension using a Pasteur pipette. Do not mix the two solutions. Centrifuge at $105,000 \times g$ for 1.5 h.

5. Invert the tube to remove the sucrose and wick off the excess solution. Add 500 μL NTE buffer to cover the pellet. This is kept overnight at 4 °C to allow the pellet to soften (*see* **Note 8**).

6. Gently resuspend the virus and transfer the supernatant into a new 1.5 mL Eppendorf tube. Centrifuge on a benchtop centrifuge at $16,000 \times g$ for 2 min. Transfer the supernatant into a new Eppendorf tube and repeat the centrifugation process.

7. Prepare a potassium tartrate-glycerol gradient ranging from 10 to 30 % in an ultra-clear tube suitable for the specific ultracentrifuge rotor. If the Beckman Ti70 rotor is used, prepare 10 mL of the gradient in a 12 mL tube. The stopcocks of the gradient maker should be opened completely (*see* **Note 9**).

8. Carefully layer the virus suspension over the top of the potassium tartrate-glycerol gradient. Ultracentrifuge the sample at $175,000 \times g$ for 2 h. The deceleration of the rotor should be set to zero.

9. To visualize the virus band, the tube is clamped such that the bottom of the tube is suspended above an inverted torch discharging a narrow beam (Fig. 2) (*see* **Note 10**).

10. The virus band is harvested using an 18 G needle connected to a 5 mL syringe. With the bevel facing up, pierce the tube about 1–2 mm below the virus band (*see* **Note 11**). Recover the entire band containing the virus fraction while pulling slowly on the plunger.

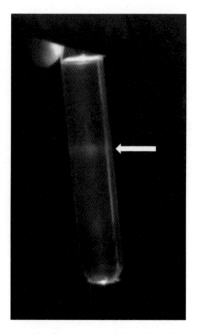

Fig. 2 The virus band (*arrowed*) can be visualized by shining a torch through the bottom of the tube, in a darkened room

11. Transfer the virus fraction to a 100 kDa MWCO centrifugal filter and conduct a buffer exchange with NTE buffer to remove the potassium tartrate. The virus is concentrated to 100–150 μL by centrifuging at $16,000 \times g$ at 4 °C (*see* **Note 12**). The final concentration of the potassium tartrate should be below 0.01 %. Use enough NTE buffer to bring the potassium tartrate down to this concentration.

3.3 Virus Quantitation

1. Run 5 μL of the reduced purified virus sample on an SDS-PAGE gel alongside twofold dilutions of BSA starting from 2 mg/mL (until 0.25 mg/mL) and a molecular weight ladder (*see* **Note 13**). The gel is stained with Coomassie Blue for 1 h at room temperature and then destained with water (Fig. 3). A 2.5 μL sample from a virus concentration of around 0.75–1 mg/mL should produce an adequately populated grid.

3.4 Cryo-EM Sample Preparation and Grid Freezing

1. Glow discharge the required number of grids (with the carbon films facing upwards) in the plasma glow discharger at 5 mA for 50 s.

2. Assemble the Vitrobot cup by fitting the ethane cup into the inner receptacle and the grid holder into the outer ring followed by the metal spindle. The metal spindle acts as a thermal connector to conduct heat away from the ethane cup (Fig. 4).

3. Up to four grid boxes can be placed into the grid box holders (*see* **Note 14**).

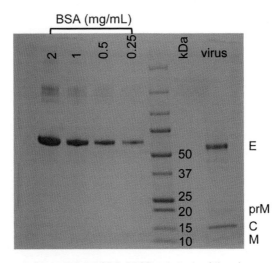

Fig. 3 Coomassie Blue-stained SDS-PAGE gel. 5 μL of the virus sample was run under reducing conditions against a BSA standard. The separated viral proteins are indicated: envelope (E), pre-membrane (prM), core (C), and membrane (M)

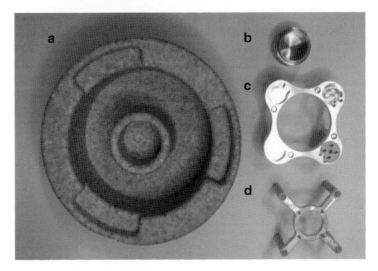

Fig. 4 Components of the Vitrobot cup (**a**). The ethane cup (**b**) should sit in the inner receptacle of the Vitrobot cup. The grid holder (**c**), holding one grid box (*lower right*), is placed over the ethane cup followed by the metal spindle (**d**)

4. Fill the humidifier reservoir in the Vitrobot with distilled water (Fig. 5).

5. Turn on the Vitrobot and program the conditions for the Vitrobot control chamber using the touch screen. At the "Console" page, set the temperature to the required value and the humidity to 100 %. At the "Options" page, select "Use foot pedal" and "Skip grid transfer." The former option allows for a hands-free control of the raising and lowering of the stage

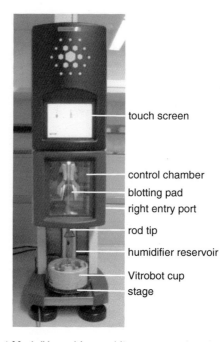

touch screen

control chamber
blotting pad
right entry port
rod tip
humidifier reservoir
Vitrobot cup
stage

Fig. 5 The Vitrobot Mark IV machine and its component parts

and blotting and plunge-freezing of the grid, while the latter option keeps the plunger holding the grid immersed in ethane while the stage is being lowered. We use the following parameters in our grid preparations: Blot force = 1; Blot total = 1; Blot time is usually between 1 and 3 s (the optimization of blotting time is required as it varies between samples) (*see* **Note 15**).

6. Secure the filter papers onto the blotting pads with the clip rings and close the door to the Vitrobot.

7. Fill both the ethane cup and the outer nitrogen ring of the Vitrobot cup with liquid nitrogen to hasten the cooling process (*see* **Note 16**).

8. After the liquid nitrogen has completely evaporated from the central ethane cup and the ethane cup is cooled down to liquid nitrogen temperature, blow ethane gas at a moderate speed into the ethane cup using a yellow tip held against the side of the cup. The gas will liquefy. Fill the cup to the brim and wait for the liquefied ethane to appear cloudy indicating that the temperature has reached −170 °C.

9. Remove the metal spindle carefully (*see* **Note 17**) to prevent further cooling.

10. Holding the Vitrobot tweezers such that the spring faces the operator, pick up a grid such that the carbon side faces the right. Slide the black clamp ring on the tweezers down so that the grid is gripped tightly (*see* **Note 18**).

11. Step on the foot pedal to lower the rod tip from the control chamber so that the Vitrobot tweezers with the grid can be attached to the rod tip. Retract the tweezers into the control chamber by stepping on the foot pedal again and place the Vitrobot cup onto the stage (*see* **Note 19**).

12. Step on the foot pedal to raise the stage with Vitrobot cup. Apply 2.5 μL of sample to the entire surface of the carbon side of the grid through the right entry port found on the side of the Vitrobot.

13. Step on the foot pedal to activate the blotting of excess liquid on the grid using the blot force and time set previously. The blotted grid will then be plunge-frozen into liquid ethane.

14. Slide the tweezers out of the rod tip while keeping the grid immersed under liquid ethane. Very swiftly, transfer the grid into the liquid nitrogen in the outer ring. While applying pressure on the tweezers, raise the clamp ring and insert the grid into the preferred position in the grid holder.

15. The grids can then be used for imaging or stored in liquid nitrogen (*see* **Note 20**).

4 Notes

1. We find that growing the cells in 13–14 T175 flasks is sufficient for seeding both cell stacks. Confluency is reached approximately 2 days later but this is dependent on the rate of growth of the cells. A two-cell stack is needed as this can be placed under a light microscope to check the cell confluency. This will give an indication of how the cells are growing in the ten-cell stack.

2. A different medium, MEM, was used for the production of immature virus as we did not have good success producing immature dengue virus in RPMI supplemented with HEPES. Note that the C6/36 cells need to be adapted to MEM before use if they have been previously cultured in RPMI.

3. The growing cells will tend to cluster at the sides of the chambers if the stacks are incubated on an uneven surface. It may be easier to place the cell stacks on a level bench surface for 20–30 min prior to placing them into the incubator to let most of the cells settle rather than attempting to balance it while it is on a shelf in the incubator.

4. The virus stock for infecting both cell stacks can be prepared in advance and frozen at –80 °C. Stocks for mature virus preparations are harvested from three T175 flasks (4 days post-infection), each containing 25 mL of media. The clarified

supernatant is mixed with enough FBS to give a final concentration of approximately 25 % (v/v) FBS. Stocks for producing immature virus preparations are harvested from 10 T175 flasks 4 days post-infection. More viruses are needed to infect cells to generate enough immature viruses as less replication cycles occur during the shortened incubation period. Each stock is sufficient to infect two- and ten-cell stacks.

5. The presence of NH_4Cl elevates the pH in the Golgi and prevents furin from cleaving prM. This keeps the virion particles produced in the immature state.

6. PEG 8000 does not readily dissolve in NTE. Heating the mixture up to around 85 °C on a magnetic stirrer will help the PEG dissolve more quickly.

7. The pellet should be visible at the bottom of the centrifuge bottle. If you are using a fixed angle rotor, some of the PEG-precipitated virus will be present down the side of the centrifuge bottle. This may not be very visible, but do remember to rinse that portion of the bottle as well to maximize the virus yield.

8. Occasionally, the pellet may be very translucent and hard to visualize. In any case, the sucrose must still be removed as soon as possible.

9. The stir bar in the gradient maker should be adjusted to spin at a moderate pace at a medium setting so that will not cause the formation of air bubbles that may compromise the linearity of the gradient.

10. It will be easier to visualize the virus band if all the lights in the room have been turned off and the torch held against the bottom of the centrifuge tube.

11. The virus will be visible as a thin, concentrated band sandwiched between two layers of slightly cloudy cellular material. To control the rate of flow of the virus into the syringe, the syringe must first be connected to the needle with the plunger pushed all the way down. Otherwise, when the centrifuge tube is pierced, the liquid within the tube will start to flow into the syringe immediately.

12. Centrifuge the sample for an initial period of 5–10 min. Use this time as a gauge to set the amount of time needed for complete buffer exchange. Depending on the viscosity of the sample, the time needed for buffer exchange will vary depending on the quality of the harvested virus.

13. The BSA standard is relatively similar in size to E protein and provides a rough estimate of the amount of E protein that is present in the virus preparation. Load equivalent volumes of BSA and virus. If the concentration of E protein is below 0.25 mg/mL, it should preferably be concentrated further

prior to freezing onto grids. We generally find that concentrations of 0.75–1 mg/mL will give a good density of virus particles for cryo-EM. The Coomassie-stained gel will also allow you to check if the sample has been contaminated with immature virus. This will be evident by the presence of large amounts of prM (19–23 kDa) protein on the gel (Fig. 3). Some dengue strains may produce larger amounts of immature virions than others. The presence of large amounts of spiky immature virus particles will be detrimental if the downstream process involves complexing with antibodies since both populations of immature virus and mature virus complexed with antibody will look spiky under the microscope.

14. The grid holder is not numbered. To differentiate between the different grid boxes, mark the rim of the Vitrobot cup with a number so that each digit corresponds to the location of each grid box. This preempts a mix up if the grids are to be transferred for storage in liquid nitrogen.

15. The blot time can be varied between 1 and 4 s. A longer blot time is sometimes required for virus samples that contain a larger amount of cellular debris.

16. The level of the liquid nitrogen must be monitored constantly and replenished should the level fall. The outer receptacle should be at least ¾ filled with liquid nitrogen so that the grid boxes always remain completely submerged.

17. You may find it easier to remove the metal spindle if you press the flat blades of the long tweezers downwards against the horizontal arms of the metal spindle. Be careful not to sweep any frost found on the arms of the metal spindle into the ethane cup.

18. For operators who are left handed, position the grid with the carbon side facing the left-hand side. In this case, application of the virus sample will be made through the side port on the left of the machine.

19. You may want to conduct a dry run to ensure that the apparatus is functioning correctly. Attach the Vitrobot tweezers (without a grid) to the rod, place the Vitrobot cup on the stage, and perform a mock plunge-freeze into the ethane cup.

20. It is necessary to screen one grid from each sample before attempting to freeze the rest of the sample. Adjustments to the blot time can be made if the ice thickness is not optimal. A very concentrated virus sample may sometimes lead to uneven ice formation on the grid and may cause particles to overlap, so a slight dilution of the sample may help and cause the ice quality to improve. However, this can lead to a smaller concentration of particles on the grid. But generally, 3–4 grids for each sample should provide sufficient numbers of particles for reconstruction.

References

1. Chang J, Liu X, Rochat RH et al (2012) Reconstructing virus structures from nanometer to near-atomic resolutions with cryo-electron microscopy and tomography. Adv Exp Med Biol 726:49–90

2. Lindenbach BD, Rice CM (2001) Flaviviridae: the viruses and their replication. In: Knipe DM, Howley PM (eds) Fields virology, vol II, 4th edn. Lippincott Williams and Wilkins, Philadelphia, PA, pp 991–1042

3. Zhang Y, Corver J, Chipman PR et al (2003) Structures of immature flavivirus particles. EMBO J 22(11):2604–2613

4. Zhang Y, Kaufmann B, Chipman PR et al (2007) Structure of immature West Nile virus. J Virol 81(11):6141–6145

5. Kuhn RJ, Zhang W, Rossmann MG et al (2002) Structure of dengue virus: implications for flavivirus organization, maturation and fusion. Cell 108:717–725

6. Stadler K, Allison SL, Schalich J et al (1997) Proteolytic activation of tick-borne encephalitis virus by furin. J Virol 71(11): 8475–8481

7. Yu IM, Zhang W, Holdaway HA et al (2008) Structure of the immature dengue virus at low pH primes proteolytic maturation. Science 319(5871):1834–1837

8. Ferreira GP, Trindade GS, Vilela JMC et al (2008) Climbing the steps of viral atomic force microscopy: visualisation of dengue virus particles. J Microsc 231(Pt1): 180–185

9. Junjhon J, Edwards TJ, Utaipat U et al (2010) Influence of pr-M cleavage on the heterogeneity of extracellular dengue virus particles. J Virol 84(16):8353–8358

10. Iancu CV, Tivol WF, Schooler JB et al (2006) Electron cryotomography sample preparation using the Vitrobot. Nat Protoc 1(6): 2813–2891

11. Dobro MJ, Melanson LA, Jensen GJ et al (2010) Plunge freezing for electron cryomicroscopy. Methods Enzymol 481:63–82

# Chapter 5

Development of a Multiplex Bead-Based Assay to Monitor Dengue Virus Seroconversion

Kaiting Ng and John E. Connolly

Abstract

Dengue virus (DENV) envelope protein is responsible for viral attachment to host cells and as such is a target of neutralizing antibody responses. However, the presence of envelope-specific antibodies against a given serotype may contribute to enhanced disease during secondary infection with another serotype. There is a need therefore for a standardized, high-throughput low-volume assay which permits the simultaneous screening of reactivity to multiple DENV serotypes. Here, we describe a method of identifying DENV serotype-specific response in exposed individuals using a multiplexed bead-based immunoassay. The ED3 domain of a specific DENV serotype is cloned into pQEAM containing 6xHIS, TEV protease, and AviTag biotinylation sites. Biotinylated ED3 proteins are expressed in *E. coli* CVB101 and purified by sequential column fractionation followed by coupling onto fluorescent avidin-coated microspheres. Methods for determining the optimum amount of biotinylated ED3 protein coupled onto the microsphere are described. The assay demonstrates both a high degree of sensitivity and specificity using well-characterized patient plasma samples. The nature of the assay permits further development to include a variety of DENV serotypes and regionally important sub-serotypes.

Key words Dengue, ED3, Bead-based immunoassay, Biotinylation

1 Introduction

Dengue virus (DENV) is a positive strand RNA flavivirus, with four co-circulating antigenically different serotypes designated, DENV1-4. Viral infection is a serious cause of morbidity and mortality with nearly half of the worldwide population living in risk areas [1]. Dengue hemorrhagic fever is characterized by severe plasma leakage, hemorrhage, and organ impairment and thought to be due to enhancement of DENV infection by the cross-reactive antibody produced in previous infections [2, 3]. This phenomenon of antibody-dependent enhancement (ADE) is highly serotype specific in that secondary infection with a different serotype is associated with severe disease [4].

Dengue virus envelope protein (E) mediates host cell attachment [5, 6]. As such, immune responses against E protein stimulate

Radhakrishnan Padmanabhan and Subhash G. Vasudevan (eds.), *Dengue: Methods and Protocols*, Methods in Molecular Biology, vol. 1138, DOI 10.1007/978-1-4939-0348-1_5, © Springer Science+Business Media, LLC 2014

production of neutralizing antibodies which can inhibit virus attachment and prevent infection [7]. Structurally, the ectodomain of E protein consists of three domains designated as EDI, EDII, and EDIII. EDII contains the fusion peptide and the receptor-binding region is located in EDIII [7, 8]. During attachment, E protein is presented as aggregate of 90 homodimers on the virus surface [9, 10]. Neutralizing epitopes within EDIII have been shown to be particularly effective at limiting infection [11, 12].

This bead-based DENV immunoassay describes a method for identifying DENV serotype-specific responses in exposed individuals. The multiplex nature of the system permits simultaneous screening of all four serotypes in a single low-volume assay. To express biotin-conjugated ED3 of SgDENV1-4 proteins for coupling onto microsphere (coated with avidin), each DENV1-4 ED3 serotype is inserted into pQEAM, a vector containing His Tag, MBP tag, 5× alanine linker, AviTag (a biotinylation tag), and TEV cleavage site for releasing ED3 from fusion protein. Each DENV1-4 ED3 is then mutated to Singapore strain (*see* Fig. 1), before expressing in *E. coli* CVB cells as a biotinylated ED3 protein as shown in Fig. 2. Each biotinylated ED3 of SgDENV1-4 protein is then coupled onto different bead regions. For bead-based DENV immunoassay, first, determine the amount (mM) of biotinylated ED3 protein coupled onto microsphere; second, test the sensitivity of the assay; and, lastly, test dengue specific serotype plasma cross-reactive to each of the SgDENV1-4 ED3-coupled beads and/or mixed SgDENV1-4 ED3-coupled beads. The nature of the Luminex platform permits further development of this assay to include a variety of DENV sub-serotypes.

2 Materials

2.1 Plasmid Construction

2.1.1 Construction of pQEAM

1. pQE-2 bacterial expression vector (Qiagen).
2. FastDigest NdeI and NotI restriction enzymes (Fermentas, or equivalent).
3. Complementary PAGE-purified oligonucleotides (AviTag-F and AviTag-R) containing AviTag, 5× alanine linker, and NdeI and NotI restriction sites (AviTag-F: 5′CCCATATGCAAGG CCTGAACGACATCTTCGAGGCACAGAAAATCGAA TGGCACGAAGCAGCAGCGGCCGCAAA, AviTag-R: 5′TT TGCGGCCGCTGCTGCTTCGTGCCATTCGATTT TCTGTGCCTCGAAGATGTCGTTCAGGCCTTG CATATGGG).
4. T4 DNA ligase (New England Biolabs or equivalent).
5. Calf intestinal alkaline phosphatase (New England Biolabs or equivalent).

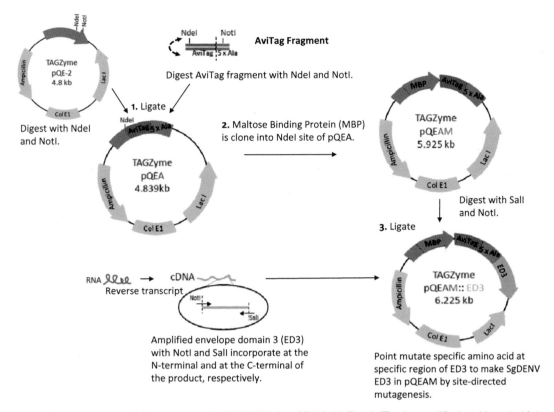

Fig. 1 Cloning strategy of Singapore strain DENV ED3 in pQEAM. (*1*) The AviTag is amplified and inserted into pQE-2 to yield pQEA. (*2*) The MBP containing TEV recognition site at its C-terminal is cloned into pQEA to yield pQEAM. The whole cloning steps are described in Subheading 2.1.1. RNAs from each DENV serotype are reverse transcribed into cDNA before amplifying ED3 of DENV. (*3*) Amplified ED3 PCR products are inserted into pQEAM to yield pQEAM::DENV ED3 before mutating specific amino acids in DENV ED3 to Singapore strain as described in Subheading 3.2

6. Forward and reverse primers (AviPCR-for and AviPCR-rev) direct towards AviTag (AviPCR-for: 5′CCATCACCATATGCAAGGCCTGAAC, AviPCR-rev: 5′GTAATTGCGGCCGCTGCTGCTT).

7. Reverse primer (pQE-R) for sequencing (pQE-R: 5′GTTCTGAGGTCATTACTGG).

8. Pfu DNA polymerase (native; Fermentas).

9. QIAquick PCR purification kit (Qiagen).

10. Plasmid prep kit (Qiagen or equivalent).

11. 0.2 mL PCR tubes.

12. PCR machine (Bio-Rad C1000 Touch™ thermal cycler, or equivalent).

13. Subcloning efficiency DH5α *Escherichia coli* (Invitrogen).

14. Nanodrop 1000 Spectrophotometer (Thermo Fisher Scientific) or any spectrophotometer.

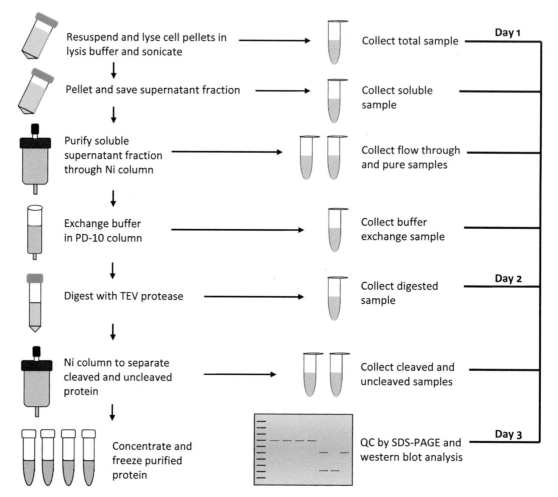

Fig. 2 Flowchart for purification of biotinylated SgDENV ED3 protein

2.1.2 Construction of
pQEAM::DENV1-4 ED3

1. RNA of dengue serotypes 1, 3, and 4.

2. Plasmid encoding DENV2 (TSV01).

3. TaqMan reverse transcription reagents (Applied Biosystems).

4. QIAquick PCR purification kit (Qiagen) or an equivalent PCR purification kit.

5. FastDigest NotI and SalI restriction enzymes (Fermentas, or equivalent).

6. Calf intestinal alkaline phosphatase (New England Biolabs or equivalent).

7. 0.2 mL PCR tubes.

8. Primers for amplifying DENV ED3 with NotI and SalI at 5′ end and 3′ end of DENV ED3, respectively:

   (a) Primers pair for targeting DENV1 ED3 (WPD1For1; forward: 5′AGCAGCGGCCGCAAAAGGGATGTCATA,

WPD1Rev1;reverse:5′GCAGTCGACTTATCCCTTCTT
GAACCAGCTTAG).

(b) Primers pair for targeting DENV2 ED3 (Den2ED3-F;
forward: 5′CCGCGGCCGCAAAAGGAATGTCATATT
CTATGTGTACAG, Den2ED3-R1; reverse: 5′GCAGT
CGACTTATCCTTTCTTAAACCAGCTGAGC).

(c) Primers pair for targeting DENV3 ED3 (VND3For1;
forward:5′AGCAGCGGCCGCAAAGGGGATGAGCTA,
VND3Rev1;reverse:5′GCAGTCGACTTATCCCTTCTT
GTACCAGTTAAT).

(d) Primers pair for targeting DENV4 ED3 (MsiaD4For1;
forward: 5′AGCAGCGGCCGCAAAGGGAATGTCATA,
MsiaD4Rev1; reverse: 5′GCAGTCGACTTACCCTTTC
CTGAACCAATGGAGT).

9. T4 DNA ligase (New England Biolabs or equivalent).

10. Reverse primer (pQE-R) and forward (AviPCR-for) for sequencing (pQE-R: 5′GTTCTGAGGTCATTACTGG, AviPCR-for:
5′CCATCACCATATGCAAGGCCTGAAC).

2.1.3 Construction of pQEAM::SgDENV3 ED3 and pQEAM::SgDENV4 ED3

1. QuikChange II site-directed mutagenesis kit (Stratagene).

2. pQEAM::DENV3 ED3 and pQEAM::DENV4 ED3 plasmids
encoding DENV3 ED3 and DENV4 ED3, respectively.

3. Two sets of mutagenic primers (125 ng) containing desired
mutation to produce envelope domain III (ED3) of Singapore
strain dengue serotype 3 (SgDENV3):

(a) First set of primers target ED3 codon mutation at L9S and
T11A (SgD3Ser-AlaF; forward: 5′GCTATGCAATGTGC
TCAAATGCCTTTGTGTTGAAG, SgD3Ser-AlaR; reverse:
5′CTTCAACACAAAGGCATTTGAGCACATTG
CATAGC).

(b) Second set of primers ED3 codon mutation at V85I
(SgD3Iso-F;forward:5′CCTTTTGGGGAAAGTAACATA
ATAATTGGAATTGGAGACAAAGC, SgD3Iso-R; reverse:
5′GCTTTGTCTCCAATTCCAATTATTATGTTACTT
TCCCCAAAAGC).

4. Forward and reverse mutagenic primers (125 ng) containing
desired mutation to produce envelope domain III (ED3) of
Singapore strain dengue serotype 4 (SgDENV4);

(a) Primers target ED3 codon mutation at T6A (SgD4Ala-F;
forward: 5′GGGAATGTCATACGCGATGTGTTCAGG
AAAG, SgD4Ala-R; reverse: 5′CTTTCCTGAACACAT
CGCGTATGACATTCCC).

5. 0.2 mL PCR tubes.

Table 1
List of primers used for priming the whole pQEAM in sequencing

Primer name	Sequence (5′ → 3′)	Length (bases)	Melting temperature (T_m) (°C)	Primer location to pQEAM
pQE-F	CGG ATA ACA ATT TCA CAC AG	20	56.3	1167-1186
pQE-R	GTT CTG AGG TCA TTA CTG G	19	58	2496-2515
pQEAM136-155	GGC TTA CCA TCT GGC CCC AG	20	66.5	136-155
pQEAM777-758	TGC ACG AGT GGG TTA CAT CG	20	62.4	777-758
pQEAM1800-1781	CGC ACC CGC GTT GTC CAC AC	20	68.6	1800-1781
pQEAM2181-2200	GCG TAT CGC GGC AAC GAT GG	20	66.5	2181-2200
pQEAM3174-3193	CGC TGG CGA TTC AGG TTC AT	20	62.4	3174-3193
pQEAM3560-3541	GCT GGC ACG ACA GGT TTC CC	20	66.5	3560-3541
pQEAM4251-4270	CGT TGC GCG AGA AGA TTG TG	20	62.4	4251-4270
pQEAM4646-4627	GGC ATG ATA GCG CCC GGA AG	20	66.5	4646-4627
pQEAM5301-5321	GGA AGC TCC CTC GTG CGC TC	20	68.6	5301-5321
pQEAM5679-5660	GGT AAC TGG CTT CAG CAG AG	20	62.4	5679-5660

6. PCR machine (Bio-Rad C1000 Touch™ thermal cycler, or equivalent).

7. Primers used for sequencing (*see* Table 1).

2.2 Preparation of Biotinylated SgDen1-4ED3 Recombinant Proteins from Escherichia coli CVB101 Competent Cells

1. *Escherichia coli* CVB101 competent cells (Avidity). Contains pBirAcm, an IPTG-inducible plasmid containing the BirA gene engineered into pACYC184. pBirAcm is maintained with chloramphenicol at 10 μg/mL.

2. pQEAM::SgDENV1 ED3 (DNA 2.0).

3. pQEAM::SgDENV3 ED3 and pQEAM::SgDENV4 ED3.

4. pQEAM::DENV2 ED3 (equivalent to Singapore strain DENV2 ED3; GenBank accession: EU448427).

5. TYH medium: 20 g/L tryptone, 10 g/L yeast extract, 11 g/L HEPES, 5 g/L NaCl, 1 g/L MgSO$_4$, adjusted to pH 7.2–7.4 with KOH; sterilized by autoclave.

6. Antibiotic stock solutions: 25 mg/mL chloramphenicol and 100 mg/mL ampicillin.

7. A sterile 20 % glucose solution.

8. 5 mM Biotin solution: 80 µL of 1 M NaOH to 12 mg of d-biotin; then top up with water to 10 mL; filter sterilize using a 0.2 µm filter.

9. Isopropyl-beta-D-thiogalactoside (IPTG): 100 mM stock solution.

10. Bacterial shaker incubator set at 37 °C.

11. Centrifuge (Sorvall Evolution RC Superspeed centrifuge, or equivalent).

12. Luria-Bertani (LB) medium: 10 g/L bacto-tryptone, 5 g/L bacto-yeast extract, 5 g/L NaCl; sterilized by autoclave.

13. LB agar (LB medium containing 1.5 % agar): sterilize by autoclave, cool to 60 °C, and add ampicillin to 100 µg/mL; pour 30 mL/Petri dish, and allow to set at room temperature for 1 h.

14. Bacterial incubator set at 37 °C.

2.3 Purification of Biotinylated SgDen1-4ED3 Recombinant Proteins

1. Lysis buffer: 20 mM NaH$_2$PO$_4$, 500 mM NaCl, 20 mM imidazole, adjusted to pH 7.4 with NaOH, 1:400 protease inhibitor cocktail set III EDTA free (Calbiochem; added fresh) and 24 U/mL benzonase nuclease (Novagen; added fresh).

2. Sonicator.

3. 0.2 µm syringe filter, 30 mL syringes.

4. Hi Trap HP column (GE Healthcare Life Sciences).

5. FPLC machine (GE Healthcare Life Sciences Äkta, or equivalent).

6. Binding buffer: 20 mM NaH$_2$PO$_4$, 500 mM NaCl, 20 mM imidazole and adjusted to pH 7.4 with NaOH.

7. Wash buffer: 20 mM NaH$_2$PO$_4$, 500 mM NaCl, 150 mM imidazole and adjusted to pH 7.4 with NaOH.

8. Elution buffer: 20 mM NaH$_2$PO$_4$, 500 mM NaCl, 500 mM imidazole and adjusted to pH 7.4 with HCl.

9. PD-10 desalting column (GE Healthcare Life Sciences).

10. 5 kDa MWCO Vivaspin (Satorius stedim or equivalent).

11. Nanodrop 1000 Spectrophotometer (Thermo Fisher Scientific) or any suitable spectrophotometer.

12. 1× PBS.

13. AcTEV protease (Invitrogen).

14. 100 % glycerol.

15. Centrifuge (Sorvall Evolution RC Superspeed centrifuge, or equivalent).

2.4 SDS-PAGE and Western Blot

1. Criterion TGX stain-free 4–12 % and gel running apparatus (Bio-Rad).

2. Laemmli sample buffer (Bio-Rad) or equivalent.

3. Beta-mercaptoethanol.

4. Heat block.

5. Running buffer: 10× Tris-Glycine-SDS (TGS) buffer (Bio-Rad).

6. Transfer buffer: 10× Tris/Glycine (Bio-Rad) or equivalent.

7. Pre-stained molecular weight markers.

8. TBST: Tris-buffered saline (TBS), 0.1 % Tween 20.

9. InstantBlue protein stains (Expedeon) or equivalent.

10. His Tag antibody horseradish peroxidase (HRP) conjugate and 5 % alkali-soluble casein (Novagen) or equivalent.

11. High-sensitivity streptavidin (HRP) conjugate (Pierce) or equivalent.

12. PVDF membrane and western blotting transfer apparatus (Bio-Rad) or equivalent.

13. ECL Prime Western Blotting detection kit (GE Healthcare Life Sciences, or equivalent).

14. Plastic container.

15. Amersham Hyperfilm ECL (GE Healthcare Life Sciences, or equivalent).

16. X-ray film developer machine (Kodak X-DMAT 2000 processor).

2.5 Coupling All Four Serotype SgDENV ED3 Recombinant Proteins to LumAvidin Microsphere

1. LumAvidin Microspheres region 53, 55, 67, 87 (Luminex).

2. Microcentrifuge (Heraeus Fresco 21 microcentrifuge, or equivalent).

3. Assay buffer: PBS, 1 % BSA, 0.05 % azide; adjust pH to 7.4.

4. SgDENV1-4 ED3-biotinylated proteins.

5. Vortex mixer (Grant bio Multi-vortex V-32 or equivalent).

6. Ultrasonic bath (Elmasonic S15H Ultrasonic bath or equivalent).

7. 1.5 mL microcentrifuge tubes.

2.6 Enumeration of Coupled Microspheres

1. Hemacytometer or cell counter.

2.7 Functional Assay of SgDENV ED3-Coupled Microspheres

1. Assay buffer: PBS, 1 % BSA, 0.05 % azide; adjust pH to 7.4.

2. SgDENV ED3-coupled microspheres: SgDENV1 ED3 microsphere, region 53; SgDENV2 ED3 microsphere, region 67; SgDENV3 ED3 microsphere, region 55; SgDENV4 ED3 microsphere, region 87.

3. Wash buffer (Millipore).

4. 4 °C in the dark.

5. Human infected with dengue specific serotype plasma.

6. Goat antihuman IgG (Fc gamma specific) phycoerythrin (PE) conjugate (eBioscience).

7. 1× PBS: filter sterilized with a 0.2 μm filter.

8. Sheath fluid (Millipore).

9. Absorbent towels.

10. Plate sealer.

11. Aluminum foils.

12. 96-well filter plates (Millipore MultiScreen$_{HTS}$-BV plate, or equivalent).

13. Vacuum manifold (Bio-Rad, or equivalent).

14. Vortex mixer (Grant bio Multi-vortex V-32, or equivalent).

15. Plate shaker (Corning LSE™ digital microplate shaker, or equivalent).

16. Flexmap 3D instrument (Luminex).

17. xPONENT software (Luminex).

3 Methods

3.1 Construction of pQEAM

Plasmid, pQEAM, was developed to obtain soluble biotinylated DENV1-4 ED3 recombinant proteins. pQEAM can also be used as a master vector for obtaining any soluble and biotinylated recombinant proteins.

1. Generate complementary pairs of AviTag-F and AviTag-R oligonucleotides:

   (a) Mix 1 μM of each AviTag-F and AviTag-R together at a 1:1 molar ratio in a PCR tube.

   (b) Add 1× ligase buffer into the mixture.

   (c) Top up the volume to 100 μL with MilliQ-H$_2$O.

   (d) Mix and quick pulse in a microcentrifuge to pull the mixture to the bottom of the tube.

   (e) Anneal AviTag-F and AviTag-R by following the conditions listed in Table 2.

Table 2
Thermocycler program for annealing complementary oligonucleotides

Step	Temperature (°C)	Time	Cycles
1	1	95 °C	5
2	70	95 °C (−1 °C/cycle)[a]	1
3		4 °C	Hold

[a]−1 °C/cycle indicates that the temperature of the heating block will decrease −1 °C per cycle

Table 3
PCR conditions for AviTag amplification

Step	Temperature (°C)	Time	Cycles
Initial denaturation	95	2 min	1
Denaturation	95	30 s	35
Annealing	61	30 s	
Extension	72	12 s	
Final extension	72	10 min	1

(f) Load 5 μL of annealed oligonucleotides (AviTag) on gel to check for double-stranded DNA (*see* **Note 1**).

(g) Purify the AviTag, using QIAquick PCR purification kit (Qiagen), as per the manufacturer's instructions.

(h) Quantify the AviTag using Nanodrop spectrophotometer or an equivalent.

2. Use AviPCR-for (forward) and AviPCR-rev (reverse) primer to amplify AviTag (*see* **Note 2**). Follow PCR condition as in Table 3. To make the PCR mix, combine the following components on ice (final concentrations are given):

(a) 0.2 μM forward primer.

(b) 0.2 μM reverse primer.

(c) 0.2 mM dNTP.

(d) 1× Pfu buffer with $MgSO_4$ (2 mM).

(e) 20 ng template.

(f) 2.5 units Pfu DNA polymerase.

(g) MilliQ-H_2O to give 50 μL per reaction.

3. Purify and quantify the AviTag PCR products following previous step in points (g) and (h) of **step 1** of Subheading 3.1.

4. Load 5 μL of PCR-purified AviTag fragment on gel to check for purity.

5. Digest 2 μg of PCR-purified AviTag with NdeI and NotI at 37 °C for 2 h and heat-inactivate enzymes at 80 °C for 5 min.

6. Linearize pQE-2 vector (*see* **Note 3**):

   (a) Digest 4 μg of pQE-2 vector with NdeI and NotI in a total volume of 80 μL and incubate at 37 °C for 1 h (*see* **Note 4**).

   (b) Heat-inactivate both enzymes at 80 °C for 5 min.

   (c) Add 1 μL of calf intestinal alkaline phosphatase (CIP) to the digested reaction mixture. Incubate at 37 °C for 1 h (*see* **Note 5**).

   (d) Perform QIAquick PCR purification to remove CIP from the reaction.

7. Load 5 μL of linearized pQE-2 (~4.8 kb); check for linearization by agarose gel electrophoresis. Include pQE-2 as an uncut plasmid control.

8. Ligate NdeI/NotI digested AviTag and dephosphorylated linearized pQE-2 together in an insert to vector molar ratio of 3:1 (*see* **Note 6** and **7**) using T4 DNA ligase (New England Biolabs) following manufacturer's protocol.

9. Heat-inactivate ligation reactions at 70 °C for 5 min, before transforming into DH5α *Escherichia coli*. Follow standard protocol for transformation.

10. Screen for potential clones (pQEA::AviTag; *see* **Note 8**) using AviPCR-for and AviPCR-rev primers for PCR reactions (*see* Table 3). Follow PCR conditions in Subheading 3.1, **step 2** (*see* **Note 9**).

11. Sequence 4 potential clones to confirm that the AviTag is inserted into pQEA plasmid and that it contains no mutation during amplification. Use pQE-R, a reverse primer, to prime from the 3′ end of multiple cloning site of pQE-2 into the insert for sequencing. Positive plasmid clone contains AviTag inserted into NdeI /NotI sites of pQE-2 and is named pQEA.

12. Insert maltose-binding protein (MBP; *see* **Note 10**) containing TEV cleavage site at its C-terminal end into the NdeI site of pQEA. The plasmid pQEA::MBP is named pQEAM.

13. Make glycerol stock for pQEAM (*see* **Note 11**).

3.2 Construction of pQEAM::DENV1-4 ED3

1. Use 1 μL of RNA of dengue serotypes 1, 3, and 4 (*see* **Note 12**) for reverse transcription to cDNA. 10 μL reaction per DENV serotype is prepared using TaqMan reverse transcription reagent (Applied Biosystems; *see* **Note 13**) following manufacturer's instructions.

2. Amplify 1 μL of cDNA from each dengue serotype 1, 3, and 4 using WPD1For1, WPD1Rev1, VND3For1, VND3Rev1, MsiaD4For1, and MsiaD4Rev1 primers, and amplify 20 ng of pTSV01 using Den2ED3-F and Den2ED3-R1 (*see* Subheading 2.1.2, **step 8** and **Note 14**).

Table 4
PCR conditions for amplifying DENV1-4 ED3

Step		Temperature (°C)	Time	Cycles
Initial denaturation		95	2 min	1
Denaturation		95	30 s	35
Annealing	DENV1	64.5	30 s	
	DENV2	64.5		
	DENV3	62.1		
	DENV4	66		
Extension		72	40 s	
Final extension		72	5 min	1

3. Prepare a PCR mix for each dengue serotype 1, 2, 3, and 4 using the following condition (*see* Table 4). Follow the protocol for PCR mix in Subheading 3.1, **step 2**.

4. Purify and quantify the DENV1-4 ED3 PCR products following previous method in points (g) and (h) of **step 1** of Subheading 3.1.

5. Load 5 μL of PCR-purified DENV1-4 ED3 products on a 1 % agarose gel and check for purity by electrophoresis using standard protocol.

6. Digest 2 μg of PCR-purified DENV1-4 ED3 with SalI and NotI at 37 °C for 2 h and heat-inactivate the enzymes at 80 °C for 5 min.

14. Linearize pQEAM vector:

    (a) Digest 4 μg of pQEAM vector with SalI and NotI in a total volume of 80 μL and incubate at 37 °C for 2 h.

    (b) Heat-inactivate both enzymes at 80 °C for 5 min.

    (c) Add 1 μL of calf intestinal alkaline phosphatase (CIP) to the digested reaction. Incubate at 37 °C for 1 h (*see* **Note 5**).

7. Verify completion of linearization by electrophoresis on an agarose gel by loading 5 μL of restriction digest of pQEAM (~5.9 kb). Load 0.5 μg of pQEAM plasmid as an uncut control.

8. Perform QIAquick PCR purification to remove CIP from the reaction.

9. Set up 4 ligation reactions by ligating each SalI/NotI digested DENV ED3 fragment from each of the four serotypes to dephosphorylated linearized pQEAM vector in an insert to vector molar ratio of 5:1 (*see* **Notes 6** and **7**) using T4 DNA ligase following the manufacturer's instructions.

10. Heat-inactivate ligation reactions at 70 °C for 5 min. Transform *Escherichia coli* DH5α following the standard protocol.

11. Screen for potential clones (pQEAM::DENV ED3 (*see* **Note 8**) using primers listed in Subheading 3.2, **step 2**, for PCR reactions (*see* Table 4 and **Note 14**). For PCR conditions, follow Subheading 3.1, **step 2**.

12. Verify two potential clones by sequencing to confirm that the DENV ED3 is inserted into pQEAM and that it contains no mutation during amplification. Use pQE-R, a reverse primer, to prime from the 3′ end of the multiple cloning site of pQEAM into the insert, and AviPCR-for sequencing. Positive clone plasmids containing DENV1 ED3, DENV2 ED3, DENV3 ED3, or DENV4 ED3 inserted into SalI/NotI sites of pQEAM are named pQEAM::DENV1 ED3, pQEAM::DENV2 ED3, pQEAM::DENV3 ED3, and pQEAM::DENV4 ED3, respectively.

13. Make glycerol stocks of pQEAM::DENV1 ED3, pQEAM::DENV2 ED3, pQEAM::DENV3 ED3, and pQEAM::DENV4 ED3 (*see* **Note 11**).

3.3 Construction of Singapore Strain ED3 of DENV3 and DENV4 in pQEAM

3.3.1 Generation of pQEAM::SgDENV3 ED3

There are three nucleotide differences when DENV3 ED3 of VN-V1015 is aligned to the Singapore strain (EU448436). Two differences are at ED3 codon L9S and T11A and the third mutation near the C-terminal region at V85I. To mutate DENV3 ED3 of VN-V1015 in pQEAM to EU448436, two separate site-directed mutagenesis reactions are performed:

1. Use SgD3Ser-AlaF and SgD3Ser-AlaR mutagenic primers and pQEAM::DENV3 ED3 as a template. Set up a reaction to mutate lysine and threonine to serine and alanine at ED3 codons 9 and 11, respectively, following QuikChange site-directed mutagenesis kit's instructions. PCR conditions and Dpn1 digestion protocol are modified from the manufacturer's protocol as follows. *See* Table 5 for PCR conditions.

2. Add 10 units DpnI into amplification reaction at time zero and a second addition of 10 units DpnI after 1 h of incubation at 37 °C (*see* **Note 15**).

3. Select 5 clones for sequencing using forward primer (pQEAM2181-2200) to identify potential positive clones (*see* Table 1).

4. Sequence the entire positive clone using the primers listed in Table 1.

5. Select a positive mutant clone as a template for PCR using SgD3Iso-F and SgD3Iso-F to mutate valine to isoleucine at ED3 codon 85. Follow the protocol in this section from **steps 1** to **4** to perform site-directed mutagenesis.

6. Positive clone is named as pQEAM::SgDENV3 ED3.

Table 5
Conditions for site-directed mutagenesis of pQEAM::DENV3 ED3 and pQEAM::DENV4 ED3

Step		Temperature (°C)	Time	Cycles
Initial denaturation		95	30 s	1
Denaturation		95	30 s	16
Annealing	DENV3	48/50	1 min	
	DENV4	50		
Extension		68	6 min 15 s	

3.3.2 Generation of pQEAM::SgDENV4 ED3

There is only one nucleotide difference when Malaysia-31586 DENV4 ED3 is aligned with Singapore strain (EU448464) DENV4 ED3.

1. Use SgD4Ala-F and SgD4Ala-R mutagenic primers and pQEAM::DENV4 ED3 as a template; set up a reaction to mutate threonine to alanine at ED3 codon 6 following QuikChange site-directed mutagenesis kit's instructions. PCR conditions and Dpn1 digestion are modified from the manufacturer protocol as follows. *See* Table 5 for PCR conditions.

2. Follow the protocol in Subheading 3.3.1 from **steps 1** to **4**.

3. Positive clone is named pQEAM::SgDENV4 ED3.

3.4 Preparation of Biotinylated ED3 of SgDENV1-4

This method describes in vivo biotinylation of protein containing AviTag in *E. coli* CVB101 strain. This is one of the most convenient methods to obtain a biotinylated protein and the number of biotin added to the protein is uniform. The overall work flow for purifying ED3 of SgDENV1-4 is shown in Fig. 2.

1. Transform clones encoding SgDENV1, SgDENV2, SgDENV3, and SgDENV4 (*see* **Note 16**) into *Escherichia coli* CVB101 competent cells (*see* Subheading 2.2, **step 1**):

   (a) Add 100 ng of the expression plasmid to 50 μL of *Escherichia coli* CVB101 competent cells, mix gently, and incubate on ice for 30 min (*see* **Note 17**).

   (b) Heat shock cells at 42 °C for 90 s; place immediately on ice to chill for 1–2 min.

   (c) Add 800 μL of LB and incubate at 37 °C for 45 min at 225 rpm in a shaker.

   (d) Plate 100 μL cells onto LB plate containing 100 μg/mL ampicillin and 10 μg/mL chloramphenicol; incubate overnight at 37 °C in an incubator.

2. Grow in 10 mL TYH medium supplemented with 100 μg/mL ampicillin and 10 μg/mL chloramphenicol from a single

colony of SgDENV1-, SgDENV2-, SgDENV3-, or SgDENV4-transformed CVB101 competent cells overnight at 37 °C at 225 rpm in a shaker incubator.

3. Transfer 5 mL of the overnight culture into 1 L of TYH medium containing 100 µg/mL ampicillin.

4. Add 20 mL of a 20 % sterile glucose solution (0.5 % final concentration) and incubate at 37 °C at 225 rpm in a shaker incubator until OD_{600} of the culture reaches 0.6.

5. Add 10 mL of 5 mM biotin solution (50 µM final concentration) (*see* **Note 18**).

6. Add 10 mL of 100 mM IPTG (1 mM final concentration) to induce for 3 h in a shaker incubator.

7. Harvest the bacterial culture in 4×250 mL centrifuge bottles by centrifugation at 6,000 rcf for 10 min.

8. Discard the supernatant and resuspend the pellet in 20 mL lysis buffer (*see* Subheading 2.3, **step 1**) per centrifuge bottle.

9. Sonicate using a large sonication probe on ice at an amplitude of 30, 5 s pulse ON, and 10 s pulse OFF for 30 min.

10. Remove cell debris by centrifugation at 20,000 rcf for 20 min at 4 °C.

11. Remove and filter the supernatant using a 0.22 µm filter.

12. Equilibrate a 5 mL HiTrap HP column with binding buffer using FPLC machine at 5 mL/min.

13. Load the filtered supernatant onto the column at 0.5 mL/min.

14. Wash the column with wash buffer until the UV reaches equilibrium and then elute at 2 mL/min with elution buffer. Collect the peak fraction. The fraction contains one of the biotinylated HisMBP-SgDENV1-4 ED3 proteins.

15. Use PD-10 desalting column to exchange the buffer to PBS.

16. Determine the protein concentration using Nanodrop™ spectrophotometer or equivalent.

17. To release HisMBP from a HisMBP-SgDENV1-4 ED3 protein, add AcTEV protease and incubate on ice or at 4 °C overnight. The TEV protease digestion mixture consists of the following (*see* **Note 19**):

    (a) 10 units of AcTEV to 100 µg of protein.

    (b) 1 mM DTT (final concentration).

    (c) 1× TEV buffer.

    (d) Add MilliQ-H_2O to final volume of 100 µL.

18. Follow Subheading 3.4, **steps 12** and **13**, for equilibration of HiTrap column and loading protein onto the column.

19. Collect the flow through containing purified biotinylated SgDENV1-4 ED3 protein (*see* **Note 20**).

20. Concentrate purified biotinylated SgDENV1-4 ED3 using a 5 kDa MWCO centrifugal filter device until volume is ~500 μL (*see* **Note 21**).

21. Determine the protein concentration using Nanodrop™ or equivalent spectrophotometer.

22. Add sterile 100 % glycerol to make 30 % glycerol stock and store at –80 °C.

3.5 SDS-PAGE and Western Blot for SgDENV1-4 ED3 Purified Proteins

Run protein fractions on SDS-PAGE to visualize purity, and perform western blot to determine the degree of biotinylation of the SgDENV1-4. Probe western blots separately with His Tag antibody HRP conjugate and streptavidin-HRP conjugate:

1. SDS-PAGE for SgDENV1-4 ED3 purified proteins:

    (a) Add 30 μL sample of each fraction with 30 μL Laemmli sample buffer containing freshly added 5 % β-mercaptoethanol.

    (b) Heat samples at 95 °C for 5 min and cool to room temperature.

    (c) Load 25 μL samples onto each 4–20 % Criterion precast gel, include a pre-stained molecular weight marker, and run at 120 V in 1× SDS running buffer until the dye has reached the bottom of the gel.

    (d) Stain one of the gels in InstantBlue protein staining buffer for 1 h.

    (e) Rinse gel in MilliQ-H_2O for 15 min.

2. Western blot for SgDENV1-4 ED3 purified proteins:

    (a) Transfer the proteins onto a PVDF membrane using standard western blotting transfer protocol.

    (b) Incubate membrane in 20 mL of 1 % alkali-soluble casein (blocking buffer) at room temperature for 1 h with shaking or overnight without shaking at 4 °C (retain blocking buffer after use).

    (c) Wash membrane four times, 15 min each time, with PBST.

    (d) Incubate membrane blot with 1: 20,000 streptavidin HRP or 1:2,000 His Tag antibody HRP diluted in 10 mL of blocking buffer at 4 °C overnight with shaking or at room temperature for 1 h with shaking.

    (e) Wash membrane four times, 15 min each time, with PBST.

    (f) Develop blot using ECL Prime Western Blotting detection kit.

3.6 SgDENV1-4 ED3 Bead-Based Immunoassay

Microsphere should be protected from prolonged exposure to light throughout this procedure.

3.6.1 Determination of the Amount of Biotin-Conjugated SgDENV1-4 ED3 Proteins to Bind onto LumAvidin Microspheres

1. Coupling of biotin-conjugated SgDENV1-4 ED3 proteins to microsphere:

   (a) Transfer 1.0×10^5 of stock microspheres to a microcentrifuge tube.

   (b) Pellet the stock microspheres by centrifugation at $\geq 8,000$ rcf for 1–2 min.

   (c) Remove supernatant and resuspend in 250 µL of assay buffer (*see* Subheading 2.5, **step 3**) by vortex and sonication for ~20 s.

   (d) Dilute the biotin-conjugated SgDENV1-4 ED3 proteins in assay buffer to a concentration of 7.5–4,000 nM in a serial twofold dilution.

   (e) Add 250 µL of each dilution to the microsphere suspension and mix immediately using a vortex.

   (f) Incubate for 30 min with mixing at room temperature.

   (g) Pellet bound microspheres by centrifugation at $\geq 8,000$ rcf for 1–2 min.

   (h) Remove the supernatant and resuspend the pelleted microsphere in 500 µL assay buffer by vortex.

   (i) Pellet bound microspheres by centrifugation at $\geq 8,000$ rcf for 1–2 min.

   (j) Repeat Subheading 3.6, **steps 8** and **9**. This is a total of two washes with assay buffer.

   (k) Remove the supernatant and resuspend the pelleted microsphere in 200 µL assay buffer by vortex and sonication for ~20 s.

   (l) Count the microsphere suspension by hemacytometer (*see* **Note 22**).

2. Bead-based SgDENV1-4 ED3 immunoassay (*see* **Note 23**):

   (a) Pre-wet the plate with 200 µL of assay buffer into each well of the plate. Shake on the plate shaker for 10 min at room temperature.

   (b) Remove assay buffer by using vacuum manifold and remove residual amount from all wells by tapping gently onto absorbent towels several times.

   (c) Add 25 µL of assay buffer to all wells.

   (d) Add 25 µL plasma from humans infected with dengue of specific serotype diluted 1:1,000 in assay buffer to sample well.

(e) Vortex and add 25 μL SgDENV1-4 ED3-coupled beads to sample well (*see* **Note 24**).

(f) Seal the plate with plate sealer. Wrap the plate with foil and incubate with agitation on a plate shaker overnight at 4 °C or 1 h at room temperature.

(g) Remove well contents by using vacuum manifold and wash 4 times with 200 μL wash buffer. For each washing, aspirate to remove well content using vacuum manifold.

(h) Add 25 μL goat antihuman IgG (Fc gamma specific) PE diluted 1:100 in PBS to all wells.

(i) Seal the plate with plate sealer. Wrap the plate with foil and incubate for 1 h at room temperature with agitation on a plate shaker.

(j) Remove well content by using vacuum manifold and wash 4 times with 200 μL wash buffer. For each washing, aspirate to remove well contents using vacuum manifold.

(k) Add 150 μL sheath fluid to all wells.

(l) Run plate on Flexmap 3D with xPONENT software.

3.6.2 Determination Sensitivity Assay

1. Add 25 μL of assay buffer to all wells.

2. Dilute the human infected with dengue specific serotype plasma in assay buffer to a concentration of 1:128,000–1:1,000 in a serial twofold dilutions.

3. Add 25 μL of each dilution to the sample wells.

4. Follow points (e)–(l) of **step 2** of Subheading 3.6.1 for sensitivity assay.

3.6.3 Determination of Cross-Reactivity of Dengue Infected Human Plasma

Cross-reactivity of dengue infected human plasma is tested in two ways: first, by testing each dengue specific serotype plasma to each of the SgDENV1-4 ED3-coupled beads and, second, by testing each dengue specific serotype plasma to a pool of SgDENV1-4 ED3-coupled beads (*see* Fig. 3).

1. Testing each dengue specific serotype plasma cross-reactive to each of the SgDENV1-4 ED3-coupled beads:

(a) Add 25 μL of assay buffer to all wells.

(b) Vortex and add 25 μL SgDENV1-4 ED3-coupled beads to sample well (*see* **Note 24**).

(c) Dengue infected human plasma was diluted 1:1,000 in assay buffer.

(d) Add 25 μL of each dengue specific serotype plasma to each of the SgDENV1-4 ED3-coupled beads.

(e) Follow protocol under points (f)–(l) of **step 2** of Subheading 3.6.1 for sensitivity assay.

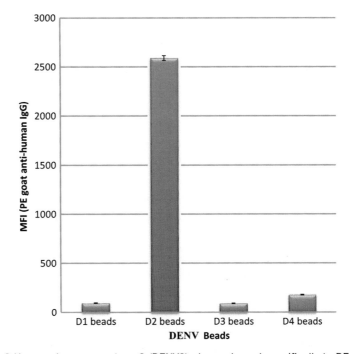

Fig. 3 Human dengue serotype 2 (DENV2) plasma bound specifically to DENV2-labeled beads. DENV2-positive plasma was diluted 1:1,000 before reacting with a mixture of beads containing ED3 of DENV1-4. Unbound antibodies were removed by washing before using goat antihuman PE for detection. Reactions were subsequently analyzed using the Flexmap 3D software. Result is presented as average mean fluorescence intensity (MFI)

2. Testing each dengue specific serotype plasma in a pool of SgDENV1-4 ED3-coupled beads:

(a) Add 25 μL of assay buffer to all wells.

(b) Vortex and add 25 μL of mixed SgDENV1-4 ED3-coupled beads to all sample wells.

(c) Add 25 μL of each dengue specific serotype plasma to mixed SgDENV1-4 ED3-coupled beads.

(d) Follow protocol under points (f)–(l) of **step 2** of Subheading 3.6.1 for sensitivity assay.

4   Notes

1. AviTag is a 73 bp DNA fragment. High percentage of agarose gel is needed to see small double-stranded DNA. In this experiment 3 % agarose is used.

2. PCR amplification is required for extending 4 bp from the recognition site to both ends of the AviTag fragment. This allows higher cleavage efficiency.

3. pQE-2 encodes an N-terminal 6× His tagged protein, an ampicillin resistance gene, a Col E1 origin of replication, a T5 promoter, lac operators, a ribosome-binding site, and a *lacI^q* repressor gene.

4. Lesser digestion time is required for plasmids than PCR products.

5. Dephosphorylate linearized vector is a very crucial method. It prevents vector self-ligation, thereby reducing vector background when the ligation reactions are performed.

6. Calculate the amount of vector (pQE-2) and insert (AviTag) to add into a ligation reaction in an insert to vector molar ratio of 3:1 using the following formula:

[Vector (ng)×insert size (bp)/ vector size (bp)]×3/1, where bp stands for base pair (number of nucleotides) and 3/1 is the insert to vector molar ratio. Ligation reaction is incubated at 16 °C overnight.

7. Include a vector control by adding only vector to the ligation reaction.

8. To screen for positive clones, randomly select at least ten colonies. Pick the selected colonies; streak onto LB agar plates containing 100 µg/mL ampicillin before dipping it into a PCR reaction.

9. AviPCR-for and AviPCR-rev primers are used previously in point (b) of **step 1** of Subheading 3.1 to amplify the annealed complementary AviTag-F and AviTag-R oligonucleotides. Positive clones will have PCR products (AviTag insert) of about 80 bp.

10. His tagged proteins typically are expressed in *E. coli* in soluble form or insoluble inclusion bodies. Protein purified from BL21 *E. coli* expressing pQEA::DENV ED3 (dengue envelop domain 3 cloned into pQEA) is known to partition in the inclusion bodies. Adding maltose-binding protein (MBP) tag to pQEA will enhance solubility of fusion protein and promote proper folding of DENV ED3 protein [13].

11. To prepare glycerol stock, pick a single colony and grow overnight at 37 °C in 10 mL of LB medium supplemented with 100 µg/mL ampicillin. Transfer 1 mL of overnight culture into 200 mL fresh LB medium and incubate at 37 °C in a shaker. When OD_{600} reaches 0.6, take some culture and add sterile 100 % glycerol to make 1 mL aliquot of 30 % glycerol stock and store at −80 °C.

12. RNAs of dengue serotypes 1, 3, and 4 are the following: DENV1, Western pacific 74; DENV3, VN/BID-V1015/2006; and DENV4, SG(EHI)D4/2641Y08. Use 140 μL of virus culture to isolate RNA and elute into 120 μL of AVE buffer, and then dilute to 1/10.

13. TaqMan Reverse Transcription Reagents may cause eye and skin irritations.

14. PCR product of DENV 1-4 ED3 is expected to be around 303 bp.

15. Additional 1 μL of DpnI is required to digest the parental supercoiled dsDNA.

16. For alignment, ED3 sequences of SgDENV1, SgDENV2, SgDENV3, and SgDENV4 are derived from the GenBank accession numbers EU069623, EU448427, EU448436, and EU448464, respectively.

17. Include competent cells that received a known amount of standard plasmid DNA and those that received no plasmid DNA as positive and negative transformation controls, respectively.

18. CVB101 cells transformed with BirA gene is expressed after IPTG induction. In the presence of biotin, BirA (biotin ligase) biotinylates a lysine side chain within AviTag in HisMBP-SgDENV ED3 protein.

19. TEV digestion mixture can be scaled up to 8 mL.

20. After TEV cleavage, the HisMBP polypeptide is released from HisMBP-SgDENV ED3 protein. The HisMBP polypeptide containing the His Tag binds to the HiTrap column, while the protein of interest (biotinylated SgDENV ED3) should elute in the flow through fraction.

21. SgDENV1-4 ED3 proteins have a molecular weight of ~15 kDa. Use centrifuge filter device with a molecular weight cut off half of the protein size for concentration.

22. Calculate total microsphere: count (1 corner of 4×4 sections) $\times$ $(1 \times 10^4) \times$ (dilution factor) $\times$ (resuspension volume in mL).

23. Background control (beads only, beads + healthy donor plasma (negative plasma), and beads + infected human dengue plasma (positive plasma)) and negative control (dengue beads + healthy donor plasma) are included in the assay.

24. During addition of beads, shake bottle intermittently to avoid settling.

References

1. Kyle JL, Harris E (2008) Global spread and persistence of dengue. Annu Rev Microbiol 62: 71–92

2. Chungue E, Poli L, Roche C, Gestas P, Glaziou P, Markoff LJ (1994) Correlation between detection of plasminogen cross-reactive antibodies and hemorrhage in dengue virus infection. J Infect Dis 170:1304–1307

3. Halstead SB (1988) Pathogenesis of dengue: challenges to molecular biology. Science 239: 476–481

4. Morens DM (1994) Antibody-dependent enhancement of infection and the pathogenesis of viral disease. Clin Infect Dis 19:500–512

5. Hung JJ, Hsieh MT, Young MJ, Kao CL, King CC, Chang W (2004) An external loop region of domain III of dengue virus type 2 envelope protein is involved in serotype-specific binding to mosquito but not mammalian cells. J Virol 78:378–388

6. Mukhopadhyay S, Kuhn RJ, Rossmann MG (2005) A structural perspective of the flavivirus life cycle. Nat Rev Microbiol 3:13–22

7. Crill WD, Roehrig JT (2001) Monoclonal antibodies that bind to domain III of dengue virus E glycoprotein are the most efficient blockers of virus adsorption to Vero cells. J Virol 75:7769–7773

8. Beasley DW, Aaskov JG (2001) Epitopes on the dengue 1 virus envelope protein recognized by neutralizing IgM monoclonal antibodies. Virology 279:447–458

9. Wahala WM, Silva AM (2011) The human antibody response to dengue virus infection. Viruses 3:2374–2395

10. Zhang Y, Corver J, Chipman PR, Zhang W, Pletnev SV, Sedlak D, Baker TS, Strauss JH, Kuhn RJ, Rossmann MG (2003) Structures of immature flavivirus particles. EMBO J 22: 2604–2613

11. Guzman MG, Hermida L, Bernardo L, Ramirez R, Guillen G (2010) Domain III of the envelope protein as a dengue vaccine target. Expert Rev Vaccines 9:137–147

12. Miller N (2010) Recent progress in dengue vaccine research and development. Curr Opin Mol Ther 12:31–38

13. Kapust RB, Waugh DS (1999) Escherichia coli maltose-binding protein is uncommonly effective at promoting the solubility of polypeptides to which it is fused. Protein Sci 8: 1668–1674

Chapter 6

Pseudo-infectious Reporter Virus Particles for Measuring Antibody-Mediated Neutralization and Enhancement of Dengue Virus Infection

Swati Mukherjee, Theodore C. Pierson, and Kimberly A. Dowd

Abstract

This chapter outlines methods for the production of dengue virus (DENV) reporter virus particles (RVPs) and their use in assays that measure antibody-mediated neutralization and enhancement of DENV infection. RVPs are pseudo-infectious virions produced by complementation of a self-replicating flavivirus replicon with the DENV structural genes in trans. RVPs harvested from transfected cells are capable of only a single round of infection and encapsidate replicon RNA that encodes a reporter gene used to enumerate infected cells. RVPs may be produced using the structural genes of different DENV serotypes, genotypes, and mutants by changing plasmids used for complementation. Further modifications are possible including generating RVPs with varying levels of uncleaved prM protein, which resemble either the immature or mature form of the virus. Neutralization potency is measured by incubating RVPs with serial dilutions of antibody, followed by infection of target cells that express DENV attachment factors. Enhancement of infection is measured similarly using Fc receptor-expressing cells capable of internalizing antibody-virus complexes.

Key words Reporter virus particles (RVP), Dengue virus, Antibody neutralization, Antibody-dependent enhancement (ADE), Replicon, Flavivirus

1 Introduction

Flaviviruses are positive-stranded RNA viruses responsible for significant morbidity and mortality across the globe. Members of this genus of significant clinical importance include yellow fever, Japanese encephalitis, tick-borne encephalitis, West Nile (WNV), and dengue (DENV) viruses. DENV is the most clinically important mosquito-borne viral pathogen worldwide and is responsible for more than 50 million human infections each year [1, 2]. While most infections are subclinical, clinical manifestations of DENV infection range from a self-limiting febrile illness to a potentially fatal hemorrhagic disease [3]. Four antigenically related groups of DENV circulate in nature, all of which are capable of causing the full spectrum of DENV disease. The envelope (E) proteins of these

Radhakrishnan Padmanabhan and Subhash G. Vasudevan (eds.), *Dengue: Methods and Protocols*, Methods in Molecular Biology, vol. 1138, DOI 10.1007/978-1-4939-0348-1_6, © Springer Science+Business Media, LLC 2014

DENV "serotypes" share roughly 60 % amino acid identity [4, 5]. This similarity is responsible for the extensive cross-reactivity of the humoral response following flavivirus infection and the complexity of serological studies of these pathogens.

Flaviviruses are small virus particles with a relatively simple genomic structure. Virions are composed of three structural proteins, a lipid envelope derived from the endoplasmic reticulum, and the viral RNA [6] (Fig. 1a). The ~11 kb viral genomic RNA encodes at least ten viral proteins that are translated as a single polyprotein that is subsequently cleaved by viral and host cell proteins. The three structural genes that form the virus particle are present at the N-terminus of the polyprotein (E, pre-membrane (prM), and capsid (C)). The nonstructural proteins responsible for viral RNA replication (as well as other critical functions) are encoded by the 3′ end of the genome [7] (Fig. 1b).

Reporter virus particles (RVP) are pseudo-infectious particles capable of a single round of replication. The backbones of this technology are replicons, which are self-replicating flavivirus RNAs containing large in-frame deletions in the structural genes. The first flavivirus replicon was constructed by Khromykh and colleagues using Kunjin virus (a strain of WNV) [8]. Replicons have been designed in many configurations. Most replicon RNAs are produced by in vitro RNA synthesis using constructs in which the subgenomic RNA is placed downstream of a bacteriophage promoter (such as T7) (Fig. 1c). Replicons have also been engineered in a "DNA-launched" format; in these constructs the replicon is placed under the transcriptional control of a eukaryotic promoter (as described in Subheading 2.1.1) (Fig. 1d). Replicons are often designed to encode reporter genes that simplify the detection and quantification of viral RNA replication. Several configurations are possible, including introduction of the reporter gene in place of the deleted structural genes or within the 3′ untranslated region (UTR) of the viral RNA (Fig. 1d; top and bottom, respectively). Subgenomic replicons have been described for several flaviviruses (reviewed in [9]) and have been extremely useful in studies of viral replication [10–14], translation [15–17], and vaccines [18–20]. Flavivirus replicons have also proven to be exceptionally valuable in high throughput screens of inhibitors of viral replication [21–24].

Flavivirus replicons may be packaged into virus particles by complementation with the structural proteins of the virus in trans, resulting in the production of pseudo-infectious RVPs. This technology was first developed by Khromykh and colleagues using Kunjin virus [25] and has since been adapted and modified for use with other viruses and via different methods of structural gene expression [26–28]. For example, RVPs have been produced by complementation of replicons with structural genes expressed via viral vectors [8, 29],

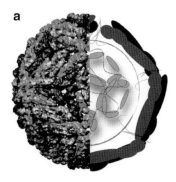

Envelope (E) protein

Pre-membrane (prM) protein

Capsid (C) protein

Lipid envelope

RNA genome (replicon RNA)

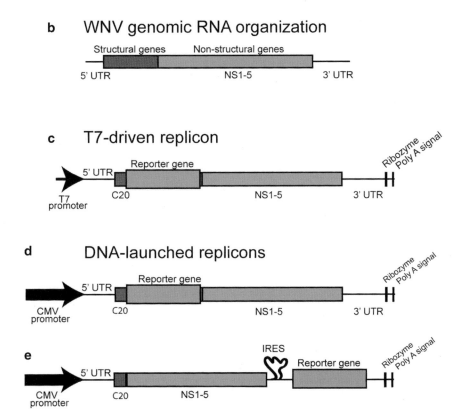

Fig. 1 Schematics of a flavivirus virion and subgenomic flavivirus replicons used to generate reporter virus particles (RVPs). (**a**) Flavivirus virions are composed of three structural proteins (CprME), a lipid membrane derived from the endoplasmic reticulum, and the viral RNA. (**b**) The ~11 kb flavivirus genome encodes the structural genes at the 5′ end and the nonstructural genes at the 3′ end. (**c, d**) Self-replicating flavivirus replicons produce intact flavivirus genomes that have large deletions within the structural gene region. Subgenomic replicons have been designed in numerous configurations. Depicted are (**c**) a replicon under the transcriptional control of the T7 bacteriophage promoter that requires in vitro RNA synthesis for production of viral RNA and (**d**) "DNA-launched" replicons under the transcriptional control of a CMV eukaryotic promoter. The location of the reporter gene is shown inserted in two possible locations that correspond to the replicons pWNII-rep-GZ and pDENV2-rep-GZ (*top*) and pWNII-rep-Ren-IB (*bottom*)

stable cell lines [30–32], and standard DNA expression vectors [26, 33]. Use of these methods has allowed for the production of RVPs with titers in excess of 10^7 infectious units/mL.

In this chapter, we will describe the production of pseudo-infectious DENV RVPs by complementation of DNA-launched WNV or DENV replicons with DENV structural genes encoded by DNA expression plasmids. The utility of this approach is its simplicity, and that it allows for the production of a large number of flavivirus variants simply by changing the expression components used during RVP production experiments.

2 Materials

2.1 DNA Plasmids

2.1.1 DNA-Launched Subgenomic Replicon, Key Components of the Replicon (See Fig. 1d)

1. DNA-launched replicons allow for the production of flavivirus RNA following the transfection of cells with plasmid DNA using standard methods [26, 34]. These vectors are designed such that the replicon RNA is placed under the transcriptional control of a viral eukaryotic promoter (e.g., CMV). Promoter placement to ensure the replicon RNA initiates at precisely the right nucleotide is critical [35–37].

2. Flavivirus RNAs are not polyadenylated. Thus, a mechanism is required to ensure the production of the flavivirus RNA with an authentic terminus following transcription by a cellular RNA polymerase in the nucleus of transfected cells. To accomplish this, our replicons encode a hepatitis delta virus (HDV) ribozyme followed by a SV40 polyadenylation signal at the 3′ end [34]. These elements facilitate export of the flavivirus replicon RNA from the nucleus and result in ribozyme-mediated cleavage to generate the authentic terminal residues of the viral genome.

3. Subgenomic replicons encode large deletions within the structural genes, preventing the production of fully infectious virus following transfection. These deletions are introduced in a manner to ensure that the single open reading frame of the viral genome remains intact, allowing translation of nonstructural genes downstream of the deletion.

4. Many replicons encode reporter genes or those that confer resistance to drugs used for selection in eukaryotic cells. When introducing such genes in place of the structural genes at the 5′ end, the 2A autoprotease of foot and mouth disease virus (FMDV) [38] is cloned downstream of the insertion. Upon translation, this protease is able to liberate the reporter or resistance gene from the flavivirus polyprotein.

2.1.2 Nomenclature and Brief Summary of Replicons Used Within This Chapter

While DENV RVPs can be generated using either a WNV or DENV subgenomic replicon, RVPs produced with a WNV replicon generally yield higher-titer stocks and allow infection to be scored at 2 days postinfection. By comparison, when DENV replicons are used, maximum signal is not achieved until roughly day 3 (*see* Subheading 3.3, **step 4**). Because functional studies with anti-flavivirus antibodies are not known to be impacted by the specific flavivirus replicon used to produce RVPs, our laboratory makes more extensive use of our WNV constructs. We have not yet studied the molecular mechanisms responsible for the higher titers achieved with WNV RNAs.

1. pWNII-rep-GZ: This DNA-launched replicon was constructed by modification of a WNV lineage II infectious clone [26, 35]. It encodes a fusion of GFP and the resistance gene for Zeocin at the 5′ end of the replicon.

2. pWNII-rep-Ren-IB: This DNA-launched replicon was constructed by modification of a WNV lineage II infectious clone [26]. It encodes Renilla luciferase at the 5′ end of the genome and a gene conferring resistance to blasticidin under translational control of the cricket paralysis virus IRES in the 3′UTR.

3. pDENV2-rep-GZ: This replicon was constructed by modification of the genome of the DENV serotype 2 strain 16681 [30]. It was constructed using the same DNA-launched configuration as pWNII-rep-GZ.

2.1.3 Mammalian Expression Vectors Encoding the DENV Structural Proteins

1. Several "standard" expression vectors encoding flavivirus structural proteins under the control of CMV or other eukaryotic promoters have been used to produce RVPs by complementation [30, 33]. The number of plasmids used to express the three structural genes may vary (*see* **Note 1**) [26]; the most straightforward approach is to encode CprME in a single expression plasmid.

2. Cell lines that express flavivirus structural genes in an inducible fashion have been described and used in RVP production experiments [30]. T-REx-293 inducible cell lines that stably express the tetracycline repressor protein (Subheading 3.1.3) were constructed by cloning DENV structural genes into the pT-REx-DEST30 plasmid (Invitrogen).

2.1.4 Mammalian Expression Vector Encoding the Human Furin Protease

This construct is used to produce populations of RVPs in which the efficiency of prM cleavage has been enhanced (prM⁻ or "mature" RVPs, described in Subheading 3.2.1). This plasmid will be referred to hereafter as pHuFurin.

2.1.5 Propagation of Flavivirus Plasmids	We have noted that many constructs that contain flavivirus sequences are very unstable when propagated in bacteria, particularly the large replicons used to produce RVPs. To reduce the potential for recombination and mutagenesis during plasmid production, all flavivirus plasmids are grown at 30 °C in bacteria strains with a genotype that minimizes recombination (e.g., Stbl2 (Invitrogen), JM109 (Sigma-Aldrich), or SURE2 (Agilent) cells).

2.2 Cell Culture

Culture media for the propagation of cell lines used within this chapter are detailed below. All cells are grown at 37 °C in a 7 % CO_2 incubator.

1. HEK-293T (ATCC), T-REx-293 (Invitrogen), and BHK-21 (ATCC) adherent cell lines are maintained in a high-glucose formulation of Dulbecco's Modified Eagle's Medium (DMEM) supplemented with 7 % fetal bovine serum (FBS) and 1× penicillin/streptomycin (PS) (complete high-glucose DMEM).

2. Raji-DCSIGNR, Raji-DCSIGN [39], and K562 (ATCC) cell lines are maintained in suspension in RPMI-1640 supplemented with 7 % FBS and 1× PS (complete RPMI). Cells should be maintained at a density of roughly 3×10^5 cells/mL (*see* **Note 2**).

2.3 Transfection Reagents

1. Low-glucose formulation of DMEM containing 25 mM HEPES and supplemented with 7 % FBS and 1 % PS (complete low-glucose DMEM) (*see* **Note 3**).

2. Lipofectamine LTX transfection reagent (Invitrogen).

3. Opti-MEM transfection medium (Invitrogen).

2.4 Buffers, Chemicals

1. 1 M NH_4Cl prepared in 1× PBS; pH adjusted to 7.4 using NaOH.

2. 32 % paraformaldehyde (PFA) solution.

3. Renilla Luciferase Assay System (Promega).

2.5 General Lab Supplies/Equipment

1. Flow cytometer with the capacity to excite and measure GFP fluorescence. Typically RVP assays are performed in our laboratory using a FACSCalibur cytometer (BD Biosciences). Use of the robotic High Throughput Sampler (HTS) option tremendously increases the ability to run the large numbers of samples required to generate detailed antibody-mediated neutralization or enhancement curves.

2. Experiments involving RVPs produced using luciferase-expressing replicons require the use of a luminometer to measure infection. Renilla and firefly luciferase require an instrument that can read emissions at 535 and 613 nm, respectively. Our laboratory uses a SpectraMax M5 luminometer (Molecular Devices) capable of reading 96-well plates.

3. Tissue culture flasks and plates, various sizes.

4. Sterile microcentrifuge tubes.

5. 0.2 μM filters and syringe units.

---

3 Methods

3.1 Production of DENV RVPs by Complementation

Flavivirus RVPs provide a quantitative tool for the study of flavivirus entry into cells and its inhibition by antibodies or small molecules [26, 30, 33, 40]. RVPs are incapable of multiple rounds of replication because the subgenomic replicon RNAs packaged by these virions do not encode the structural gene products required to produce an infectious virion. Several methods for the production of RVPs are possible, each with distinct experimental advantages. This chapter is focused on the production of RVPs for use in studies of antibody-mediated neutralization and enhancement. In this context, the study of multiple antigenic variants is desirable [41, 42]. The methods detailed below principally focus on strategies that allow for rapid mutagenesis and exchange of the structural gene components. Alternative methods to produce flavivirus RVPs may yield higher titers.

3.1.1 Co-transfection of Cells with Replicon and Structural Gene Plasmids

This straightforward method describes the production of RVPs by co-transfection of cells with plasmids encoding a flavivirus replicon and the DENV structural genes. To date, RVPs have been produced in multiple cell types using this method including Vero, BHK-21, and HEK-293T (*see* **Note 4**). The efficiency of this method is governed in part by the transfection efficiency of the RVP producer cell type, which may be optimized directly. Due to ease of transfection and relatively high levels of recombinant protein expression, our laboratory typically uses HEK-293T cells for RVP transfection experiments as detailed below.

1. Plating cells for transfection:
   The day before the transfection, plate 1×10^6 cells in each well of a 6-well plate in a volume of 2 mL per well. The use of a low-glucose formulation of DMEM (Subheading 2.3, **item 1**) slows the rate of media acidification (which can inactivate pH-sensitive DENV RVPs) and is used throughout the transfection protocol. Plating density should be adjusted to achieve ~80–90 % confluency on the day of transfection.

2. Lipid-mediated transfection:
   Pre-plated HEK-293T cells are transfected using Lipofectamine LTX according to the manufacturer's instructions. The volumes provided below are appropriate for transfecting a single well of a 6-well plate and can be scaled to accommodate changes in the surface area of the tissue culture plate/flask.

(a) Mix 1 μg replicon plasmid and 3 μg pDENV-CprME expression construct to a total volume of 500 μL with Opti-MEM reduced serum medium in a sterile microcentrifuge tube. Use of an equal amount of an empty expression vector in place of the CprME construct provides a useful negative control for each experiment.

(b) Add 12 μL Lipofectamine LTX reagent. This corresponds to a 1:3 ratio of μg DNA:μL Lipofectamine.

(c) Briefly vortex the lipid-DNA complex, gently spin the tube to collect contents and incubate for 30 min at room temperature (*see* **Note 5**).

(d) Gently add lipid-DNA complexes to cells dropwise, being careful not to disrupt the cell sheet. Gently swirl the plate before returning to the 37 °C incubator.

(e) After 4 h, replace the media with 2 mL of low-glucose complete DMEM (*see* **Note 6**).

(f) Transfected cells may then be further incubated at 37 °C or another temperature of the investigator's choice (Subheading 3.1.4).

3. Harvesting RVPs:
Culture media containing RVPs should be harvested and filtered through a 0.2 μM filter, aliquoted, and stored at –80 °C until use (*see* **Note 7**). Factors that determine the optimum time for harvesting are discussed in Subheading 3.1.4.

3.1.2 Transfecting a Stable Replicon-Expressing Cell Line with Structural Gene Plasmids

Cell lines that stably propagate flavivirus replicons have been described and may be employed to produce RVPs by complementation [29, 30, 43]. This reduces the complexity of RVP production procedures by eliminating the requirement to transfect the replicon plasmid (*see* **Note 8**).

1. Methods for the production of cell lines that harbor flavivirus replicons have been described previously [26, 30]. Briefly, RVPs are produced using a replicon that encodes a eukaryotic selectable marker (such as Zeocin in pWNII-rep-G:Z; Subheading 2.1.2). Cells (such as HEK-293T or BHK-21) are infected with RVPs, incubated for 48 h to allow for reporter gene expression, and then maintained in the presence of the selecting drug.

2. In order to generate RVPs, cell lines containing the replicon are transfected with 3 μg of a plasmid encoding the DENV structural genes. The transfection is performed identical to that described in Subheading 3.1.1, except 9 μL of Lipofectamine LTX is used (instead of 12 μL) in order to maintain a 1:3 ratio of μg DNA:μL Lipofectamine, accounting for the absence of the replicon plasmid at the transfection step.

3. RVPs are harvested as described above in Subheading 3.1.1.

3.1.3 Use of Stable Cell Lines to Generate DENV RVPs

Cell lines that may be induced to produce RVPs have been described [30]. The utility of this approach is that RVPs may be produced at a relatively large scale simply by plating large numbers of cells, and does not require the use of expensive transfection reagents. In our experience, a downside of this approach is the difficulty of propagating cells that harbor both a replicon and an inducible structural gene cassette.

1. Cell lines that may be induced to produce RVPs are constructed in two steps. First, a cell line that constitutively expresses the tetracycline repressor protein (such as the T-Rex 293 cell line) is infected with RVPs to produce a cell line that stably propagates a flavivirus replicon (as described in Subheading 3.1.2). After selection of a stable replicon-expressing line, this cell line is transfected with a DNA expression construct in which the structural genes are under the control of a tetracycline-inducible promoter, followed by additional antibiotic selection. Our only experience with this approach exploits the Invitrogen T-REx system.

2. RVP production is induced by the addition of an appropriate concentration of tetracycline (this should be determined empirically).

3. RVPs are harvested as described above in Subheading 3.1.1.

3.1.4 Determining the Optimal Kinetics for Harvesting RVPs

The kinetics of infectious RVP release from cells is governed by several factors. The most significant factor is the temperature at which RVP producing cells are cultured [30]. For example, WNV RVP production peaks at day 2 when cells are cultured at 37 °C and on day 3 when incubated at 30 °C. Flaviviruses may be differentially stable in solution and thus decay at different rates in a temperature-dependent fashion. DENV RVP production is most efficient at lower temperature (30 °C) due in part to increased virus particle stability at this temperature [30]. Structural gene variants may alter the rate of RVP production and/or stability; the optimal kinetics for production should be determined empirically by serial harvest of RVP-containing supernatants (*see* **Note 9**).

3.2 Manipulation of the Maturation State of DENV RVPs

Populations of flaviviruses released from infected cells are heterogeneous with respect to the efficiency of the virion maturation process [44]. Virus stocks may contain significant levels of uncleaved prM protein. While the stoichiometry of prM cleavage required for the production of an infectious virus has not yet been determined, several lines of evidence suggest partially mature prM-containing viruses are infectious [44]. The transfection protocols in Subheading 3.1 describe production of a "standard" preparation of RVPs composed of immature, mature, and partially mature RVPs in unknown proportions. Modifications to this protocol to enhance or decrease the efficiency of the maturation process are possible and are described below. These result in considerably more homogeneous populations of immature and mature forms of the virus.

3.2.1 Generating "Mature" or prM⁻ DENV RVPs

The mature form of DENV is characterized by complete cleavage of prM during viral egress by the cellular serine protease furin [45]. The efficiency of maturation may be enhanced by co-transfection of a plasmid encoding human furin (pHuFurin) into RVP producing cells [39, 43].

1. Mix the following plasmids with Opti-MEM to a final volume of 500 μL.

   1 μg replicon,

   3 μg pDENV-CprME, and

   1 μg pHuFurin.

2. Add 15 μL Lipofectamine LTX reagent for a 1:3 ratio of μg DNA:μL Lipofectamine.

3. Perform transfection and harvest RVPs as described in Subheading 3.1.1.

3.2.2 Generating "Immature" or prM⁺ DENV RVPs

Cleavage of prM by furin-like proteases occurs during viral egress through the trans-Golgi network and is pH dependent due to a requirement for conformational changes in the virion associated with exposure of the furin cleavage site [46]. Furin-mediated cleavage can be inhibited by treating cells with the weak base ammonium chloride (NH_4Cl), which raises the pH of the intracellular compartments and prevents cleavage of prM by furin [43, 47].

1. Transfect cells as described in Subheading 3.1.1, with the following modifications.

2. Replace media at 4 h post-transfection with complete low-glucose DMEM supplemented with 20 mM NH_4Cl.

3. After 1 h, replace the media again with complete low-glucose DMEM containing 20 mM NH_4Cl. Replacing the media at this point ensures that any RVPs produced shortly after transfection (before the NH_4Cl was present or had taken effect) are removed.

4. The kinetics of prM⁺ RVP production is generally faster than standard or mature preparations and peaks at much lower titers. Earlier time points, such as 1 day post-transfection, may need to be harvested. If harvesting on multiple days post-transfection (recommended), continue to replace with media containing NH_4Cl (*see* **Notes 9** and **10**).

3.2.3 Measuring the Efficiency of prM Cleavage on RVPs

The effectiveness of manipulations that change the efficiency of prM cleavage may be assayed biochemically and functionally. Biochemically, decreases in the efficiency of prM cleavage can be assayed by Western blot and manifest as an increase in the level of uncleaved prM in NH_4Cl preparations relative to a standard

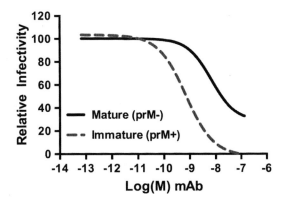

Fig. 2 Assessing the efficiency of prM cleavage on DENV RVPs. The extent of prM cleavage can be altered by producing DENV RVPs in the presence of NH_4Cl or exogenous furin, resulting in populations of virus particles that resemble immature or mature forms of the virus, respectively. Biochemically, the efficiency of cleavage can be measured by Western blot analysis with an antibody specific for the M peptide of prM (the portion that remains within the virus particle after cleavage is complete). Alternately, the maturation state of a RVP stock can be assessed functionally. Depicted are the results of a neutralization assay performed with DENV RVPs and an antibody sensitive to the maturation state of the virus. The epitope for this particular antibody is not readily accessible on the mature form of the virus, resulting in decreased neutralization potency

preparation produced in parallel. The opposite results are expected when producing RVPs in cells overexpressing exogenous furin. Differences in the maturation state between RVP preparations can also be assayed functionally by performing neutralization assays with certain "maturation-sensitive" antibodies. This particular class of flavivirus-specific antibodies has been shown to recognize epitopes that are more readily accessible on the immature form of the virus [43, 48]. Using such antibodies, RVPs produced in the presence of NH_4Cl will be more sensitive to neutralization as compared to RVPs produced in the presence of furin (Fig. 2).

3.3 Determining the Infectious Titer of RVP Preparations

RVP titer may be determined on any DENV-permissive cellular substrate selected by the investigator. Our laboratory routinely uses Raji B-cell lines that stably express the flavivirus attachment factors DC-SIGN or DC-SIGNR (*see* **Note 11**) [30, 39].

1. Using complete RPMI-1640 media as a diluent, perform two-fold dilutions of an RVP stock across a 96-well plate. Generally, we assay each dilution in duplicate. To avoid artifacts associated with edge effects, only the inner 60 wells of the plate are utilized (*see* **Note 12**) [49].

2. To the diluted RVPs, add 5×10^4–1×10^5 Raji-DCSIGN (or DC-SIGNR) cells per well to a total volume of 200 μL (100 μL RVP, 100 μL cells).

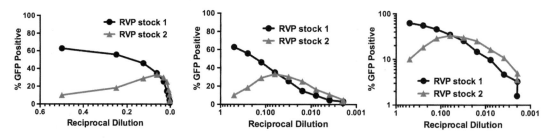

Fig. 3 Titering of DENV RVPs. Twofold dilutions of DENV RVPs were incubated with Raji-DCSIGNR cells and the % GFP-positive infected cells assessed by cytometry; the same results are shown plotted three separate ways. Depicted are the titer results of two RVP stocks that illustrate often-observed patterns of infectivity. RVP stock 1 initially follows a nonlinear titration pattern, such that a twofold dilution of virus results in a less than 50 % reduction in infected cells. RVP stock 2 contains an unknown factor that is sometimes produced in HEK-293T cells and is inhibiting infectivity until sufficiently diluted out

3. Fill the unused outer wells with media and then incubate the plate at 37 °C in a 7 % CO_2 incubator.

4. After an appropriate incubation (48 h if using a WNV replicon or 72 h if using a DENV replicon (*see* **Note 13**)), stop the infection by adding 32 % paraformaldehyde (PFA) to a final concentration of 1.2 %. To achieve maximal cell dispersion, vigorously resuspend the cells (by pipetting up and down) both prior to and during the addition of PFA (*see* **Note 14**).

5. Vigorously resuspend the cells immediately prior to analysis by flow cytometry to detect the number of GFP-positive cells (*see* **Note 15**). Use of the HTS on a BD FACSCalibur allows for the analysis of the inner 60 wells of a 96-well plate in roughly 45 min (with settings set to measure a volume of 100 μL/well).

6. Alternatively, the results of infection can be scored using luminometry. The luminescence data is collected as an endpoint measurement with a 2-s delay followed by a 10-s measurement period.

3.4 Calculating DENV RVP Titer

The relationship between the amount of RVPs added to cells and the number of infected cells is typically nonlinear for reasons that are not yet clear (Fig. 3). This complicates determining the infectious titer of an RVP stock using a single dilution of virus particles. This section includes a description of commonly observed nonlinear characteristics of RVP titer data (many of which have also been observed with fully infectious flavivirus preparations), followed by a discussion of how to best approximate the infectious titer of an RVP preparation.

1. As shown in Fig. 3 (RVP stock 1), the percentage of GFP-positive cells observed over serial dilutions of a particular RVP stock may be nonlinear, such that a twofold dilution in RVPs does not always result in a 50 % reduction in the percentage of

infected cells. Generally, this phenomenon appears in dilutions at the "high" end of the titration curve, i.e., the wells that contain the largest amounts of RVPs.

2. In some, but not all experiments, the percentage of infected cells may actually increase with increasing RVP dilutions. Eventually, this relationship is reversed, and the titration curve follows the expected pattern in which infectivity decreases as the RVP stock is further diluted (Fig. 3; RVP stock 2). This phenomenon is thought to be caused by an unknown inhibitory factor that is released from RVP producer cells after transfection. Inhibition at the "high" end of the titration curve results in infectivity results that are not representative of the actual amount of RVPs present at that dilution. In the example provided in Fig. 3, a 1:4 dilution of RVPs yields the same percentage of infected cells as a 1:128 dilution, despite the 32-fold difference in the amount of RVPs present. A better estimate of the infectivity of this stock is obtained by analysis at the "low" end of the titration curve, at which point the unknown inhibitory factor has been sufficiently diluted to a concentration at which it no longer reduces infection (*see* **Note 16**).

3. Calculation of RVP infectious titer. It is best to calculate the RVP titer by averaging three to four data points that span the most linear portion of the titration curve using the following formula:

$$Infectious\ units/mL\ (IU/mL) = (proportion\ of\ infected\ cells \times number\ of\ cells\ in\ the\ experiment)\ /\\ mL\ of\ virus - containing\ supernatant).$$

The most linear portion is found at the "low" end of the titration curve (the section of the curve that has the most dilute RVPs, yet robust data that is significantly above background). The examples described above and illustrated in Fig. 3 underscore the necessity of measuring 10 twofold dilutions when determining the titer. Fewer dilutions would often result in uninformative portions of the titration curve, resulting in a miscalculation of the actual infectious titer.

3.5 Use of DENV RVPs to Measure Antibody-Mediated Neutralization of Infection

Once titered, DENV RVP stocks are ready for use in antibody neutralization assays. This section will provide a detailed protocol and discussion of the graphing and analysis of results. Additionally, the importance of satisfying the laws of mass action in the assay will be discussed.

3.5.1 Performing a Standard Neutralization Assay Using Raji-DCSIGNR Cells

Neutralization assays are generally performed in duplicate or triplicate and utilize the inner ten wells of each row of a 96-well plate. All dilutions should be made in complete RPMI-1640 media.

1. Perform nine serial dilutions of antibody and aliquot 100 μL of each dilution into adjacent wells of a 96-well plate. To the tenth well, add 100 μL media as a control.

2. Dilute an RVP stock to a concentration that will not violate mass action (*see* Subheading 3.5.4). Add 100 μL of diluted RVPs to each well.

3. Add media to the unused outer wells of the plate.

4. Incubate antibody-virus complexes for 1 h at 37 °C to allow the binding reaction to reach equilibrium (*see* **Note 17**).

5. Add between 5×10^4 and 1×10^5 Raji-DCSIGN or Raji-DCSIGNR cells per well for a total volume of 300 μL (100 μL antibody, 100 μL RVP, 100 μL cells).

6. Incubate at 37 °C for 48 h if using RVPs produced with a WNV replicon or 72 h for RVPs produced with a DENV replicon (*see* **Note 13**).

7. Stop infection by adding PFA to a final concentration of 1.2 % (*see* **Note 14**) and determine the number of GFP-positive cells/well on a cytometer.

3.5.2 Graphing of Results from DENV RVP Neutralization Assays

Precise and reproducible measurements of the neutralization potency, such as the EC_{50}, can be calculated from RVP neutralization assay results.

1. Using an available scientific graphing program (our lab uses Prism (GraphPad Software)), plot the percent GFP-positive cells on the *y*-axis against log-transformed antibody concentrations on the *x*-axis.

2. Examine your results. Most dose-response neutralization curves will have a sigmoidal shape. Using graphing software, fit the data with a curve using the option for nonlinear regression, allowing for a variable slope.

3.5.3 Analysis of Nonlinear Regression Curves Fit to Neutralization Data

While graphing software will fit a curve to most any data set, care should be taken to visually inspect your results. Some tips on assessing the quality of dose-response neutralization curves are discussed below. Investigators should also utilize the statistical outputs of the graphing software, such as the 95 % confidence interval of the fitted curve, to assess the robustness of the data.

1. Best-fitting regression curves will contain a plateau region at the "top," such that the percentage of infected cells in the presence of the lowest concentrations of antibody is equivalent to that of the no-antibody control well. When the results are normalized such that the infectivity in the absence of antibody (the control well) is set to 100 %, the curve will plateau at 100 (Fig. 4, left panel). An incomplete data set is shown in the right panel of Fig. 4; the antibody dilution at the lowest concentration is still capable of neutralizing the RVPs, as compared to the control well. Experiments that yield such data should be repeated, altering the antibody dilution scheme such that lower concentrations are reached.

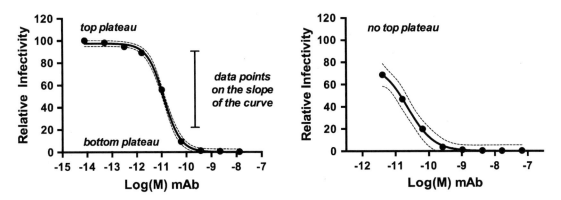

Fig. 4 Nonlinear regression analysis of RVP neutralization curves. Results from two neutralization assays were normalized such that the infectivity in the absence of antibody was set to 100 %. The data was then fit with nonlinear regression curves. The *left panel* illustrates a robust data set that has plateau regions at the *top* and *bottom* and data points on the slope of the curve. In contrast, the antibody dilutions used to generate the data set in the *right panel* did not capture the entire dose-response curve. The *dotted lines* represent the 95 % confidence intervals of the regression curves and demonstrate the effect that the quality of the data can have on the accuracy of the analysis

2. Best-fitting regression curves will also plateau at the "bottom" of the curve, such that the concentration of antibody has reached saturation and neutralization is complete. For the example curve shown in Fig. 4, this plateau corresponds to 100 % neutralization.

3. The quality of data in the "middle" of the curve also requires attention, as this portion contains the data points that will be critical in calculation of the EC_{50} value of an antibody. The Hill slope is a measure of the steepness of this portion of the curve and can vary significantly among antibodies (hence, the importance of choosing the option of fitting a line with a variable slope (Subheading 3.5.2)). For antibodies that yield a very steep Hill slope (Fig. 4, left panel), it may be useful to alter the antibody dilutions in the assay in such a way that maximizes the number of data points in this region (e.g., performing threefold instead of fourfold dilutions). Conversely, neutralization assays performed with antibodies that yield shallow Hill slopes (Fig. 5, left panel) may need to be altered such that a broader range of antibody concentrations is covered.

4. Additional consideration must be given when neutralization assays are performed with antibodies sensitive to the maturation state of the virus (discussed in Subheading 3.2.3). When a maturation-sensitive antibody is used to neutralize a "mature" (prM$^-$) or standard RVP preparation, the resulting curve may show evidence of a resistant fraction. The resistant fraction is composed of viral particles that cannot be neutralized, even in the presence of saturating concentrations of antibody,

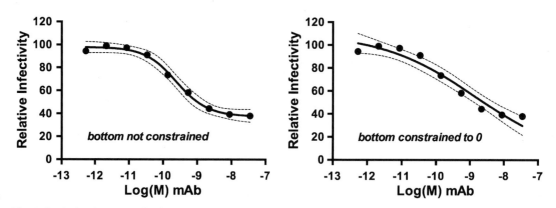

Fig. 5 Analysis of a neutralization assay with a resistant fraction. "Mature" (prM⁻) RVPs were incubated with an antibody sensitive to the maturation state of the virus. At saturating concentrations of antibody, a resistant fraction indicates the presence of virions that cannot be neutralized because the number of available epitopes is below the threshold required for neutralization. The same data set is shown analyzed two ways. The *left panel* does not place any constraint on the bottom of the curve. In the *right panel*, the curve was generated by constraining the bottom to zero, based on the incorrect assumption that saturating antibody concentrations would result in complete neutralization of infectivity. The *dotted lines* represent the 95 % confidence intervals of the regression curves

because the number of accessible epitopes is below the threshold required for neutralization [50]. A neutralization curve with a resistant fraction is shown in Fig. 5. The same data set is shown analyzed two ways; in the left panel, the regression analysis is performed with no constraint on the bottom of the curve, while in the right panel the bottom was constrained to zero. The analysis in the left panel clearly has the best-fit curve and underscores the importance of understanding the biology of flavivirus-antibody interactions when analyzing assay results.

5. Not all neutralization assays can be designed to fit the criteria outlined above. The concentration of antibody required to reach saturating levels may not be possible due to volume constraints of the assay or the need to conserve antibody samples. In addition, RVP mutagenesis studies designed to identify resistant variants will by design not fit the "ideal" curve. Such results are by no means uninformative. Investigators must analyze results taking such caveats into consideration.

3.5.4 Compliance with the "Percentage Law"

The goal of neutralization assays is to identify the conditions that enable an antibody to block virus infection. In this context, the most useful results are those that are dependent solely on the concentration of antibody in the experiment and its affinity for the viral antigen. The percentage law (which is an extension of the assumptions of the law of mass action) states that the result of a neutralization assay should not be affected by virus input (antigen concentration), as discussed in [51]. The most reproducible

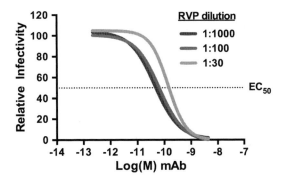

Fig. 6 Violating mass action in a neutralization assay. A high-titer RVP stock was used in a neutralization assay at either a 1:30, 1:100, or 1:1,000 dilution. The two latter dilutions satisfy the percentage law, resulting in identical neutralization curves despite the tenfold difference in RVP input. At a 1:30 dilution, however, the high amount of antigen (RVP) violates mass action, resulting in a neutralization curve that provides a false value of the EC_{50} concentration for this particular antibody

results are those that comply with this rule. The relatively high specific infectivity of flavivirus RVPs and existence of sensitive and specific methods to detect infection allow experiments to be carried out with a minimal amount of antigen (RVP) input, the important criteria for satisfying the percentage law. To validate compliance with the "percentage law":

1. Perform a series of neutralization assays in which the concentration of RVPs is varied over a range of dilutions (Fig. 6). If the level of antigen is significantly high in a particular RVP stock, lower dilutions of virus (i.e., more RVPs) will yield dose-response curves in which the EC_{50} is shifted towards higher antibody concentrations (Fig. 6, RVP dilution 1:30). This shift indicates that mass action has been violated; the results no longer reflect the characteristics of the antibody, but are a measure of the amount of antigen in the well. One must identify the range of RVP concentrations where the dose-response curve is unaffected by increases or decreases in RVP input.

2. It is important to remember that all DENV RVP stocks will contain some level of subviral particles (SVPs) [52], which although functionally invisible (noninfectious) may be a likely culprit when mass action is violated. When expressing DENV structural genes from separate plasmids, special care must be taken to minimize the production of SVPs (*see* **Note 1**).

3.6 Antibody-Dependent Enhancement (ADE) Assay with DENV RVPs

In addition to infecting cells through poorly understood virus attachment-endocytosis mechanisms, DENV can infect certain cell types expressing Fc or complement receptors in the presence of DENV-reactive antibodies. This phenomenon is referred to as ADE and

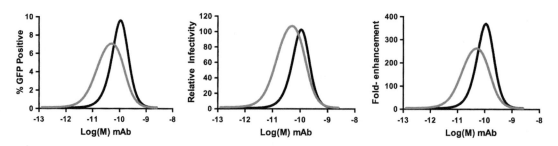

Fig. 7 Antibody-dependent enhancement (ADE) curves. Example ADE curves are shown analyzed three independent ways. In the *left panel*, the raw data (% GFP-positive infected cells) is plotted against increasing antibody concentration. In the *middle panel*, the results have been normalized such that the peak infectivity was set to 100 %. Finally, in the *right panel* the results are shown as fold enhancement of infectivity as compared to the infectivity in the absence of antibody

involves uptake of antibody-virus complexes by Fc receptors [53]. An increase in the efficiency of virus infection is observed at sub-neutralizing concentrations of the DENV-specific antibody [54].

3.6.1 Performing a Standard ADE Assay

ADE assays are identical to neutralization assays described in Subheading 3.5 with the exception of the cell type used. Our lab generally uses K562 cells, which express the Fcγ-RII receptor (CD32A). 5×10^4–1×10^5 K562 cells are added per well.

3.6.2 Analysis of ADE Assays

1. Using a scientific graphing program, plot the percent GFP-positive cells on the *y*-axis against log-transformed antibody concentrations on the *x*-axis. For antibodies that are capable of complete neutralization of DENV, the results typically resemble a bell-shaped curve (Fig. 7, left panel).

2. In addition to the raw data (percent infected cells), results from ADE assays can be expressed after normalizing to a reference data point. One option is to normalize the results such that the data point that resulted in the highest infectivity is set to 100 % (Fig. 7, middle panel). Alternately, the results can be expressed as the fold enhancement in infectivity compared to the low level of infectivity observed in the control well in the absence of antibody (Fig. 7, right panel). The ideal method for presenting the data is up to the investigator's discretion and will depend on the experiment.

4   Notes

1. DENV RVPs can be generated with a single plasmid encoding CprME or using two distinct plasmids encoding C and prME. The advantage of the two-plasmid approach is a reduced chance of DNA recombination following transfection and reduced plasmid size (which simplifies mutagenesis).

However, care should be taken to limit the production of large amounts of noninfectious subviral particles. Subviral particles (SVP) can be produced by transfection with only prME and are formed even in the absence of viral proteins encoded by the replicon [52]. To limit the number of cells transfected solely with the prME expressing construct (which would make SVPs), the ratio of plasmids used during transfection should be adjusted. For example, the transfection protocol described in Subheading 3.1.1 should be modified as follows:

1 μg WNVII-rep-GZ.

2.8 μg pDENV-C.

0.2 μg pDENV-prME.

2. Healthy Raji B-cell lines form small clumps and settle to the bottom of the flask. Care should be taken to resuspend cells before splitting or using in experiments.

3. The manufacturer's instructions for Lipofectamine LTX suggest transfecting cells in the absence of antibiotics, as they have been shown to interfere in some cell types. Using the protocols describes within, we have found no difference in the efficiency of RVP production in HEK-293T cells when transfections were performed in the presence or absence of PS.

4. Investigators should be wary of producing DENV RVPs in cell lines that mount a robust interferon response. It has been demonstrated that lipid-mediated transfection of DNA triggers the production of interferon [55]; this could subsequently inhibit replication of the transfected replicon RNA due to the inherent interferon sensitivity of flaviviruses.

5. We incubate DNA-lipid transfection complexes for a minimum of 30 min. In our hands, longer incubations (up to 4 h) have no significant effect on transfection efficiency.

6. HEK-293T cell monolayers are easily disturbed by excessive movement of the tissue culture plate or forceful addition of medium. Extra care must be taken to keep the monolayer in place, as this allows harvesting of RVPs on multiple days (each time involving the removal and replacement of media). In these instances, we find it best to tilt the plate, and add media dropwise against the side of tissue culture vessel.

7. The ideal method for producing and storing RVPs is to avoid repetitively freezing and thawing a stock of RVPs. This is simplified by aliquoting RVPs into experimentally appropriate volumes. However, an appropriate volume may be difficult to determine for new types of experiments or when using new structural gene variants for which the expected titer is unknown. Using the protocols described within, we have found that infectivity of RVP stocks remains stable for numerous (~4) thaws.

8. In our hands, the use of cell lines that stably maintain replicon RNA to produce DENV RVPs confers no significant advantage in RVP titers achieved as compared to the co-transfection protocol described in Subheading 3.1.1. While the use of stable cell lines simplifies the transfection protocol, our HEK-293T cell lines that constitutively propagate flavivirus replicons adhere less robustly to tissue culture plastic and can be difficult to work with, and our BHK cell lines are less transfectable.

9. In our hands, optimal DENV RVP production occurs in HEK-293T cells that are cultured at 30 °C. In general, RVPs of consistently high titer can be harvested at 48, 72, and 96 h post-transfection. Incubation at 37 °C yields comparable titers, but not for the same length of time; RVP titers at this temperature often peak and quickly decline within 72 h post-transfection. One exception to this rule is the generation of "immature" (prM$^+$) DENV RVPs; these should be produced at 37 °C as the level of uncleaved prM retained on the particles is reduced when 30 °C is used.

10. prM$^+$ RVPs are rather difficult to make due to their low titers and the detrimental effect that long-term incubation in the presence of NH$_4$Cl has on 293T cells. Alternatively one can use furin inhibitors to reduce the efficiency of furin cleavage; in our hands use of the furin inhibitor Dec-RVKR-CMK (Enzo Life Sciences) poses less detrimental effects on the HEK-293T producer cells. Scaling up (transfection of more wells/flasks, harvesting daily) while producing prM$^+$ RVPs is often necessary.

11. Raji-DCSIGN or Raji-DCSIGNR cells are preferentially used in our lab for titer analysis and neutralization assays due to their high permissiveness to DENV RVP infection. Other cell types of the investigator's choice may be employed. Efficient infection has been observed with Vero cells.

12. The outer wells of 96-well plates are not used in our experiments to avoid irregularities associated with the incompletely understood "edge effect." Because this "edge effect" may not be completely abrogated by excluding the outer wells, experiments should be designed to minimize any effect the positioning of samples on the plate might have on the results or analysis.

13. When using a WNV replicon, GFP expression peaks at ~30 h and remains stable until ~50 h postinfection. Our DENV2 replicon expresses GFP with reduced kinetics. As a practical matter, experiments using RVPs generated with WNV or DENV replicons can be stopped any time 2 or 3 days postinfection, respectively.

14. Raji-DCSIGNR cells form clumps that settle to the bottom of the well during an experiment. We have found it necessary to break up these clumps before the addition of PFA, as well as thoroughly mix in the PFA during addition. Not doing so will result in significantly fewer events read on the cytometer.

15. The number of total events (cells) to collect per well is up to the researcher's discretion. Generally, our lab attempts to collect ~80,000 events/well.

16. In our experience, high levels of the inhibitory factor are more often observed when DENV RVPs are harvested from 293T cells incubated at 37 °C as opposed to 30 °C after transfection.

17. Through the process of structural dynamics, flavivirus virions explore an ensemble of conformations at equilibrium [33]. Binding of some DENV-specific antibodies is particularly temperature sensitive, requiring incubation at 37 °C for binding to fully occur [56].

References

1. Gubler DJ, Kuno G, Markoff L (2007) Flaviviruses. In: Knipe DM, Howley PM (eds) Fields virology. Lippincott-Raven, Philadelphia, pp 1154–1227

2. Monath TP (1994) Dengue: the risk to developed and developing countries. Proc Natl Acad Sci U S A 91:2395–2400

3. Halstead SB (2007) Dengue. Lancet 370: 1644–1652

4. Rico-Hesse R (1990) Molecular evolution and distribution of dengue viruses type 1 and 2 in nature. Virology 174:479–493

5. Holmes EC, Twiddy SS (2003) The origin, emergence and evolutionary genetics of dengue virus. Infect Genet Evol 3:19–28

6. Mukhopadhyay S, Kuhn RJ, Rossmann MG (2005) A structural perspective of the flavivirus life cycle. Nat Rev Microbiol 3:13–22

7. Lindenbach BD, Rice CM (2003) Molecular biology of flaviviruses. Adv Virus Res 59:23–61

8. Khromykh AA, Westaway EG (1997) Subgenomic replicons of the flavivirus Kunjin: construction and applications. J Virol 71: 1497–1505

9. Khromykh AA (2000) Replicon-based vectors of positive strand RNA viruses. Curr Opin Mol Therapeut 2:555–569

10. Hoenninger VM, Rouha H, Orlinger KK et al (2008) Analysis of the effects of alterations in the tick-borne encephalitis virus 3′-noncoding region on translation and RNA replication using reporter replicons. Virology 377:419–430

11. Chen Y-L, Yin Z, Duraiswamy J et al (2010) Inhibition of dengue virus RNA synthesis by an adenosine nucleoside. Antimicrob Agents Chemother 54(7):2932–2939

12. Suzuki R, Fayzulin R, Frolov I et al (2008) Identification of mutated cyclization sequences that permit efficient replication of West Nile virus genomes: use in safer propagation of a novel vaccine candidate. J Virol 82(14): 6942–6951

13. Liu WJ, Sedlak PL, Kondratieva N et al (2002) Complementation analysis of the flavivirus Kunjin NS3 and NS5 proteins defines the minimal regions essential for formation of a replication complex and shows a requirement of NS3 in cis for virus assembly. J Virol 76: 10766–10775

14. Khromykh AA, Meka H, Guyatt KJ et al (2001) Essential role of cyclization sequences in flavivirus RNA replication. J Virol 75: 6719–6728

15. Manzano M, Reichert ED, Polo S et al (2011) Identification of cis-acting elements in the 3′-untranslated region of the dengue virus type 2 RNA that modulate translation and replication. J Biol Chem 286: 22521–22534

16. Tilgner M, Deas TS, Shi P-Y (2005) The flavivirus-conserved penta-nucleotide in the 3′ stem-loop of the West Nile virus genome requires a specific sequence and structure for RNA synthesis, but not for viral translation. Virology 331:375–386

17. Holden KL, Stein DA, Pierson TC et al (2006) Inhibition of dengue virus translation and RNA synthesis by a morpholino oligomer targeted to the top of the terminal 3′ stem-loop structure. Virology 344:439–452

18. Chang DC, Liu WJ, Anraku I et al (2008) Single-round infectious particles enhance immunogenicity of a DNA vaccine against West Nile virus. Nat Biotechnol 26:571–577

19. Harvey TJ, Anraku I, Linedale R et al (2003) Kunjin virus replicon vectors for human immunodeficiency virus vaccine development. J Virol 77:7796–7803

20. Suzuki R, Winkelmann ER, Mason PW (2009) Construction and characterization of a single-cycle chimeric flavivirus vaccine candidate that protects mice against lethal challenge with dengue virus type 2. J Virol 83(4):1870–1880

21. Patkar CG, Larsen M, Owston M et al (2009) Identification of inhibitors of yellow fever virus replication using a replicon-based high-throughput assay. Antimicrob Agents Chemother 53:4103–4114

22. Puig-Basagoiti F, Qing M, Dong H et al (2009) Identification and characterization of inhibitors of West Nile virus. Antiviral Res 83:71–79

23. Ishikawa T, Widman DG, Bourne N et al (2008) Construction and evaluation of a chimeric pseudoinfectious virus vaccine to prevent Japanese encephalitis. Vaccine 26:2772–2781

24. Lo MK, Tilgner M, Shi PY (2003) Potential high-throughput assay for screening inhibitors of West Nile virus replication. J Virol 77(23):12901–12906

25. Khromykh AA, Varnavski AN, Westaway EG (1998) Encapsidation of the flavivirus kunjin replicon RNA by using a complementation system providing Kunjin virus structural proteins in trans. J Virol 72:5967–5977

26. Pierson TC, Sánchez MD, Puffer BA et al (2006) A rapid and quantitative assay for measuring antibody-mediated neutralization of West Nile virus infection. Virology 346:53–65

27. Jones CT, Patkar CG, Kuhn RJ (2005) Construction and applications of yellow fever virus replicons. Virology 331:247–259

28. Scholle F, Girard YA, Zhao Q et al (2004) trans-Packaged West Nile virus-like particles: infectious properties in vitro and in infected mosquito vectors. J Virol 78:11605–11614

29. Puig-Basagoiti F, Deas TS, Ren P et al (2005) High-throughput assays using a luciferase-expressing replicon, virus-like particles, and full-length virus for West Nile virus drug discovery. Antimicrob Agents Chemother 49:4980–4988

30. Ansarah-Sobrinho C, Nelson S, Jost CA et al (2008) Temperature-dependent production of pseudoinfectious dengue reporter virus particles by complementation. Virology 381:67–74

31. Harvey TJ, Liu WJ, Wang XJ et al (2004) Tetracycline-inducible packaging cell line for production of flavivirus replicon particles. J Virol 78:531–538

32. Mattia K, Puffer BA, Williams KL et al (2011) Dengue reporter virus particles for measuring neutralizing antibodies against each of the four dengue serotypes. PloS One 6:e27252

33. Dowd KA, Jost CA, Durbin AP et al (2011) A dynamic landscape for antibody binding modulates antibody-mediated neutralization of West Nile virus. PLoS Pathog 7:e1002111

34. Varnavski AN, Young PR, Khromykh AA (2000) Stable high-level expression of heterologous genes in vitro and in vivo by noncytopathic DNA-based Kunjin virus replicon vectors. J Virol 74:4394–4403

35. Pierson TC, Diamond MS, Ahmed AA et al (2005) An infectious West Nile virus that expresses a GFP reporter gene. Virology 334:28–40

36. Yamshchikov V, Mishin V, Cominelli F (2001) A new strategy in design of + RNA virus infectious clones enabling their stable propagation in E. coli. Virology 281:272–280

37. Mishin VP, Cominelli F, Yamshchikov VF (2001) A "minimal" approach in design of flavivirus infectious DNA. Virus Res 81:113–123

38. Varnavski AN, Khromykh AA (1999) Noncytopathic flavivirus replicon RNA-based system for expression and delivery of heterologous genes. Virology 255:366–375

39. Davis CW, Nguyen H-Y, Hanna SL et al (2006) West Nile virus discriminates between DC-SIGN and DC-SIGNR for cellular attachment and infection. J Virol 80:1290–1301

40. Qing M, Liu W, Yuan Z et al (2010) A high-throughput assay using dengue-1 virus-like particles for drug discovery. Antiviral Res 86:163–171

41. Sukupolvi-Petty S, Austin SK, Engle M et al (2010) Structure and function analysis of therapeutic monoclonal antibodies against dengue virus type 2. J Virol 84:9227–9239

42. Nelson S, Poddar S, Lin T-Y et al (2009) Protonation of individual histidine residues is not required for the pH-dependent entry of west nile virus: evaluation of the "histidine switch" hypothesis. J Virol 83:12631–12635

43. Nelson S, Jost CA, Xu Q et al (2008) Maturation of West Nile virus modulates sensitivity to antibody-mediated neutralization. PLoS Pathog 4:e1000060

44. Pierson TC, Diamond MS (2012) Degrees of maturity: the complex structure and biology of flaviviruses. Curr Opin Virol 2:168–175

45. Kuhn RJ, Zhang W, Rossmann MG et al (2002) Structure of dengue virus: implications for flavivirus organization, maturation, and fusion. Cell 108:717–725

46. Yu IM, Zhang W, Holdaway HA et al (2008) Structure of the immature dengue virus at low pH primes proteolytic maturation. Science 319:1834–1837

47. Gollins SW, Porterfield JS (1986) The uncoating and infectivity of the flavivirus West Nile on interaction with cells: effects of pH and ammonium chloride. J Gen Virol 67(Pt 9):1941–1950

48. Cherrier MV, Kaufmann B, Nybakken GE et al (2009) Structural basis for the preferential recognition of immature flaviviruses by a fusion-loop antibody. EMBO J 28:3269–3276

49. Lundholt BK, Scudder KM, Pagliaro L (2003) A simple technique for reducing edge effect in cell-based assays. J Biomol Screen 8:566–570

50. Pierson TC, Xu Q, Nelson S et al (2007) The stoichiometry of antibody-mediated neutralization and enhancement of West Nile virus infection. Cell Host Microbe 1:135–145

51. Klasse PJ, Sattentau QJ (2001) Mechanisms of virus neutralization by antibody. Curr Top Microbiol Immunol 260:87–108

52. Wang P-G, Kudelko M, Lo J et al (2009) Efficient assembly and secretion of recombinant subviral particles of the four dengue serotypes using native prM and E proteins. PloS One 4:e8325

53. Halstead SB (2003) Neutralization and antibody-dependent enhancement of dengue viruses. Adv Virus Res 60:421–467

54. Morens DM, Halstead SB, Marchette NJ (1987) Profiles of antibody-dependent enhancement of dengue virus type 2 infection. Microb Pathog 3:231–237

55. Muñoz-Jordan JL, Sánchez-Burgos GG, Laurent-Rolle M et al (2003) Inhibition of interferon signaling by dengue virus. Proc Natl Acad Sci U S A 100:14333–14338

56. Lok S-M, Kostyuchenko V, Nybakken GE et al (2008) Binding of a neutralizing antibody to dengue virus alters the arrangement of surface glycoproteins. Nat Struct Mol Biol 15:312–317

Chapter 7

Cell-Based Flavivirus Infection (CFI) Assay for the Evaluation of Dengue Antiviral Candidates Using High-Content Imaging

Kah Hin Tan, Kitti Chan Wing Ki, Satoru Watanabe, Subhash G. Vasudevan, and Manoj Krishnan

Abstract

Large-scale screening of antiviral compounds that target dengue virus life cycle requires a robust cell-based assay that is rapid, easy to conduct, and sensitive enough to be able to assess viral infectivity and cell viability so that antiviral efficacy can be measured. In this chapter we describe a method that uses high-content imaging to evaluate the in vitro antiviral efficacy in a modification to the cell-based flavivirus immunodetection (CFI) assay that was described previously in Wang et al. (Antimicrob Agents Chemother 53(5):1823–1831, 2009).

Key words Dengue HTS assay, Antiviral testing, High-content imaging, 4G2 antibody, Dose-response curve, EC50

1 Introduction

Dengue virus (DENV), a mosquito-borne flavivirus, is a major human health risk for nearly 2.5 billion people worldwide with an estimated annual global infection incidence of nearly 100 million [1, 2]. The concerns for spread of dengue is exacerbated by the fact that there are 4 antigenically distinct serotypes of dengue viruses (DENV 1-4) which give rise to secondary dengue infections that cause severe dengue diseases such as dengue hemorrhagic fever (DHF) and dengue shock syndrome (DSS). Currently, there are no approved vaccines available that can confer simultaneous protection against all four serotypes [3]. In addition, there are no effective antiviral drugs available for treatment of dengue. These facts underscore the importance for screening of potential novel candidates for in vitro anti-dengue activity.

Radhakrishnan Padmanabhan and Subhash G. Vasudevan (eds.), *Dengue: Methods and Protocols*, Methods in Molecular Biology, vol. 1138, DOI 10.1007/978-1-4939-0348-1_7, © Springer Science+Business Media, LLC 2014

There are several in vitro cell-based assays developed for screening of potential antiviral candidates. The gold standard for assessing inhibitors against dengue virus is a focus-forming assay, which is also known as plaque-reduction neutralization assay (PRNT). It allows the virus-test compounds' interaction in a multiwell dish and then measure for the test compounds' effect on the virus infectivity in the target cells. A semisolid medium has to be overlaid onto the cells to restrict the spread of progeny virus [4]. Each plaque or foci formed represent a plaque-forming virus unit (pfu) and is quantitated by crystal violet staining. The assay has been refined with the use of antibody staining prior/during/after infection for better sensitivity. The number of plaques formed could be compared back to the dilutions to determine the total plaque-forming virus unit and percent reduction of plaque numbers by test compounds. Despite being indicative of the effect of compound on virus replicative capacity, PRNT method is both labor-intensive and time-consuming, requiring almost a week to complete the total experiment. Numerous washing steps and manual quantitation steps, which involve an element of human subjectivity, are some of the additional limitations that make the PRNT method unsuitable for high-throughput screening of potential antivirals.

Alternatively, virus-induced cytopathic effect (CPE) can be quantified as a measure of inhibition of viral infection. Various CPE assays involving probes have been developed for the use of high-throughput screening in recent years [5]. For example, the Cell Titer-Glo® Luminescent Cell Viability Assay kit that can measure the virus-induced cytotoxicity can be used to quantify CPE. This assay depends on the principle that the ATP present in actively metabolizing intact viable cells would release a strong luminescence signal upon addition of luciferase enzyme and substrate [5, 6]. However, during high virus-induced CPE, the cells will be less viable, and hence the amount of ATP available for luciferase reaction would be minimal, leading to dampened luminescence signal. A restoration of cell viability upon treatment (gain in luminescence signal) as compared to the infection-induced CPE positive control is indicative of the antiviral activity. Although the methodology of these refined CPE assays is highly sensitive [6] the fact that CPE may not be obvious for several viruses, and other factors such as cost of the kits used make it a less popular approach for preliminary screening.

Taking all together, an ideal dengue virological cell-based assay for preliminary high-throughput compound screening should be rapid, easy to conduct, and yet sensitive enough to assess viral infectivity and cell viability. In this perspective, the cell-based flavivirus immunodetection (CFI) assay that was developed and described previously in Wang et al. [7] meets many of the prerequisites for a high-throughput assay. The method uses the principle of an indirect enzyme-linked immunosorbent assay (ELISA) and the readout is based on colorimetric analysis that gives a net optical density reading. However, colorimetric readout is not always accurate and dynamic range is not always consistent, which may lead to inconsistency and inaccuracy of the EC_{50} (effective concentration at which 50 % inhibition) value.

This chapter describes a modified CFI assay adapted from Wang et al. [7] by employing a new method of analysis using high-content fluorescence imaging, in conjunction with the conventional assay technique. In this method, fixed cells are probed with Alexa Fluor 488 (green fluorescent)-conjugated secondary antibody targeting a viral protein detecting primary antibody and DAPI (4′,6-diamidino-2-phenylindole) for nuclei staining. Fluorescent cells are captured using high-content imaging at 4× magnification in order to maximize the sample area covered in each well in a minimal period of time. Green fluorescent cells and nuclei are analyzed and quantified using MetaXpress software (Molecular Devices Corporation). The data are plotted into sigmoidal curve and EC50 values calculated using GraphPad Prism 5. This approach provides an inexpensive, more visual and reliable readout in a high-throughput manner, allowing rapid assessment of potential antiviral candidates.

2   Materials

2.1   Cell Seeding on a 96-Well Plate

1. RPMI 1640 medium.
2. 0.25 % trypsin-EDTA.
3. Hemocytometer.
4. Trypan blue.
5. Fetal bovine serum, heat-inactivated.
6. Penicillin/streptomycin.
7. 96-well plate (flat bottom, tissue culture treated).
8. Multichannel pipettes.

2.2   Test Compound and Virus Preparation

1. Dimethyl sulfoxide, DMSO.
2. Water bath, 37 °C.
3. Vortex.
4. Melsept SF, formaldehyde-free surface disinfectant.
5. Serum-free RPMI 1640 medium.
6. Celgosivir HCl, a positive control drug for dengue virus inhibition [8].
7. Test compounds.

2.3 Virus Infection and Treatment	1. Aspirator and vacuum (VACUSIP and multichannel adapter).
	2. Incubator, 5 % CO_2, 37 °C.
	3. Shaker.

2.4 Fixation and Probing	1. Phosphate buffered saline, PBS.
	2. Methanol, kept cold at –20 °C.
	3. Freezer, –20 °C.
	4. Tween-20.
	5. Fridge, 4 °C.
	6. 4G2 antibody, mouse monoclonal, anti-dengue E-protein (produced in-house).
	7. Fetal bovine serum, heat-inactivated.

2.5 High-Content Screening	1. Alexa Fluor 488-conjugated anti-mouse antibody.
	2. DAPI.
	3. ImageXpress Micro high-content fluorescence microscope.

2.6 Calorimetric Reading	1. HRP-conjugated antibody, anti-mouse.
	2. 3,3′,5,5′-Tetramethylbenzidine substrate.
	3. 1 M sulfuric acid H_2SO_4.
	4. Tecan Safire II plate reader.

2.7 Data Analysis	1. MetaXpress.
	2. Microsoft Excel.
	3. GraphPad Prism 5.

3 Methods

The fluorescence imaging-based CFI assay described in this protocol utilizes detection of dengue envelope (E) antigen that is expressed in the infected cells. Viral infectivity can be detected through high-content imaging following an indirect immunostaining using primary antibody (in-house-produced anti-dengue E-protein antibody, 4G2) and fluorochrome-conjugated secondary antibody. Alternatively, horseradish peroxidase (HRP)-conjugated secondary antibodies can be used for colorimetric (optical density) readout as described previously [7].

3.1 Cell Seeding on a 96-Well Plate

This procedure describes the preparation of monolayer (*see* **Note 1**) of adherent cells (e.g., BHK21; baby hamster kidney cells) for drugs screening against viral infection, using the high-content imaging-based CFI assay.

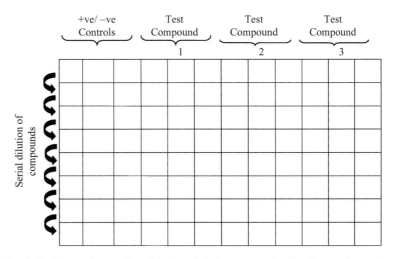

Fig. 1 Serial dilution profile of test antiviral compounds. The figure shows the plate layout for control and test compounds. Test compounds and controls are tested from 100 to 200 μM in fourfold dilutions to 6–12 nM

1. Dislodge healthy cells in trypsin-EDTA for <5 min (*see* **Note 2**). Add complete growth medium to resuspend cells and transfer into a centrifugal tube.

2. Centrifuge cells at $900 \times g$ for 3 min at room temperature (*see* **Note 3**).

3. Discard medium and resuspend cells in fresh growth medium.

4. Count cell number and seed cells at 1.3×10^4 cells/well/100 μL in 96-well plate (Fig. 1) (*see* **Note 4**).

5. Incubate cells for 24 h at 37 °C in 5 % CO_2 incubator (*see* **Note 5**).

3.2 Preparation of Negative/Positive Controls, Test Compound, and Virus

Test compound should be dissolved in a compatible solvent, generally DMSO, to a high-concentration stock and subsequently diluted in a serum-free medium (SFM) to testing concentration.

1. Dissolve solid test compound in a compatible solvent to a 20 mM stock.

2. Dilute test compounds with SFM to 2× test concentration (*see* **Note 6**) and serially dilute (4× serial dilution) eight times (*see* **Note 7**).

3. The solvent used to dissolve compounds (generally DMSO) should be used as negative controls, identically diluted as in **step 2**.

4. A positive control compound should be used and prepared identically as in **step 2**.

5. Thaw virus stock at 37 °C in the water bath and completely by warming between palms.

6. Dilute virus stock with SFM to give multiplicity of infection (M.O.I.) 0.3 (*see* **Note 8**).

7. Mix working test compounds with working virus inoculums (or 4 % FBS medium) at ratio 1:1 in a sample block.

3.3 Virus Infection and Treatment

Potent antiviral compounds may target host targets or virus and can also affect any stage of infection starting with binding to the host cells surface. The effect of the time of treatment of compounds on viral infection can provide a preliminary understanding of the stage of infection where the tested compounds act. The three general compound treatment approaches in relation to viral infection are pretreatment, co-incubation, and posttreatment.

Pretreatment

Seeding	Drug	Virus Infection	Maintaining	Fixation
(24 hours)	Treatment (1 hour)	(1 hour)	(48 hours)	

Day 1	Day 2	Day 3	Day 4

1. Dilute test compounds, negative and positive controls with 4 % FBS medium at ratio 1:1 (final 2 % FBS and 1× test compound concentration).

2. Aspirate medium from the wells.

3. Add 1× test compounds, negative and positive controls into respective wells (in triplicate) at 50 μL/well.

4. Incubate cells with test compound inoculums for an hour at 37 °C in 5 % CO_2 incubator.

5. Dilute virus stock with SFM to give M.O.I. 0.3 per 50 μL.

6. Aspirate medium containing test compounds, negative and positive controls from the wells.

7. Add in diluted virus into respective wells at 50 μL/well.

8. Incubate cells with virus inoculums for an hour at 37 °C in 5 % CO_2 incubator.

9. Aspirate medium containing virus inoculums.

10. Add in 4 % FBS medium into the wells at 100 μL/well.

11. Incubate cells at 37 °C in 5 % CO_2 incubator for 48 h.

Co-treatment

Seeding	Drug Treatment & Virus	Test Compound	Fixation
(24 hours)	Infection (1 hour)	(48 hours)	

Day 1	Day 2	Day 3	Day 4

1. Aspirate medium from the wells.
2. Add the mixture of virus with test compounds, negative or positive controls at 2× working concentration into respective wells, in triplicate at 50 μL/well.
3. Incubate cells with inoculums at 37 °C in 5 % CO_2 incubator for an hour.
4. Dilute test compounds, negative and positive controls with 4 % FBS medium at ratio 1:1 (final 2 % FBS and 1× test compound working concentration).
5. Aspirate inoculums from the wells and add in 100 μL/well of the 1× test compounds.
6. Incubate cells with test compounds, negative and positive controls in maintaining medium at 37 °C in 5 % CO_2 incubator for 48 h.

Posttreatment

Seeding	Virus Infection	Test Compound	Fixation
(24 hours)	(1 hour)	(48 hours)	

Day 1	Day 2	Day 3	Day 4

1. Dilute virus stock with SFM to give M.O.I. 0.3 in 50 μL.
2. Aspirate medium from the wells.
3. Add in diluted virus into respective wells at 50 μL/well.
4. Incubate cells with virus inoculums for an hour at 37 °C in 5 % CO_2 incubator.
5. Dilute test compounds, negative and positive controls with 4 % FBS medium at ratio 1:1 (final 2 % FBS and 1× test compound working concentration).

6. Aspirate virus inoculums from the wells and add in 100 μL/ well of the diluted test compounds, in triplicate.

7. Incubate cells with test compounds, negative and positive controls in maintaining medium at 37 °C in 5 % CO_2 incubator for 48 h.

3.4 Fixation, Immunostaining, and High-Content Imaging (See Note 9)

1. Aspirate medium from wells.

2. Gently add 100 μL of PBS into each wells, swirl gently, and aspirate.

3. Add 100 μL of ice-cold methanol into each well and keep plate in −20 °C for 15 min.

4. Aspirate methanol and wash cells thrice for 5 min each with PBS-T (PBS + 0.1 % Tween-20).

5. Aspirate washing buffer and add in 100 μL blocking buffer PBS-FT (PBS-T + 1 % FBS).

6. Gently swirl plate and incubate at room temperature for an hour. Blocking can be kept overnight at 4 °C if required.

7. Aspirate blocking buffer and add in primary antibody (4G2; anti-dengue E-protein antibody; diluted in PBS-FT) at 50 μL/ well. Swirl gently and incubate for 3 h at room temperature. Incubation can be kept overnight at 4 °C.

8. Aspirate primary antibody and wash cells thrice with 100 μL PBS-T per well for 5 min each.

9. Aspirate washing buffer and add in secondary antibody (Alexa Fluor 488 anti-mouse antibody; diluted 1,000× in PBS-FT) at 50 μL per well. Incubate for an hour at room temperature.

10. Aspirate antibody and wash cells thrice with 100 μL PBS-T per well for 5 min each.

11. Add DAPI (3.3 mg/mL; diluted in water) at 50 μL per well.

12. Perform high-content fluorescence microscopy by imaging at channels detecting both DAPI (Fig. 2a) and Alexa Fluor 488 (Fig. 2b) at 4Å~ magnification.

3.5 Data Analysis

1. Determine both the total number of cells in each image (indicated by DAPI staining) and also the percentage of cells that are positive for green fluorescence (indicating percentage of viral E-protein expressing cells), using the high-content screening image data analysis software MetaXpress.

2. Generate high-content screening data in spreadsheet and plot dose-response curve (drug dose vs. percent infectivity) using GraphPad Prism, non-regressive, sigmoidal curve.

3. Generate optical density from calorimetric reading and plot dose-response curve using GraphPad Prism, non-regressive, sigmoidal curve. Figure 3 shows a typical dose-response curve obtained for the positive control (Celgosivir).

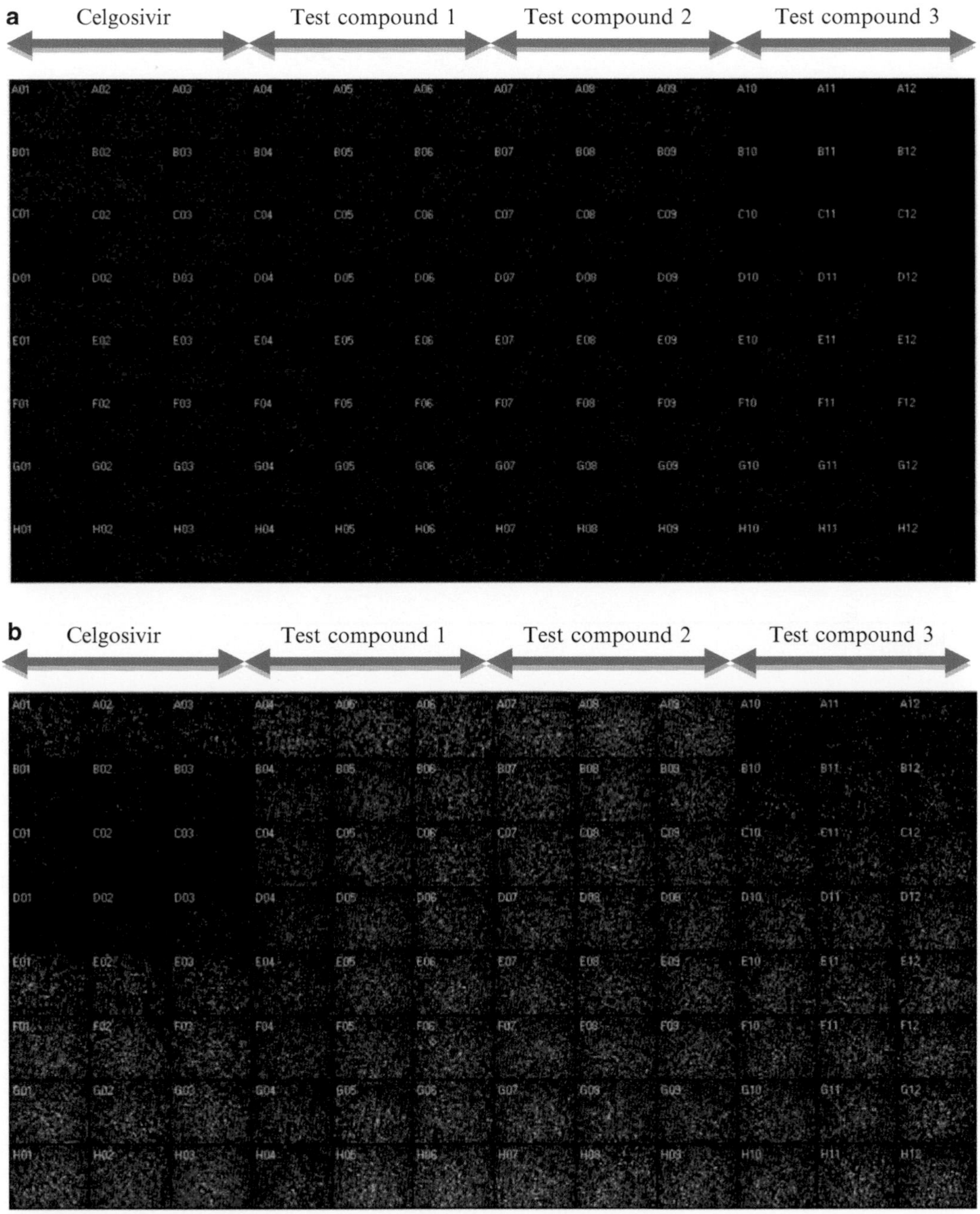

Fig. 2 Images showing the DAPI staining that reveal the viability of cells and the gradient of green fluorescence (infected cells). Celgosivir HCl (host target antiviral) is included as a positive control which has been previously shown to work prior and after dengue infection without affecting the cell viability (Rathore et al. [8]). Negative control (DMSO only) did not show any significant effect on the cells' viability nor virus infectivity (data not shown here)

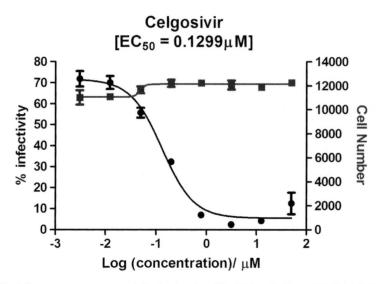

Fig. 3 Dose-response curve plotted using GraphPad Prism 5. Almost 70 % inhibition achieved for the positive control, Celgosivir HCl

4 Notes

1. Monolayer of cells is crucial for a successful and reliable CFI assay, taking into account that stacking cells may be dislodged during washing steps or yield less clear images.

2. Cells should not be left in trypsin-EDTA for more than 5 min. The following can enhance dislodging of cells:

    (a) Prior gentle rinse with small volume (e.g., 1–2 mL) of trypsin-EDTA.

    (b) Warm trypsin-EDTA to 37 °C before use.

    (c) Incubate cells with trypsin-EDTA at 37 °C.

    (d) A gentle knock on the side of the flask, when cells are observed to be round under the microscope.

3. Room temperature refers to 25 °C.

4. Seeding of cells should be done using a multichannel pipette. Cells should be dispensed smoothly into the wells with tips slightly slanted touching the wall of wells in order to get an even distribution of cells.

5. Cells should be incubated for 24 h to achieve monolayer. Alternatively, seeding cell number can be increased to 1.5×10^4 cells/well/100 μL if incubation time is shortened to about 18 h.

6. Test compound is to be mixed with virus inoculum at ratio 1:1. Hence, working stock should be prepared 2× concentration (e.g., 25 μL of 400 μM working test compound to be mixed with 25 μL virus inoculum, final test concentration is 200 μM).

7. Suggested testing concentration range for an unknown test compound is between 200 µM and 12 nM. However, testing concentration can be expended or narrowed for better accuracy if prior information is available.

8. Direct dilution of virus stock to M.O.I. 0.3 should be avoided especially when the virus titer is high (i.e., 10^8pfu/mL). Serial dilution is recommended for better accuracy and infection.

9. ImageXpress Micro high-content microscope reads from the bottom of the plate; therefore, touching the plate base should be avoided so that the readout is clear. Wipe clean the plate base with 30 % ethanol before proceeding to high-content reading.

Acknowledgements

This work was supported by the National Medical Research Council, Singapore (www.nmrc.gov.sg), under grant NMRC/1315/2011 to SGV and a Duke-NUS Graduate Medical School start-up grant to MK and SGV.

References

1. Bhatt S, Gething PW, Brady OJ, Messina JP, Farlow AW, Moyes CL, Drake JM, Brownstein JS, Hoen AG, Sankoh O, Myers MF, George DB, Jaenisch T, Wint GR, Simmons CP, Scott TW, Farrar JJ, Hay SI (2013) The global distribution and burden of dengue. Nature 496(7446):504–507

2. World Health Organization (2009) Dengue guidelines for diagnosis, treatment, prevention and control. http://www.who.int/rpc/guidelines/9789241547871/en/

3. Whitehead SS, Blaney JE, Durbin AP, Murphy BR (2007) Prospects for a dengue virus vaccine. Nat Rev Microbiol 5(7):518–528

4. Roehrig JT, Hombach J, Barrett AD (2008) Guidelines for plaque-reduction neutralization testing of human antibodies to dengue viruses. Viral Immunol 21(2):123–132

5. Green N, Ott RD, Isaacs RJ, Fang H (2008) Cell-based assays to identify inhibitors of viral disease. Expert Opin Drug Discov 3(6):671–676

6. Noah JW, Severson W, Noah DL, Rasmussen L, White EL, Jonsson CB (2007) A cell-based luminescence assay is effective for high-throughput screening of potential influenza antivirals. Antiviral Res 73(1):50–59

7. Wang QY, Patel SJ, Vangrevelinghe E, Xu HY, Rao R, Jaber D, Schul W, Gu F, Heudi O, Ma NL, Poh MK, Phong WY, Keller TH, Jacoby E, Vasudevan SG (2009) A small-molecule dengue virus entry inhibitor. Antimicrob Agents Chemother 53(5):1823–1831

8. Rathore AP, Paradkar PN, Watanabe S, Tan KH, Sung C, Connolly JE, Low J, Ooi EE, Vasudevan SG (2011) Celgosivir treatment misfolds dengue virus NS1 protein, induces cellular pro-survival genes and protects against lethal challenge mouse model. Antiviral Res 92(3):453–460

Part II

Reverse Genetic Systems to Study Virus Replication and Evolution

Chapter 8

Development and Application of Dengue Virus Reverse Genetic Systems

Andrew D. Davidson

Abstract

The development of dengue virus "reverse genetic" systems based on full-length cDNA clones corresponding to the viral RNA genome has been an important technological platform for advancing dengue virus research. Mutations can be introduced into the genome to study their effect on virus replication and pathogenesis while attenuated or chimeric viruses can be constructed that are potential vaccine candidates. The deletion of the virus structural genes has led to the production of noninfectious, but replication competent viral subgenomes (termed replicons) that have been used to study viral replication and are useful for the screening of antiviral compounds. This article describes the development of dengue virus reverse genetic systems and protocols to manipulate the viral genome, recover infectious virus, and produce replicon-containing cell lines.

Key words Dengue virus, Dengue virus reverse genetics, Dengue virus replicon, Dengue virus infectious clone, Flavivirus, Flavivirus reverse genetics

1 Introduction

The construction of full-length cDNA clones corresponding to the dengue virus (DENV) RNA genome has been an important technological platform for advancing DENV research. Transcription from a promoter engineered upstream of a full-length DENV cDNA clone produces DENV RNA transcripts, which when introduced into a permissive target cell lead to a productive infection. In a process termed "reverse genetics," the DENV genome may be manipulated via the full-length cDNA clone. Specific mutations may be introduced into the viral genome to study their effects on viral replication and pathogenesis, while entire genes may be either substituted or deleted to produce, respectively, novel chimeric viruses and noninfectious viral subgenomes (termed replicons) capable of intracellular replication. DENV reverse genetic systems have been extensively used to develop live attenuated vaccines and identify antiviral compounds (*see* ref. 1) that are urgently needed to stop the spread of dengue.

Radhakrishnan Padmanabhan and Subhash G. Vasudevan (eds.), *Dengue: Methods and Protocols*, Methods in Molecular Biology, vol. 1138, DOI 10.1007/978-1-4939-0348-1_8, © Springer Science+Business Media, LLC 2014

Typically, full-length viral cDNA clones have been constructed in plasmid vectors. However, the construction of DENV full-length cDNA clones that can be stably propagated using *Escherichia coli* plasmid vector systems has proven difficult due to toxicity problems arising from the viral cDNA. This may lead to rearrangement, deletion, or insertion of foreign sequences into the DENV cDNA clone. A number of strategies have been used to circumvent this problem, including the use of low-copy-number bacterial plasmids [2, 3], yeast shuttle vectors [4, 5], and bacmid vectors [6]. In addition, the viral genome has been engineered to make it more stable in high-copy-number plasmids [7].

We have produced an infectious cDNA clone corresponding to the DENV-2 strain New Guinea C in the low-copy-number vector pWSK29 [8] (termed pDVWS601 [9, 10]). Transcription of the cDNA clone in vitro, using T7 RNA polymerase, results in the production of RNA transcripts that can be introduced into permissive cells to recover infectious virus (termed v601). The DENV-2 NGC reverse genetic system and DENV replicon systems derived from it [11–13] have been used extensively to investigate DENV replication and pathogenesis and for DENV vaccine and antiviral compound development. In this article the techniques and strategies for the development and manipulation of DENV full-length clones and replicons are described, based on the approach used for the DENV-2 NGC system. A detailed description is provided for the methods used to handle the low-copy-number plasmid infectious clone system, to produce in vitro RNA transcripts and recover virus from them and the production of cell lines stably expressing DENV replicons. It is expected that these methods are applicable not only to the DENV-2 NGC infectious clone system but also to the further development and application of infectious clones for other DENV strains.

2   Materials

Many of the protocols used in the Methods are based on standard molecular biology techniques and have not been described in detail in this article. Representative commercial kits used in the Methods are described below but unless specified can often be replaced with similar kits from another supplier.

2.1   Full-Length DENV Infectious cDNA and Replicon Clone Construction and Manipulation

1. Culture fluid containing infectious DENV.

2. QIAamp Viral RNA Mini Kit (QIAGEN).

3. DENV-specific oligonucleotide primers.

4. SuperScript™ One-Step RT-PCR for Long Templates (Invitrogen).

5. Low-copy-number plasmid vector: pWSK29 (GenBank: AF016889.1, (8)) which has been modified to remove the T7 promoter [9].

6. cDNA library efficiency (>10^8 cfu/µg of pUC19) competent *E. coli* strain DH5α cells (Bioline).

7. OL-PCR: DENV-specific mutagenic and flanking oligonucleotide primers.

2.2 Plasmid Clone Isolation and Propagation

1. Plasmid preparation kits: QIAprep Spin Miniprep Kit, QIAGEN Plasmid Midi Kit (QIAGEN; *see* **Note 1**).

2.3 Production of In Vitro Transcripts Corresponding to the DENV Genome

1. T7 RNA transcription: mMESSAGE mMACHINE® T7 kit (Ambion-Invitrogen) or RiboMAX™ Large Scale RNA Production System-T7 (Promega) and 40 mM $m^7G(5')$ ppp($5'$)G RNA cap structure analogue (New England Biolabs).

2. RNeasy MinElute Cleanup Kit (QIAGEN).

3. Equipment and reagents for performing formaldehyde agarose gel electrophoresis (*see* ref. 14).

2.4 Virus Recovery and the Production of Cell Lines Containing DENV Replicons

1. Culture cells: baby hamster kidney (BHK-21) cells, *Aedes albopictus* C6/36 cells.

2. TransMessenger™ Transfection Reagent (QIAGEN).

3. RNA extraction and RT-PCR analysis: *see* Subheading 2.1.

4. Indirect immunofluorescence assay: primary antibodies against DENV structural and nonstructural proteins.

5. Alexa Fluor® labeled anti-mouse and anti-rabbit IgG secondary antibodies (Invitrogen).

3   Methods

3.1 Full-Length DENV Infectious cDNA Clone Construction

Traditionally, DENV full-length infectious cDNA clones have been constructed by preparing overlapping cDNA clones corresponding to the genome of the DENV strain of interest and joining them in a stepwise cloning strategy using unique restriction sites. During the cloning process, additional sequences are added to cDNA clones corresponding to the 5′ and 3′ ends of the viral genome to facilitate transcription of a DENV RNA genome that includes a minimal number of nonviral nucleotides (*see* Fig. 1). The production of DENV RNA transcripts can either occur in vivo after transfection of a plasmid in which the full-length DENV cDNA clone is fused downstream of a promoter recognized by the cellular RNA polymerase II (e.g., a cytomegalovirus (CMV) promoter) or more commonly in vitro using a promoter recognized by a bacteriophage RNA polymerase (T7/SP6) (*see* Fig. 2a, b). Typically, a unique restriction enzyme site is engineered at the 3′ end of the viral genome to allow the production of in vitro RNA transcripts containing 2–3 nonviral nucleotides after cleavage of the cDNA clone (*see* Fig. 2c). However, an authentic 3′ end can be produced

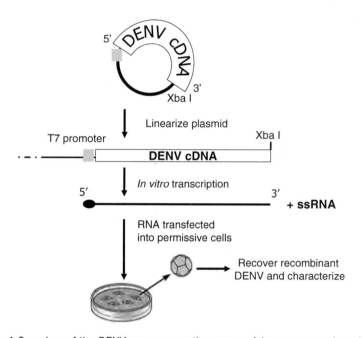

Fig. 1 Overview of the DENV reverse genetic process. A low-copy-number plasmid vector containing a full-length cDNA copy of the DENV RNA genome is linearized at a unique *Xba* I site engineered at the 3′ end of the genome. The linearized plasmid is transcribed in vitro using a bacteriophage T7 RNA polymerase promoter engineered adjacent to the 5′ end of the full-length clone to produce RNA transcripts that correspond to the DENV genome. The RNA transcripts are transfected into cells permissive for DENV replication to recover recombinant virus that can then be characterized

by engineering a hepatitis delta virus ribozyme sequence at the 3′ end of the viral genome (*see* Fig. 2d). An overview of the methods used to produce full-length clones containing these features is described below.

1. Harvest the culture supernatant from DENV infected cells and extract the DENV genomic RNA using a QIAamp Viral RNA Mini Kit following the manufacturer's protocol.

2. Using oligonucleotide primers specific to the DENV genome of interest, amplify the genome as a series of overlapping cDNA clones of ~3–4 kb in size, using a one-step RT-PCR kit (*see* **Note 2**). The cDNA clone corresponding to the 5′ terminal region of the genome should be simultaneously fused to either a bacteriophage T7 RNA polymerase promoter using an oligonucleotide primer containing the promoter sequence (*see* Fig. 2a) or the CMV promoter [15] by OL-PCR mutagenesis [16, 17] (*see* Fig. 2b). The cDNA clone representing the 3′ terminal region of the genome should be simultaneously fused to a sequence containing a unique restriction site (*see* Fig. 2c) or the hepatitis delta ribozyme sequence using OL-PCR mutagenesis (*see* Fig. 2d and **Note 3**).

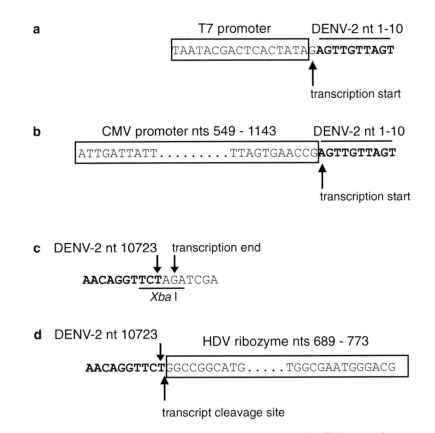

Fig. 2 Modification of the 5′ and 3′ terminal cDNA clones for RNA transcript production. Sequences engineered at the 5′ and 3′ ends of the full-length cDNA clone to facilitate transcription are shown. (**a**) The sequence of a fusion between the bacteriophage T7 RNA polymerase promoter and the 5′ terminal end of the DENV-2 genome. An extra G is added before the DENV-2 sequence to facilitate transcription from the T7 promoter. It is assumed the extra G is replaced by a cap structure analogue during in vitro transcription. (**b**) A fusion between the human cytomegalovirus (CMV) major immediate-early promoter (nts 549-1143; GenBank: M60321) and the 5′ terminal end of the DENV-2 which results in the production of infectious transcripts by cellular RNA pol II. (**c**) A sequence showing a fusion between the 3′ terminal end of the DENV-2 genome and a *Xba* I restriction enzyme site. In vitro transcription of a full-length cDNA clone linearized with *Xba* I results in the addition of 2 nonviral nucleotides to the viral genome. (**d**) A fusion between the 3′ terminal end of the DENV-2 genome and a human hepatitis delta virus ribozyme (nts 689-773; GenBank: M28267). Transcripts containing this sequence will be self-cleaved by the ribozyme to produce an authentic viral 3′ end

3. Purify the overlapping cDNA fragments by gel electrophoresis followed by gel extraction using a commercial kit and clone the fragments into a suitable vector using a commercially available blunt-end PCR product cloning system (e.g., a Zero Blunt® TOPO® PCR cloning kit) following the manufacturer's protocol.

4. Propagate and screen the resulting bacterial colonies to identify plasmid cDNA clones representing the viral genome using standard molecular biology techniques (*see* ref. 14). Sequence the cDNA clones to ensure that no mutations have been introduced into the cDNA fragments during the cloning process and that the sequence of the genome conforms to the published sequence.

5. The cDNA fragments are then joined in a series of stepwise subcloning reactions using unique restriction sites that are either naturally occurring or have been intentionally engineered into the cDNA fragments during the cloning process (*see* **Note 4**). If the intermediate plasmid subclones are stable, a high-copy-number plasmid can be used for this purpose. However, once plasmid instability is encountered or in the final stages of clone assembly, the fragments should be joined in the low-copy-number vector pWSK29 which has been modified to avoid duplication of the appropriate RNA polymerase promoter (*see* **Note 5**) [9]. cDNA clones housed in the low-copy-number vector should be handled as described below (*see* Subheading 3.3). In the final stages of clone construction, cDNA library efficiency competent *E. coli* strain DH5α cells should be used for plasmid transformation to increase the recovery rate of full-length cDNA clones in the low-copy-number vector.

6. Once constructed, the full-length DENV cDNA clone should be tested to ensure it can be used to recover infectious virus as described below (*see* Subheadings 3.3–3.5). A comparison of the phenotypic properties of the natural and recovered recombinant viruses should be done to ensure that any additional sequences introduced into the viral genome do not effect infectivity (*see* **Note 6**).

3.2 Genetic Manipulation of the Full-Length DENV Infectious cDNA Clone

Once a full-length cDNA clone corresponding to the viral genome of interest has been constructed, it may then be manipulated to alter the DENV genome either (a) to introduce site-specific mutations and/or nucleotide deletions/insertions to examine their effects on virus replication and/or pathogenesis or (b) to create reporter replicons or viruses expressing heterologous genes. Both of these types of manipulations are described below.

3.2.1 The Introduction of Site-Specific Mutations into the Full-Length DENV cDNA Clone

1. This process involves the exchange of a subgenomic fragment from the full-length cDNA clone with a corresponding PCR-derived fragment containing the mutation(s) of interest using unique restriction sites in the full-length cDNA clone as follows.

2. Digest the DENV full-length cDNA clone with restriction enzymes that have unique cleavage sites surrounding the region to be mutated using standard procedures (*see* **Note 7**).

Typically enough vector is prepared for 10–20 ligation reactions (1–2 μg) such that gene cassettes containing sets of mutations can be introduced between the unique restriction sites. The restriction digests may be performed as a double digest if the enzyme buffers are compatible or sequentially after purification of the plasmid DNA between steps using a commercially available PCR cleanup column. It is recommended that the digested plasmid is treated with calf intestinal phosphatase after digestion to ensure there is no self-ligation of the plasmid backbone.

3. The vector backbone is then purified by agarose gel electrophoresis followed by gel extraction using a commercially available kit and the integrity and concentration of the vector determined by agarose gel electrophoresis.

4. A subgenomic cDNA fragment containing nucleotide substitutions and/or insertions/deletions in the region of the viral genome of interest is then prepared by OL-PCR using a high-fidelity thermostable DNA polymerase [16]. The OL-PCR product is produced using mutagenic and flanking oligonucleotide primers so that it overlaps the unique restriction enzyme sites in the viral genome used to prepare the vector backbone. The fragment is digested using the same unique restriction sites as described above for the vector and the resulting DNA fragment(s) purified by gel extraction (*see* Subheading 3.2, **step 3**).

5. The vector and insert DNA fragments are ligated using standard procedures and an aliquot of the ligation reaction used to transform cDNA library efficiency competent ($>1 \times 10^8$ cfu/μg of pUC19) *E. coli* strain DH5α cells. The transformation mix is plated on LB agar plates containing 100 μg/mL ampicillin and grown overnight at 37 °C.

6. Two types of colonies may be present, small and large. Analysis of the large colonies should be avoided as they generally are the result of plasmid rearrangement. A number of individual colonies should be selected and used to prepare plasmid miniprep DNA (*see* Subheading 3.3, **step 1**). The plasmid should be checked for rearrangement by *Eco* RI digestion. Typically, the successful introduction of specific mutations into the full-length cDNA clone is verified by sequence analysis. As the amount of the low-copy-number plasmid is often limiting, a fragment spanning the PCR-derived region of the clone containing the mutation(s) is amplified from the low-copy-number plasmid by PCR using standard techniques and sequenced to verify that only the mutations of interest are present.

7. Once full-length cDNA clones containing the mutations of interest have been generated, the effects of the mutations on virus recovery and/or replication can be analyzed (*see* Subheadings 3.3–3.5).

3.2.2 The Production
of DENV Replicons

1. A number of strategies have been used to produce DENV replicons; however, we have found the substitution of the genes encoding the DENV structural proteins with genes encoding reporter proteins or selectable markers to be the most robust in terms of replicon replication. Following this strategy, the DENV full-length cDNA clone is modified to remove the DENV capsid, pre-membrane, and envelope (C, prM, and E) genes except for the sequences encoding the first 27 amino acids of the C protein (C_{27}, including the first methionine) which contains a cyclization sequence required for genome replication [18–20] and the last 24 amino acids of the E protein (E_{24}), which acts as a signal sequence for insertion of NS1 into the endoplasmic reticulum [21, 22] (*see* Fig. 3). The full-length cDNA clone is cleaved with unique restriction enzymes flanking the C, prM, and E genes to remove these sequences which are then replaced with an OL-PCR fragment that retains the C_{27} and E_{24} coding sequences separated by a linker containing a selection of unique restriction enzyme sites (*see* Subheading 3.2.1) for further manipulation of the replicon plasmid.

2. Once the replicon plasmid is established, gene cassettes can be introduced into the replicon using the unique restriction sites

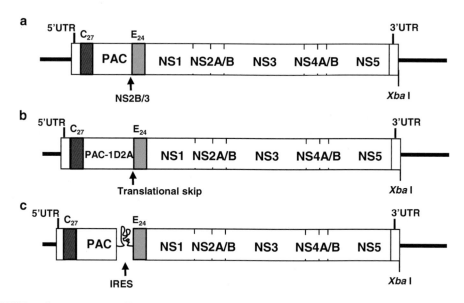

Fig. 3 DENV replicon constructs. Three alternatives for the construction of selectable or reporter replicons are shown. In all cases the structural genes, except for sequences encoding the first 27 amino acids of the C protein and last 24 amino acids of the E protein, are substituted with a heterologous gene while retaining the remainder of the DENV-2 genome sequence. (**a**) Translation of a selectable marker (such as the *PAC* gene product, conferring resistance to puromycin) is separated from the remainder of the viral polyprotein using an introduced NS2B/3 cleavage site. (**b**) The foot-and-mouth-virus 1D2A peptide is fused to the end of the *PAC* gene product promoting a ribosomal skip that removes the fusion protein from the remainder of the viral polyprotein. (**c**) Translation of the *PAC* gene product is terminated at an introduced stop codon. Translation of the viral polyprotein occurs using an internal ribosome entry site (IRES) engineered upstream of the polyprotein sequence

in the linker region. We have used three strategies to express reporter proteins or proteins coding for selectable markers from the replicon backbone. In all strategies the heterologous protein(s) is(are) translated using the first methionine codon of the C gene sequence, resulting in a fusion protein which has an N-terminal C_{27} extension and any additional amino acids encoded by the linker sequence. However, to ensure faithful replicon replication, the C-terminus of the protein of interest is separated from the remainder of the virus polyprotein by one of three methods (*see* Fig. 3) as described in **steps 3–5** below.

3. A sequence encoding a peptide cleavage site for the NS2B/3 protease is engineered at the C-terminus of the protein of interest. We have successfully used the NS4B/5 cleavage sequence for this purpose. If more than one heterologous protein is to be expressed, multiple NS2B/3 protease cleavage sites can be used to separate the heterologous proteins from each other and from the remainder of the virus polyprotein.

4. The heterologous gene of interest is followed by a sequence encoding the 1D2A peptide derived from foot-and-mouth disease virus. The inclusion of the 1D2A peptide results in a translation "skip" [23] which "cleaves" the 1D2A fusion protein from the remainder of the DENV polyprotein [11, 12].

5. The heterologous gene(s) of interest is(are) terminated with a translation stop codon. An internal ribosome entry site is then engineered downstream of the heterologous gene to reinitiate translation of the DENV polyprotein [13].

6. Once a replicon plasmid clone has been produced, it is then propagated and used to produce replicon expressing cells as described below (*see* Subheadings 3.3, 3.4, and 3.6).

3.3 Plasmid Clone Isolation and Propagation

1. To screen bacterial colonies for low-copy-number plasmids containing the sequences of interest, 5 mL of LB medium containing 100 μg/mL ampicillin is inoculated with an individual colony and incubated no longer than 16 h at 37 °C with rotation at 250 rpm. Bacteria are harvested from the 5 mL overnight culture by centrifugation at $13,000 \times g$ for 3 min. All further steps are performed entirely as described in the QIAprep Spin Miniprep protocol (*see* **Note 1**). DNA is eluted from the column with 50 μL of 10 mM Tris–HCl (pH 8.5) and 10 μL used for restriction enzyme analysis.

2. Plasmids are initially screened for integrity using the enzyme *Eco* RI which reveals a characteristic plasmid rearrangement (*see* Fig. 4a and **Note 8**). If the plasmid is intact, the presence of introduced gene cassettes and/or mutations is confirmed by PCR amplification and sequencing of any newly introduced sequences to confirm that only the desired mutations have been introduced into the DENV cDNA clone.

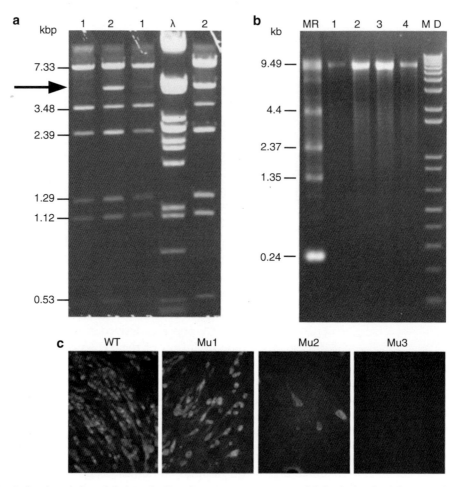

Fig. 4 Analysis of experimental steps in the virus recovery process. (**a**) Analysis of minipreparations of the DENV full-length cDNA clone plasmid pDVWS601 and derivatives by *Eco* RI digestion and agarose gel electrophoresis. *Lanes* labeled 1 and 2 show the intact and rearranged plasmids, respectively. The sizes of the expected *Eco* RI fragments are shown in kbp with a characteristic band indicative of plasmid rearrangement arrowed. A *Pst* I digested λ DNA marker is shown. (**b**) Formaldehyde agarose gel electrophoresis of RNA transcripts produced by in vitro transcription of *Xba* I linearized pDVWSK601 derivatives (*lanes 1–4*). The quality of the transcripts is suitable for virus recovery. The size (kb) of RNA size markers (MR) is shown. DNA markers are shown for comparison (MD). (**c**) Analysis of DENV replication in cells transfected with in vitro RNA transcripts at 48 h post-transfection. DENV replication was detected by indirect immunofluorescence assay using an antibody against the DENV E protein. The assay shows the difference in replication between the wild-type virus (WT), a DENV mutant that has decreased replication (Mu1), a DENV mutant that is defective in cell to cell spread (Mu2), and RNA transcripts with carrying a mutation that abolishes virus replication (Mu3)

3. Once the plasmids of interest have been successfully produced, they are then propagated using a plasmid midi- or maxiprepa-ration. For the analysis of the effects of introduced mutations on viral recovery, a midipreparation is sufficient, whereas if the plasmid is to be used for further cloning steps, a plasmid maxi-preparation is done. Due to the very-low-copy number of the

plasmid vector, standard midiprep and maxiprep procedures have been modified as described below.

4. Plasmid midipreparation: A QIAprep Spin Miniprep Kit is used for plasmid isolation as follows: 250 mL of LB medium containing 100 µg/mL ampicillin is inoculated with a freshly grown single bacterial colony (*see* **Note 9**) and incubated for 16 h at 37 °C with shaking at 250 rpm. The OD600 is measured, and the bacteria harvested by centrifugation at 6,000 × *g* for 10 min at 4 °C. If the OD is >1.0, the volumes of buffers used in the procedure are doubled. The pellet is resuspended in 3 mL of chilled buffer P1. 3 mL of buffer P2 is then added to the sample, mixed well by inversion, and incubated for 5 min at room temperature. Finally, 5 mL of buffer N3 is added and the samples mixed again by inversion. The sample is then centrifuged for 10 min at 13,000 × *g*. The supernatant is transferred to two QIAprep Spin columns on a vacuum manifold and the DNA bound to the column by applying the vacuum. All further steps are performed as described in the QIAprep Spin Miniprep protocol. A typical plasmid yield is 10–15 µg.

5. Plasmid maxipreparation: A QIAGEN Plasmid Midi Kit (Qiagen) is used for plasmid isolation as follows. 4 × 500 mL of pre-warmed LB medium containing 100 µg/mL ampicillin is inoculated with a single colony (200 µL of sterile medium is inoculated with a freshly grown single colony, vortexed, and distributed among the flasks containing 4 × 500 mL of medium) and incubated for 16 h at 37 °C and 250 rpm. 10 mL of culture is then removed and used to prepare a plasmid minipreparation as described above (*see* Subheading 3.3, **step 1**). The remainder of the bacteria are harvested by centrifugation at 6,000 × *g* for 10 min at 4 °C. The bacterial pellets are then frozen at −70 °C in the centrifuge buckets until the integrity of the plasmid is confirmed by *Eco* RI digestion of the plasmid minipreparation sample (*see* **Note 10**). Once the plasmid integrity is confirmed, the pellets are resuspended in a total of 100 mL of chilled buffer P1. Then 100 mL of buffer P2 is added to the sample, mixed by inversion, and incubated for 5 min at room temperature. Finally, 100 mL of chilled buffer P3 is added, the sample mixed again by inversion and then incubated on ice for 30 min. After centrifugation for 30 min at 10,500 × *g* at 4 °C, the supernatant is filtered through a pre-wetted Whatman 1 filter. The DNA in the filtrate is precipitated by the addition of 0.7 volumes of isopropanol and collected by centrifugation at 9,500 × *g* for 30 min at 4 °C. The pellet is redissolved in 0.5 mL of TE buffer (pH 8) and buffer QBT added to a final volume of 5 mL. All further purification steps are performed entirely as described in the QIAGEN plasmid purification handbook following the

protocol for very-low-copy plasmid purification using a QIAGEN-tip 100. The final DNA pellet is redissolved in 80 μL of H_2O or TE buffer. A typical plasmid yield is 60–80 μg for pDVWS601 and 80–120 μg for replicon plasmids derived from pDVWS601.

3.4 Production of In Vitro Transcripts Corresponding to the DENV Genome or Replicon

1. Linearize 2–5 μg of purified pDVWS601 (and/or derivatives) with *Xba* I. Check the concentration of the DNA and the efficiency of the *Xba* I digest by agarose gel electrophoresis.

2. Purify the linearized plasmid by phenol:chloroform:isoamylalcohol (25:24:1) extraction followed by extraction with chloroform:isoamylalcohol (24:1). Precipitate the DNA by the addition of 0.1 volumes of 3 M sodium acetate (pH 5.2) and 2.5 volumes of 100 % ethanol. Incubate on ice for at least 30 min or leave indefinitely at –20 °C.

3. Collect the precipitate by centrifugation at $13,000 \times g$ for 20 min at 4 °C. Wash the pellet with 70 % ethanol and collect the DNA by centrifugation. Air-dry the DNA pellet and resuspend in sterile, nuclease-free H_2O at an approximate concentration of 0.25 μg/μL. Determine the concentration of the DNA by gel electrophoresis before in vitro transcription.

4. Transcribe the *Xba* I linearized pDVWS601 plasmid in vitro using a T7 RNA transcription kit (either a mMESSAGE mMACHINE® T7 kit or a RiboMAX™ Large Scale RNA Production System-T7). The manufacturer's recommendations should be followed with the following exceptions: (a) a smaller scale reaction is usually sufficient to produce enough RNA for transfection experiments and (b) the ratio of RNA cap structure analogue to rGTP must be balanced to produce the optimum amount of capped RNA transcripts. When using the mMESSAGE mMACHINE kit, we have found the following mixture to be optimal in a 10 μL reaction: 0.5 μg of DNA template in 2.4 μL nuclease-free H_2O, 5 μL 2× NTP/CAP, 0.6 μL rGTP, 1 μL 10× reaction buffer, and 1 μL of T7 enzyme mix. While for the RiboMAX™ T7 kit, the following mix is used in a 20 μL reaction: 1–2 μg DNA template in 6 μL of nuclease-free H_2O, 4 μL 5× transcription buffer, 6 μL 25 mM rNTP mix (7.5 mM each of rATP, rCTP, and UTP and 2.5 mM rGTP), 2 μL 10 mM $m^7G(5')ppp(5')G$ and 2 μL of T7 enzyme mix. The transcription reaction is incubated at 37 °C for 2 h. The DNA template is then removed by the addition of 2 U of RNase free DNase I followed by incubation at 37 °C for 15 min.

5. The in vitro RNA transcripts are then purified from the reaction mix using a RNeasy MinElute Cleanup Kit following the recommendations of the manufacturer. The RNA is eluted in a final volume of 30 μL. An aliquot (~1–3 μL) of the RNA sample

is analyzed by formaldehyde RNA agarose gel electrophoresis (*see* ref. 14) to ensure the RNA transcripts are predominantly full length (*see* Fig. 4b). The remainder of the RNA is precipitated by the addition of 0.1 volumes of 3 M sodium acetate (pH 5.2) and 3 volumes of ethanol. The RNA precipitate is collected by centrifugation and washed with 70 % ethanol as described above (*see* Subheading 3.4, **step 3**). The pellet is air-dried under sterile conditions and resuspended in 20 μL of nuclease-free H_2O. The concentration of the RNA is determined spectrophotometrically by A_{260} reading. The RNA is stored at –70 °C until required for transfection.

3.5 *Virus Recovery*

1. 24 h prior to RNA transfection, seed BHK-21 cells in 2.5 mL of the appropriate media into triplicate wells (per RNA sample) of a 6-well plate, each containing a sterile 13 mm diameter glass coverslip. Repeat the process if required using *Aedes albopictus* C6/36 cells (*see* **Note 11**). Incubate the cells under their usual growth conditions such that they achieve 60–80 % confluency at the time of transfection.

2. Transfect duplicate wells of cells with 2 μg of DENV RNA transcript per well using the TransMessenger™ Transfection reagent following the manufacturer's protocol (*see* **Note 12**). Use the remaining well of cells as a control. Remove the transfection mix after 3 h, wash the cells twice with PBS, and replace with 2.5 mL of fresh media. Incubate the transfected BHK-21 and C6/36 cells at 37 and 28 °C, respectively, under appropriate growth conditions.

3. After 48–72 h (*see* **Note 13**), harvest the virus and process the cells as follows.

4. Remove 140 μL of culture supernatant from each transfected and control cell culture supernatant and transfer to a sterile microfuge tube. Store at –70 °C for subsequent RNA extraction and RT-PCR analysis.

5. Remove the remaining culture supernatant and use to set up a Passage 1 infection on C6/36 cells as described below (*see* Subheading 3.5, **step 7**).

6. Add 4 mL of PBS to each well of cells and wash the coverslips gently; transfer the coverslips to a fresh tray. Fix the cells on the coverslips by incubation in ice cold methanol:acetone (50:50) at –20 °C for 5 min. Air-dry the fixed cells and either store at –70 °C or directly assay for evidence of viral replication by immunofluorescence assay using an antibody recognizing the DENV E protein (*see* Fig. 4c and **Note 14**).

7. Passage 1 C6/36 cell infection: To produce virus stocks for further work, the culture supernatant from the transfected cells is used to infect a 75–85 % confluent C6/36 cell monolayer in

a 75 cm² T flask as follows. Wash the cell monolayer with 20 mL of PBS and remove completely. Transfer the transfected cell culture supernatant to the cells. The volume should be at least 2 mL. Incubate at 28 °C for 60–90 min. Add 13 mL (15 mL in total) of maintenance media to the cells and incubate at 28 °C for 4–6 days.

8. If there is visible cytopathic effect (CPE) (*see* **Note 15**), harvest the culture supernatant when there is 80 % CPE; otherwise, harvest after 5–6 days if the cells still appear healthy. First remove 140 μL from each cell culture supernatant and transfer to a sterile microfuge tube. Store at –70 °C for subsequent RNA extraction and RT-PCR analysis. Harvest the remainder of the culture supernatant and clarify by centrifugation at 1,000 × *g* for 10 min. Use 2 mL of the culture supernatant to set up a Passage 2 infection on a fresh round of C6/36 cells as described in **step 7** above (*see* **Note 16**). Store the remainder of the culture supernatant in 1.5–2 mL aliquots at –70 °C as required.

9. The Passage 2 C6/36 cell culture supernatants should be harvested after 4–6 days as described in **step 8**. The aliquoted culture supernatant samples (~8 × 1.5 mL) are then used as working stocks for determining the virus titer and further virus characterization. At this stage a 140 μL aliquot is removed from each culture supernatant sample, the RNA extracted, and the virus analyzed by RT-PCR and sequencing (*see* **Note 17**).

3.6 Production of Cell Lines Containing DENV Replicons

1. 24 h prior to RNA transfection, seed BHK-21 cells or another cell line of interest (*see* **Note 18**) in 2.5 mL of the appropriate media into triplicate wells (per RNA sample) of a 6-well plate. Incubate the cells at their usual growth conditions such that they achieve 50 % confluency at the time of transfection.

2. Transfect duplicate wells of cells with 2 μg of DENV replicon RNA transcript per well as described above (*see* Subheading 3.5, **step 2**). Use the remaining well of cells as a control. Remove the transfection mix after 3 h, wash the cells twice with PBS, and replace with 2.5 mL of fresh media.

3. Incubate the transfected BHK-21 cells at 37 °C under appropriate growth conditions until they are confluent (~24 h), detach the cells with half strength trypsin, collect by centrifugation, resuspend in 10 mL of media, and transfer to a 10 cm diameter petri dish.

4. After 16–24 h add puromycin to the media to a final concentration of 3.5 μg/mL or another selective drug of choice (*see* **Note 19**).

5. Change the media every 3–4 days until all of the control cells are dead and colonies of puromycin resistant cells develop from

the population of replicon RNA transfected cells. The colonies can then either be detached with trypsin and reseeded to produce a single mixed cell population or single colonies can be isolated, transferred to a 24-well culture dish, and expanded.

6. The cells can then be examined for replicon expression by indirect immunofluorescence using antibodies against DENV nonstructural proteins (e.g., NS1 or NS5) and cellular RNA extraction followed by RT-PCR.

7. Once established, the replicon expressing cell population(s) is subjected to routine passaging using the media and drug concentrations described above and can be cryopreserved using standard procedures.

4 Notes

1. Although we have used a QIAprep Spin Miniprep for this purpose, other miniprep kits are also suitable.

2. If the sequence of the DENV strain of interest is known, strain-specific primers can be designed so that a minimum number of overlapping cDNA fragments [3, 4] representing the genome can be produced and joined using naturally occurring restriction sites. If the sequence is not available, it may be necessary to use DENV consensus primers to produce a series of smaller fragments which can then be used to determine the viral genome sequence. The RT-PCR fragments can then either be joined or larger RT-PCR fragments prepared using the determined sequence to minimize the number of cloning steps required to produce the full-length cDNA clone. The one-step RT-PCR kit used for this process should contain a high-fidelity thermostable DNA polymerase to ensure a minimum number of mutations are introduced into the cDNA clones.

3. Most of our studies have been done with a full-length DENV cDNA clone that is under the control of a T7 RNA polymerase promoter and is linearized with *Xba* I prior to the production of in vitro RNA transcripts. However, we have modified the clone so that it is under the control of the CMV promoter and has a HDV ribozyme sequence adjacent to the 3′ terminus. Plasmids containing this clone can be used to recover infectious virus directly after transfection of the plasmid into permissive cells. The sequences used to construct the plasmid-based infectious cDNA clone are shown in Fig. 2.

4. During the cloning process, it is desirable to introduce silent mutations into the full-length cDNA clone that result in the production of unique restriction sites. Unique restriction sites are useful for manipulation of the full-length cDNA clone in further procedures.

5. If a T7 RNA polymerase promoter is fused to the 5′ end of the DENV genomic cDNA clone, the original T7 RNA polymerase promoter in pWSK29 must be removed to avoid duplication.

6. The phenotypic properties of the natural virus that we have compared to the recombinant virus include growth analysis in a range of mammalian and mosquito cell culture lines, plaque size, temperature sensitivity, and neurovirulence in a suckling mouse model. This comparison is required to ensure that any additional nucleotides added to the 5′ and 3′ ends of the genome or silent mutations introduced into the viral genome do not have unforeseen consequences on the viral lifecycle.

7. The rapid introduction of mutations into the full-length cDNA clone is facilitated by the use of convenient unique restriction enzyme sites. Ideally mutations are introduced into the cDNA clone using PCR-derived fragments 1–2 kbp in size. Larger fragments are more difficult to prepare and take longer to sequence. A number of silent mutations specifying unique restriction enzyme sites have been engineered into the DENV-2 NGC clone pDVWS601 for this purpose.

8. We have found that the full-length DENV-2 cDNA clone (and derived replicons) constructed in pWSK29 can undergo rearrangement if not propagated using good microbiological practice. The rearrangement is characteristic and easily visualized by *Eco* RI digestion. Aside from this rearrangement, no other changes to the clone have been detected when grown in bacteria.

9. It is not recommended to use glycerol cultures to grow bacterial colonies for plasmid preparations. Instead, retransform the full-length clone of interest and use a freshly grown colony as the starting point for plasmid preparations. Avoid the use of starter cultures as the inoculum for plasmid midi- and maxipreparations as their use increases the chances of plasmid rearrangement.

10. This procedure is followed to avoid performing the more time-consuming maxiprep procedure on a clone that has rearranged.

11. We routinely test the recovery of mutant viruses in both mammalian BHK-21 and insect C6/36 cells at 37 and 28 °C, respectively. For some mutants this assay reveals cell-type-specific or temperature-sensitive specific defects. However, if the object is solely to recover the wild-type virus, then RNA transfection is only done using BHK-21 cells.

12. Less RNA transcript can be used for transfection (i.e., 1 μg/well); however, there is a corresponding decrease in the amount of virus recovered.

13. Incubation for 72 h will produce more virus, but there will be pronounced cytopathic effect if the virus replicates.

This may be important if good immunofluorescence images are required. If this is the case, a 48-h incubation is recommended.

14. In our experience, if the virus is capable of even low-level replication, it will be possible to detect the E protein in transfected cells by immunofluorescence assay at 72 h post-transfection.

15. In most cases CPE will not be visible on the 1st passage. If CPE appears very early, remove the culture supernatant and store at 4 °C until all of the culture supernatants are harvested.

16. Although not strictly necessary, we have found that by passaging the recombinant viruses twice in C6/36 cells, the titer is increased to a level that is suitable for further characterization of the viruses. We routinely use C6/36 cells for virus propagation as the virus replicates to higher titer and has a glycosylation pattern that may affect infection of mammalian cells.

17. We typically sequence any region containing introduced mutations to ensure the mutations have not undergone reversion. If the mutation has undergone reversion, the sequence of earlier passage viruses is analyzed to determine the stability of the introduced mutation. Depending on the use of the viruses, the entire genome of mutant viruses can be sequenced.

18. The ease of producing replicon-containing cell lines is cell type specific. The replicons can most easily be established in BHK-21 cells that are easy to transfect and are deficient in the interferon response. Therefore, BHK-21 cells are used as a control when establishing replicons in other cell types. We have established replicons in Vero cells and human HEK-293, A549, K562, and THP-1 cells.

19. Depending on the selectable marker used, other antibiotics may be required for the selection of replicon-containing cells. We routinely use a replicon-conferring puromycin resistance as the selection is rapid. The puromycin concentration used varies for different cell types (typically from 2 to 10 μg/mL).

References

1. Ward R, Davidson AD (2008) Reverse genetics and the study of dengue virus. Future Virol 3:279–290

2. Kinney RM, Butrapet S, Chang GJ, Tsuchiya KR, Roehrig JT, Bhamarapravati N, Gubler DJ (1997) Construction of infectious cDNA clones for dengue 2 virus: strain 16681 and its attenuated vaccine derivative, strain PDK-53. Virology 230:300–308

3. Lai CJ, Zhao BT, Hori H, Bray M (1991) Infectious RNA transcribed from stably cloned full-length cDNA of dengue type 4 virus. Proc Natl Acad Sci U S A 88:5139–5143

4. Polo S, Ketner G, Levis R, Falgout B (1997) Infectious RNA transcripts from full-length dengue virus type 2 cDNA clones made in yeast. J Virol 71:5366–5374

5. Puri B, Polo S, Hayes CG, Falgout B (2000) Construction of a full length infectious clone for dengue-1 virus Western Pacific,74 strain. Virus Gene 20:57–63

6. Suzuki R, de Borba L, Duarte dos Santos CN, Mason PW (2007) Construction of an infectious cDNA clone for a Brazilian prototype strain of dengue virus type 1: characterization of a temperature-sensitive mutation in NS1. Virology 362:374–383

7. Pu SY, Wu RH, Yang CC, Jao TM, Tsai MH, Wang JC, Lin HM, Chao YS, Yueh A (2011) Successful propagation of flavivirus infectious

cDNAs by a novel method to reduce the cryptic bacterial promoter activity of virus genomes. J Virol 85:2927–2941

8. Wang RF, Kushner SR (1991) Construction of versatile low-copy-number vectors for cloning, sequencing and gene expression in Escherichia coli. Gene 100:195–199

9. Gualano RC, Pryor MJ, Cauchi MR, Wright PJ, Davidson AD (1998) Identification of a major determinant of mouse neurovirulence of dengue virus type 2 using stably cloned genomic-length cDNA. J Gen Virol 79(Pt 3):437–446

10. Pryor MJ, Carr JM, Hocking H, Davidson AD, Li P, Wright PJ (2001) Replication of dengue virus type 2 in human monocyte-derived macrophages: comparisons of isolates and recombinant viruses with substitutions at amino acid 390 in the envelope glycoprotein. Am J Trop Med Hyg 65:427–434

11. Jones M, Davidson A, Hibbert L, Gruenwald P, Schlaak J, Ball S, Foster GR, Jacobs M (2005) Dengue virus inhibits alpha interferon signaling by reducing STAT2 expression. J Virol 79:5414–5420

12. Masse N, Davidson A, Ferron F, Alvarez K, Jacobs M, Romette JL, Canard B, Guillemot JC (2010) Dengue virus replicons: production of an interserotypic chimera and cell lines from different species, and establishment of a cell-based fluorescent assay to screen inhibitors, validated by the evaluation of ribavirin's activity. Antiviral Res 86:296–305

13. Ng CY, Gu F, Phong WY, Chen YL, Lim SP, Davidson A, Vasudevan SG (2007) Construction and characterization of a stable subgenomic dengue virus type 2 replicon system for antiviral compound and siRNA testing. Antiviral Res 76:222–231

14. Ausubel FM, Brent R, Kingston RE, Moore DD, Seidman JG, Smith JA, Struhl K (eds) (2003) Current Protocols in Molecular Biology, John Wiley & Sons, Inc.

15. Khromykh AA, Varnavski AN, Sedlak PL, Westaway EG (2001) Coupling between replication and packaging of flavivirus RNA: evidence derived from the use of DNA-based full-length cDNA clones of Kunjin virus. J Virol 75:4633–4640

16. Ho SN, Hunt HD, Horton RM, Pullen JK, Pease LR (1989) Site-directed mutagenesis by overlap extension using the polymerase chain reaction. Gene 77:51–59

17. Horton RM, Cai ZL, Ho SN, Pease LR (1990) Gene splicing by overlap extension: tailor-made genes using the polymerase chain reaction. BioTechniques 8:528–535

18. Alvarez DE, Lodeiro MF, Luduena SJ, Pietrasanta LI, Gamarnik AV (2005) Long-range RNA–RNA interactions circularize the dengue virus genome. J Virol 79: 6631–6643

19. Khromykh AA, Meka H, Guyatt KJ, Westaway EG (2001) Essential role of cyclization sequences in flavivirus RNA replication. J Virol 75:6719–6728

20. You S, Falgout B, Markoff L, Padmanabhan R (2001) In vitro RNA synthesis from exogenous dengue viral RNA templates requires long range interactions between 5′- and 3′-terminal regions that influence RNA structure. J Biol Chem 276:15581–15591

21. Falgout B, Chanock R, Lai CJ (1989) Proper processing of dengue virus nonstructural glycoprotein NS1 requires the N-terminal hydrophobic signal sequence and the downstream nonstructural protein NS2a. J Virol 63: 1852–1860

22. Rice CM, Lenches EM, Eddy SR, Shin SJ, Sheets RL, Strauss JH (1985) Nucleotide sequence of yellow fever virus: implications for flavivirus gene expression and evolution. Science 229:726–733

23. Doronina VA, Wu C, de Felipe P, Sachs MS, Ryan MD, Brown JD (2008) Site-specific release of nascent chains from ribosomes at a sense codon. Mol Cell Biol 28:4227–4239

Chapter 9

Construction of Self-Replicating Subgenomic Dengue Virus 4 (DENV4) Replicon

Sofia L. Alcaraz-Estrada, Rosa del Angel, and Radhakrishnan Padmanabhan

Abstract

Dengue virus serotypes 1–4 are members of mosquito-borne flavivirus genus of *Flaviviridae* family that encode one long open reading frame (ORF) that is translated to a polyprotein. Both host and virally encoded proteases function in the processing of the polyprotein by co-translational and posttranslational mechanisms to yield 10 mature proteins prior to viral RNA replication. To study *cis-* and *trans*-acting factors involved in viral RNA replication, many groups [1–8] have constructed cDNAs encoding West Nile virus (WNV), DENV, or yellow fever virus reporter replicon RNAs. The replicon plasmids constructed in our laboratory for WNV [9] and the DENV4 replicon described here are arranged in the order of 5′-untranslated region (UTR), the N-terminal coding sequence of capsid (C), *Renilla luciferase* (*Rluc*) reporter gene with a translation termination codon, and an internal ribosome entry site (IRES) element from encephalomyocarditis virus (EMCV) for cap-independent translation of the downstream ORF that codes for a polyprotein precursor, $C_{ter}E$-NS1-NS2A-NS2B-NS3-NS4A-NS4B-NS5, followed by the 3′-UTR. In the second DENV4 replicon, the *Rluc* gene is fused sequentially downstream to the 20 amino acid (aa) FMDV 2A protease coding sequence, neomycin resistance gene (Neo^r), a termination codon, and the EMCV leader followed by the same polyprotein coding sequence and 3′-UTR as in the first replicon. The first replicon is useful to study by transient transfection experiments the *cis*-acting elements and trans-acting factors involved in viral RNA replication. The second DENV4 replicon is used to establish a stable monkey kidney (Vero) cell line by transfection of replicon RNA and selection in the presence of the G418, an analog of neomycin. This replicon is useful for screening and identifying antiviral compounds that are potential inhibitors of viral replication.

Key words Dengue virus serotype 4, *Renilla* luciferase reporter replicon, Overlap extension PCR, Neomycin resistance selectable marker, Foot-and-mouth disease virus 2A, Cap-independent translation, Internal ribosome entry site, Polyprotein precursor, Antiviral small-molecule screening, Bicistronic expression

1 Introduction

Positive-sense flavivirus life cycle has the following steps: entry via receptosome-mediated endocytosis, uncoating, release of the viral RNA into the cytoplasm, translation of the viral RNA in the endoplasmic reticulum (ER), polyprotein processing, viral RNA

Radhakrishnan Padmanabhan and Subhash G. Vasudevan (eds.), *Dengue: Methods and Protocols*, Methods in Molecular Biology, vol. 1138, DOI 10.1007/978-1-4939-0348-1_9, © Springer Science+Business Media, LLC 2014

replication, and assembly and release of the mature virions. To study the mechanisms involved in translation and RNA replication in the absence of those involved in entry, assembly, and release of virions, the self-replicating subgenomic replicon plasmid constructs have been very useful. Two types of replicon constructs are described. In the first type, the replicon cDNA plasmid encodes all the *cis*-acting elements and *trans*-acting factors essential for translation and replication along with the gene encoding a readily quantifiable reporter (e.g., *Rluc* gene). The wild type (WT) and mutant replicon RNAs are transiently expressed in mammalian cells to study the effects of mutations in translation and replication. The second type of replicon plasmid construct encodes a selectable marker that allows isolation of a stable mammalian cell line expressing the replicating subgenomic RNA in cultured cells. This chapter describes the construction of both types of replicon plasmids and characterization of the functionality of replicon RNAs in mammalian cells.

The yeast-*E. coli* shuttle plasmid vector, pRS424, is used for cloning the cDNA encoding the replicon RNAs. The vector plasmid contains elements essential for replication and antibiotic selection (Ori and Amp^r) in *E. coli* Stbl2 strain and in yeast *S. cerevisiae* YPH857 strain (2 μ Ori, *TRP1*) [10]. The cDNA encoding the replicon RNA in mammalian cells consists of the following in the order from the 5′-end: 5′-UTR-$C_{25codons}$-Rluc*-IRES-$E_{37codons}$-NSorf-3′-UTR (* indicates translation termination codon). The N-terminal region consisting of 25 codons of C gene fused in frame with the *Rluc* gene contains an essential *cis*-acting RNA element, 5′-cyclization sequence (5′CS), that interacts with the complementary sequence within the 3′-UTR (3′CS1), and this functional interaction is essential for RNA synthesis in vitro as shown using mini-genome RNA template [11]. The 5′CS and 3′CS1 along with the 5′-upstream AUG region (UAR) and 3′-UAR located within the 5′- and 3′-UTRS are required for cyclization of the RNA, a prerequisite for viral RNA replication [3, 12].

The translation of the *Rluc* gene is under cap-dependent translational control for which the 5′ cap and 5′- and 3′-UTRs as well as the cHP element [13] within the 5′ coding region play an essential role. However, translation of the ORF encoding the polyprotein starting with 37 C-terminal codons of E protein gene is essential for proper processing of the E-NS1 junction and folding of the polyprotein in the ER [14] (Fig. 1).

Fig. 1 (continued) is used as the template. (**b**) 1B1. PCR2 is performed by mixing fragments 1 and 2 from 1A1 to 1A2, respectively, along with primers A and D. Fragment 4 is produced. 1B2. Another PCR is performed by mixing fragments 2 and 3 from 1A2 to 1A3, respectively, to produce fragment 5. (**c**) PCR-3 is performed by mixing fragments 4 and 5 as well as primers A and F to produce the DENV4 replicon clone encoding the ORF 5′-UTR-$C_{N-ter-Rluc}$- FMDV 2A-Neo^{r*} (* translation termination codon) followed by IRES and the C-terminal 37 codons of E gene fused to N-terminal 8 codons of NS1 (fragment 6). DENV4 cDNA is digested with NheI enzyme (a site located 240 nt from 5′-end) as indicated by underlined sequence prior to yeast recombination.

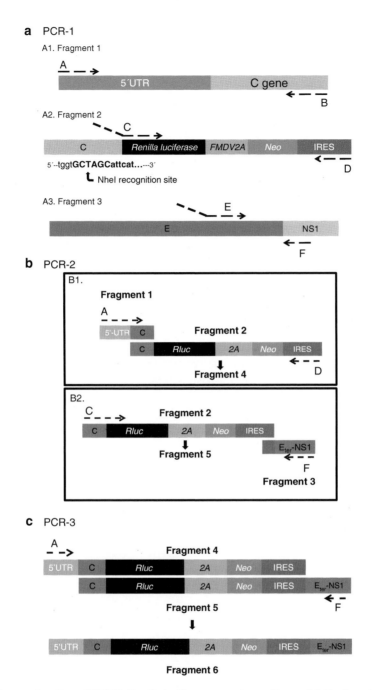

Fig. 1 Strategy for construction of DENV4 *Renilla* luciferase reporter replicons. (**a**) Strategy for construction of DENV4 *Renilla* luciferase reporter replicon encoding neomycin resistance gene (Neoʳ) as a selectable marker. 1A1. The primers A (forward primer located in 5′-UTR nucleotides 1–22) and B (reverse primer from codon 11–18 of C gene) for PCR1 to generate fragment 1 are shown in Table 1. The template is DENV4 cDNA clone in pRS424 yeast-*E. coli* shuttle vector [19]. The length of fragment 1 is 165 bp. The region amplified is not drawn to scale. 1A2. Fragment 2 is generated by PCR using the primers C (forward primer located at nt 42–64 in the C gene) and D (reverse primer located from the 3′-end sequence of IRES element) and DENV2 replicon cDNA (E. D. Reichert and Padmanabhan, unpublished results) as the template. Fragment 2 is 2,406 bp in length. The details of methods used are described in the text. 1A3. Fragment 3, 158 bp in length, is generated by PCR. The forward primer E is a hybrid which fuses the 3′-end 23 nt sequence of IRES with 27 nt, the first 9 codons of 37-codon region from the carboxy-terminal of E. The first codon ATT of 9 codons is mutated to initiator codon ATG downstream from IRES. The reverse F is from the N-terminal coding region of NS1. pRS424 DENV4 cDNA

The second replicon plasmid is designed for stable expression of the replicon RNA in mammalian cells. The replicon plasmid consists of the following in the order from the 5′-end: 5′-UTR-$C_{25codons}$-$Rluc$-FMDV2A-Neo^r*-IRES-$E_{37codons}$-NSorf-3′-UTR. Neo^r codes for neomycin resistance gene and * represents translation termination codon (Fig. 1). Mammalian cells expressing the replicon encoding Neo^r gene are selectable by treatment with neomycin analog, G418. FMDV 2A protease is derived from the foot-and-mouth disease virus. When the 20 aa sequence of FMDV 2A is inserted in frame between two genes (in this case, $Rluc$ and Neo^r), the translation of $Rluc$ is terminated at the FMDV peptide. Translation is reinitiated at the first AUG codon of the Neo^r gene and terminated at the translation termination codon (*).

Previous studies have revealed that some flavivirus genomes are difficult to clone as a full-length cDNAs using commonly used plasmid vectors and propagated in *E. coli* [15–17]. Polo et al. described a strategy to clone the full-length cDNA for DENV2 (New Guinea C strain) in *E. coli* using yeast-*E. coli* shuttle vector, pRS424, and in vivo homologous recombination in yeast [18]. In this study, we describe this approach to generate DENV4 replicons.

2   Materials

1. General chemicals and reagents: Use molecular biology grade chemicals and reagents. Use Milli-Q water or equivalent for preparation of all solutions. Store all reagents at room temperature unless indicated otherwise.

2. Plasmid clones: The plasmid clone encoding the full-length cDNA of DENV4 in the pRS424 vector is a gift from Dr. Robin Levis [19]. The cDNAs encoding DENV2 replicons (New Guinea C strain) are from Padmanabhan's laboratory (Reichert and Padmanabhan, unpublished results).

3. *Renilla luciferase* (*Rluc*) assay kit: The kit is purchased from Promega (Madison, WI), and it contains all the reagents necessary to measure *Rluc* activity using a luminometer.

4. PCR reagents: Taq polymerase, 10× PCR buffer, and mixture of four ribonucleoside triphosphates (NTP; 250 µM of each) are required. In some cases, 10× Standard Taq reaction buffer (100 mM Tris-HCl, pH 8.3, 500 mM KCl, 15 mM $MgCl_2$) is used.

5. Reagents for ligation: 10× ligation buffer (500 mM Tris-HCl, pH 7.5, 100 mM $MgCl_2$, 5 mM ATP, 100 mM dithiothreitol (*see* **Note 1**)); 10 mM ATP (*see* **Note 2**), 10 mg/mL acetylated bovine serum albumin (BSA) (*see* **Note 3**), and T4 DNA ligase are required.

6. Materials for bacterial transformation: *E. coli* DH5α or Stbl2 competent cells are required.

7. Restriction enzymes: Restriction enzymes and their respective 10× buffers are from suitable sources.

8. Materials for in vivo recombination in yeast: Yeast strain: *Saccharomyces cerevisiae* YPH857 (*Matαura3-52 lys2-801 ade2-101 his3Δ200 trp1Δ63 leu2Δ1 cyh2^R*) [20] is used. Solutions I, II, and III of the S.c. EasyComp™ Transformation Kit are from Invitrogen (K5050-01).

9. Reagents for plasmid isolation from yeast: Solutions 1, 2, and 3 and Zymolyase (4 U/μL) are components of Zymoprep yeast plasmid miniprep II from Zymo Research (Cat. No. D2004).

10. Materials and solutions for in vitro transcription: AmpliScribe SP6 High Yield Transcription Kit that includes 10× reaction buffer (final concentration, 1×; *see* **Note 4**) is from Epicentre Biotechnologies. If the cap analog, m7G(5′)ppp(5′)G, is supplied by the manufacturer in a lyophilized form, add H_2O to prepare a stock solution of 40 mM (*see* **Note 5**). Final NTP concentrations for the production of capped DENV4 Neo^r replicon RNA are 5 mM each of ATP, CTP, and UTP and 1 mM of GTP. Make up a stock reaction mixture containing 25 mM each of ATP, CTP, UTP and 5 mM of GTP by adding 5 μL each of 100 mM stock solutions of ATP, CTP, and UTP and 1 μL of 100 mM GTP into 4 μL of H_2O to produce 20 μL of the mixture (*see* **Note 6**).

11. Materials for mammalian cell electroporation: Use 1× phosphate-buffered saline (PBS) for cell electroporation. For the preparation of 1× PBS buffer, add the following: 8 g sodium chloride, 0.2 g potassium chloride, 1.44 g Na_2HPO_4, and 0.24 g KH_2PO_4 in 800 mL of H_2O. Adjust the pH to 7.4 with HCl. Add H_2O to 1 L. Autoclave and store at room temperature. The Gene Pulser Xcell™ Eukaryotic System and sterile electroporation cuvettes (5 × 0.4 cm) are from Bio-Rad.

12. Mammalian cell culture: For the preparation of complete medium, mix the following: Dulbecco's Modified Eagle Medium (DMEM), 10 % heat-inactivated fetal bovine serum (FBS), 4.5 g/L glucose, 2 mM glutamine, and 1× of penicillin-streptomycin. Heat inactivation of FBS: incubate FBS for 30 min at 56 °C, mixing every 5 min. Make 50 mL aliquots and store at –20 °C. Use 0.25 % trypsin-EDTA (1×) to dissociate adherent cells from the cell culture flasks or plates.

13. Mammalian cell cloning: Bel-Art cell culture paper cloning disks (Scienceware^R is trademark of Bel-Art products) 5 mm size (available from Sigma-Aldrich-cat. No. Z374458-100EA) and a sterile fine-point forceps. G418: 50 mg/mL stock solution (*see* **Note 7**).

3   Methods

1. General methods: Follow established protocols and regulations in the Institutional Biosafety Committee manual for handling biohazard materials such as recombinant DNA and plasmid-transformed *E. coli* waste disposal.

2. For the PCR amplification of the fragments, use about 1 μg of plasmid DNA template and 10 μM of oligonucleotides primers (e.g., A, B, C, D, E, F, and G in Table 1).

3. Screening colonies by PCR: For screening the yeast colonies, the colony is picked with a sterile inoculation loop and PCR is performed with primers F and G.

4. Ligation of DNA fragments: Use 0.5 μL T4 DNA ligase (200 cohesive end units) in a total volume of 10 μL.

5. Bacterial transformation: Prepare SOC medium by mixing 2 g of tryptone, 0.5 g yeast extract, 0.25 mL of 1 M KCl, 1 mL of 1 M NaCl, 1 mL of 1 M MgCl$_2$, and 98 mL water. Sterilized by autoclave. After it is cooled down, add 1 mL of 1 M MgSO$_4$ and 1 mL of 2 M glucose. Filter through a 0.22 μm filter and store in 5 mL aliquots.

6. Restriction enzyme digestions: Follow incubation conditions for restriction enzyme digestions recommended by the manufacturer. For the digestion of the DENV4 cDNA clone in pRS424 with NheI, 40 μg of plasmid DNA is used in the presence of 1× manufacturer-supplied buffer (10 mM Tris-HCl, pH 7.9, 50 mM NaCl, 10 mM MgCl$_2$, and 1 mM dithiothreitol) and 100 μg/mL bovine serum albumin. For the digestion of the pGEM-T easy clone containing the DENV4-Luc-Neo^r insert with EcoRI, 40 μg of plasmid DNA is used in the presence of 1× manufacturer-supplied buffer for EcoRI (10 mM Tris-HCl, pH 7.5, 50 mM NaCl, 10 mM MgCl$_2$, and

Table 1
Primers used for PCR

Name	Sequence
A	5′-AGTTGTTAGTCTGTGTGGACCG-3′
B	5′-CGCGGTTTCTCTCGCGTTTCAGCATATTGAAAGG-3′
C	5′-GCTGAAACGCGAGAGAAACCGCG-3′
D	5′ GGTATTATCGTGTTTTTCAAAGGAAAACCACG-3′
E	5′-CCTTTGAAAAACACGATAATACCATGGGGTTCTTAGTGTTGTGGATTGGC-3′
F	5′-CCATGACACCACACAACCCATGTC-3′
G	5′-CTTCTTGACGAGCATTCCTA-3′

0.025 % of Triton X-100). To linearize the pRS424 DENV4-Luc-Neo^r replicon for in vitro transcription, Sac II is used: about 40 μg plasmid DNA is used in the presence of 1× manufacturer-supplied buffer (20 mM Tris-acetate, pH 7.9, 50 mM potassium acetate, 10 mM magnesium acetate, and 1 mM dithiothreitol).

7. Preparation of competent yeast cells: The following solution is used to make the yeast culture medium: Prepare 10× dextrose solution by dissolving 200 g of dextrose (D-glucose) in 800 mL of water, bring the volume up to 1,000 mL, and sterilize by filtration using 0.45 μm filter. To make 1 L of yeast extract/peptone/dextrose medium (YPD), mix in 900 mL of deionized Milli-Q water, 10 g of yeast extract, 20 g of peptone, and 0.1 g adenine, and sterilize in an autoclave. After the solution is cooled down, add 100 mL of sterile 10× dextrose (D-glucose). Store the medium at room temperature. Use this solution within approximately 2 months.

8. Cloning by in vivo yeast homologous recombination: To prepare the amino acid dropout mixture, combine 1.5 g of valine, 0.2 g of arginine, 0.3 g of isoleucine, 0.2 g methionine, 0.5 g of phenylalanine, 0.3 g of lysine, 1.0 g of glutamic acid, 0.2 g adenine sulfate, 3.7 g of serine, 0.2 g of uracil, 0.3 g of tyrosine, 0.3 g leucine, 1.0 g of aspartic acid, 0.2 of histidine, and 2.0 g of threonine (*see* **Note 8**). For the preparation of the synthetic dextrose solid media without tryptophan (SD-trp), add 20 g of agar to 800 mL water, and autoclave. In a separate flask mix 20 g of dextrose, 6.7 g of yeast nitrogen base with ammonium sulfate, and 1.2 g of the amino acid dropout mix, and bring it to a volume of 200 mL with water and sterilize by filtration. After the sterilized solutions cool down, mix them and dispense into Petri dishes. Store the Petri dishes at 4 °C.

9. Yeast replicon clone culture: Prepare SD-trp liquid media by mixing 20 g of dextrose, 6.7 g of yeast nitrogen base with ammonium sulfate, and 1.2 g of the amino acid dropout mix, and bring it to a volume of 1,000 mL with water. Sterilize by filtration. Store at 4 °C.

3.1 Construction of the DENV4-Luc-Neo^r Replicon by PCR

3.1.1 PCR-1

1. Follow standard PCR protocols: For initial PCR, the reaction mixture (50 μL) consists of the following components in an appropriate microcentrifuge tube (0.2 or 0.5 mL capacity, depending on the capacity of the heating block in the thermal cycler used for PCR): H$_2$O (37.5 μL), 10× PCR buffer (5 μL), 2.5 mM each of dNTP mixture (4 μL; the final concentration of each dNTP is 200 μM), 10 μM primer A (Table 1; Fig. 1, A1) (1 μL), 10 μM primer B (1 μL), 500 ng of pRS424 DENV4 cDNA as the DNA template (0.5 μL; 1 μg), and 5 U Taq polymerase (1 μL). Mix well and then spin briefly in a microcentrifuge.

2. Perform PCR-1 using the following cycle profiles: Initial denaturation at 94 °C for 3 min, followed by 30–35 cycles of denaturation at 94 °C for 1 min, annealing at 55 °C for 2 min, and extension at 72 °C for 4 min (depending on product length), and final extension incubated at 72 °C for 2 min to produce fragment 1 (Fig. 1, A1).

3. Perform two additional PCR reactions in parallel using conditions described above (**steps 1** and **2** under Subheading 3.1.1): In one, primers C, D (Table 1) and the DENV2 replicon cDNA (5′-UTR-$C_{25codons}$-$Rluc$-FMDV2A-Neo^r*-IRES-$E_{74codons}$-NSorf-3′-UTR; Reichert and Padmanabhan, unpublished results) as the template; in the other, perform PCR using primers E, F and pRS424 DENV4 cDNA as the template. These reactions produce PCR fragments 2 and 3, respectively (Fig. 1, A2 and A3). After synthesis, maintain the samples at 16 °C (*see* **Note 9**) until processed further.

4. Purify the PCR fragments using standard agarose gel electrophoresis and recovery using a kit from Zymo Research (Irvine, CA) following the manufacturer's recommendations. Use a minimum volume for elution (5–10 μL) of the PCR DNA fragment adsorbed to the matrix.

3.1.2 PCR-2

1. In a 50 μL reaction volume, mix the following components: 10× PCR buffer (5 μL), 2.5 mM each of dNTP mix (4 μL), all of the recovered fragment 1 (10 μL), all of the recovered fragment 2 from the **step 4** under Subheading 3.1.1 (10 μL), 10 μM primer A (1 μL), 10 μM primer D (Table 1) (1 μL), and 5 U Taq polymerase (1 μL). Make up the reaction volume to 50 μL with H_2O and mix well. Perform PCR following conditions described in **steps 1** and **2** under Subheading 3.1.1. Fragment 4 is obtained (Fig. 1, B1).

2. Perform another PCR reaction in parallel using gel-purified fragments 2 and 3 mixed with primers C and F and conditions described for PCR-1. Fragment 5 is obtained.

3. Purify the final PCR products by gel electrophoresis as described in **step 4** under Subheading 3.1.1. Elute the DNA fragments 4 and 5 in minimum volumes for each fragment (10 μL).

3.1.3 PCR-3

1. Assemble the following components in a microcentrifuge tube: 10× PCR buffer (5 μL), 2.5 mM each of dNTP mix (4 μL), all of the recovered fragments 4 and 5 (10 μL each), 10 μM primer A (1 μL), 10 μM primer F (Table 1) (1 μL), and 5 U Taq polymerase (1 μL); add H_2O to a volume of 50 μL and mix.

2. Perform PCR following conditions described in **steps 1** and **2** under Subheading 3.1.1.

3. Purify the final PCR product by gel electrophoresis as described in **step 4** of Subheading 3.1.1. Clone the PCR product into the pGEM-T Easy[R] vector following the manufacturer's instructions. Select a few clones and verify their sequences to make sure that there are no mutations (*see* **Note 10**).

3.2 Construction of the DENV4-Luc Replicon (without Neo[r] gene) by PCR

This replicon is constructed also by overlap PCR as described under Subheading 3.1 with the difference that for PCR-2 the template used was a DENV2 replicon without the Neo[R] gene (Erin D. Reichert and R. Padmanabhan, unpublished results) (*see* Fig. 2).

3.3 Cloning of the PCR Fragment into the pGEM-T Easy Vector

1. Purify the final PCR product by gel electrophoresis as described in **step 4** under Subheading 3.1.1. Elute the product in minimum volume (10 µL) in a 1.5 mL Eppendorf tube and dry in a vacuum centrifuge (*see* **Note 11**). Resuspend the purified PCR product in 6.5 µL of H_2O.

2. Perform the ligation reaction by adding 6.5 µL of the DNA from the above step, 10× ligation buffer (1 µL), 10 mM ATP (1 µL), 10 mg/mL acetylated BSA (1 µL), 200 cohesive end units of T4 DNA ligase (0.5 µL). Incubate the ligation reaction overnight at room temperature. Use the whole ligase reaction mixture for transformation of competent *E. coli* DH5α cells.

3. Grow selected bacterial colonies in LB broth + ampicillin (100 µg/mL). Plasmid DNAs from colonies containing the pGEM-T Easy DENV4-*Rluc*-Neo[r] insert are prepared on a mini scale following standard recombinant DNA methods.

4. Verify the sequences of the insert DNA encoding the *Renilla* luciferase gene fused to the part of capsid gene (25 codons) at the 5′-end and the EMCV leader sequence at the 3′-end followed by 37 amino acid coding sequence of the E gene in the DENV4 replicon to ensure that there are no mutations due to the PCR amplifications.

5. Purify pGEM-T Easy DENV4-*Rluc*-Neo[r] DNA from a medium scale (250 mL) culture of the transformed *E. coli* cells using standard recombinant DNA methods.

3.4 Cloning of the DENV4 Replicon

1. DNA preparation: For the restriction enzyme digestion of the pRS424 DENV4 cDNA plasmid, mix the following components: H_2O (4 µL), 40 µg of the plasmid (80 µL), 10× buffer (10 µL), 10 mg/mL BSA (1 µL) and 50 U of NheI (5 µL) (*see* **Note 12**). Incubate overnight at 37 °C.

2. Purify the restriction digest of DENV4 cDNA following standard protocols involving phenol/chloroform/isoamyl alcohol extraction and precipitation with ethanol. Do not resuspend in water at this step (*see* **Note 13**).

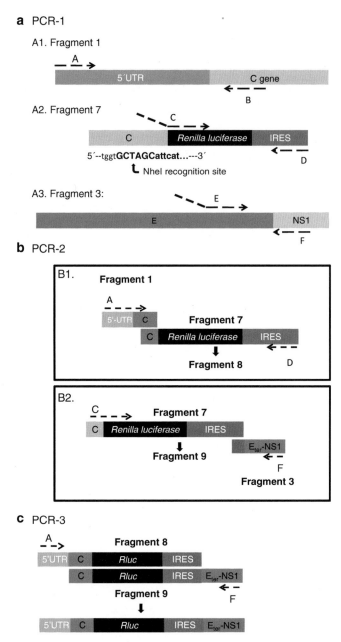

Fig. 2 Strategy used for construction of DENV4 Renilla luciferase reporter replicon without a selectable marker. Sequences of primers used for PCR are indicated in Table 1 (*see* also Fig. S1 in ref. 19). The pRS424 DENV4 cDNA clone and DENV2 replicon cDNA without FMDV 2A and neomycin resistance gene (pRS424DENV2LuclresDen2) (E. D. Reichert and Padmanabhan, unpublished results) are used for construction of the DENV4 replicon. Conditions are same as in Fig. 1 for PCR amplifications to get fragment 7 using primers C and D and pRS424DENV2LuclresDen2 as the template. Additional PCRs are performed to fuse fragments 1 with 7 and 7 with fragment 3 to yield fragments 8 and 9. To obtain the final product (fragment 10), fragments 8 and 9 are used for the final PCR using primers A and F

3. In parallel, digest the DENV4-*Luc*-Neo^r plasmid containing the PCR fragment cloned into the pGEM-T Easy^R vector with EcoRI (*see* **Note 14**). For the reaction, mix the following: H$_2$O (5 μL), 40 μg of the plasmid (80 μL), 10× buffer for EcoRI (10 μL) and 50 U of EcoRI (5 μL). Incubate overnight at 37 °C.

4. Purify the restriction enzyme digested DNA by gel electrophoresis as described in **step 4** under Subheading 3.1.1. Elute the fragment in 10 μL with water.

5. Mix the eluted DNA from **step 4** with the precipitated DENV4 cDNA in **step 2** under the Subheading 3.4. Resuspend the two components very well.

3.5 Preparation of Competent S. cerevisiae Cells

1. The solutions used to make the competent *S. cerevisiae* cells are provided in the S.c. EasyComp™ Transformation Kit (Invitrogen) with the exception of the YPD medium, and all solutions should be equilibrated to room temperature before starting.

2. For the preparation of the competent *S. cerevisiae* cells, inoculate one colony of the yeast into 10 mL of YPD medium and incubate overnight at 30 °C in a shaking incubator (250 rpm).

3. Determine the OD$_{600}$ of the overnight culture; should be between 3.0 and 5.0.

4. Dilute to an OD$_{600}$ of 0.2–0.4 in a total volume of 10 mL of YPD.

5. Grow the cells at 30 °C in a shaking incubator (250 rpm) until the OD$_{600}$ reaches 0.6–1.0.

6. Pellet the cells by centrifugation at 500×*g* (1,500 rpm) for 5 min at room temperature and discard the supernatant.

7. Resuspend the cell pellet in 10 mL of Solution I.

8. Pellet the cells by centrifugation at 500×*g* (1,500 rpm) for 5 min at room temperature. Discard the supernatant.

9. Resuspend the cell pellet in 1 mL of Solution II.

10. Aliquot 100 μL each in 1.5 mL microcentrifuge tubes and store at −80 °C until used.

3.6 Cloning by In Vivo Yeast Homologous Recombination

1. Equilibrate the Solution III of the S.c. EasyComp™ Transformation Kit and SD-trp plates to room temperature. Thaw on ice an aliquot of competent yeast cells from **step 10** under Subheading 3.5.

2. Mix the DNA fragments prepared in **step 5** under the Subheading 3.4 with the competent yeast cells on ice.

3. Add 500 μL of the Solution III from S.c. EasyComp™ Transformation Kit.

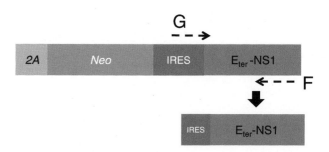

Fig. 3 Screening of yeast colonies by PCR. To select positive clones among the yeast colonies, PCR is performed using primers G and F and single, well-isolated colonies of yeast as the source of DNA. The expected size of PCR fragment is 565 bp

4. Incubate at 30 °C for 1 h and mix the contents by inverting the tube every 15 min.

5. Plate the entire transformation mixture (proximally 600 μL) on to a SD-trp plate.

6. Incubate in a 30 °C incubator for 24 h or until small colonies appear.

7. Screen the colonies for the desired insert by PCR using the primers F and G (Fig. 3; *see* Table 1 for primer sequences). The PCR reaction mixture (10 μL) consists of the following components: H_2O (8.65 μL), 10× PCR buffer (1 μL), 2.5 mM each of dNTP mixture (0.05 μL; the final concentration of each dNTP is 12.5 μM), 10 μM primer F (0.125 μL), 10 μM primer G (0.125 μL) (final concentration of each primer is 0.125 μM), 0.25 U Taq polymerase (0.05 μL), and a yeast colony as the source of DNA template. Mix well and spin briefly in a microfuge. Perform the PCR using the following cycle profiles: initial denaturation at 94 °C for 6 min, (*see* **Note 15**), followed by 30–35 cycles of denaturation at 94 °C for 1 min, annealing at 55 °C for 2 min, and extension at 72 °C for 30 s min (depending on product length), and final extension, incubated at 72 °C for 2 min.

3.7 Yeast Culture and Plasmid Extraction

All solutions and columns used in this step are provided in the Zymoprep yeast plasmid miniprep II kit (Zymo Research).

1. Inoculate 5 mL of SD-trp medium with a positive colony and incubate overnight at 30 °C in a shaking incubator (250 rpm).

2. Pellet the cells by centrifugation at $600 \times g$ for 2 min. Discard the supernatant.

3. Resuspend the pellet with 200 μL of Solution 1. Add 3 μL of Zymolyase provided in the kit (*see* **Note 16**).

4. Incubate 37 °C for 20 min.

5. Add 200 μL of the Solution 2 and mix.

6. Add 400 μL of the Solution 3 and mix.

7. Centrifuge at maximum speed in an Eppendorf centrifuge for 3 min.

8. Transfer the supernatant to a Zymo-Spin I Column™ and spin for 30 s.

9. Discard the flow-through and add 550 μL of wash buffer. Spin for 2 min. Discard the wash buffer.

10. Place the column in a 1.5 mL microfuge tube, elute with 10 μL of water and spin for 30 s. The eluate is concentrated to 5 μL using a vacuum centrifuge.

3.8 Transformation of Stbl2 Cells with Plasmid DNA Isolated from Yeast

1. *E. coli* **Stbl2** competent cells are used for transformation by the extracted DNA from yeast. Thaw an aliquot (50 μL) of Stbl2 competent cells on ice. Mix the yeast DNA with the competent cells and incubate in ice for 30 min. Incubate at 42 °C for 25 s. Add 1 mL of SOC medium (*see* Subheading 3, **step 5** under Methods). Incubate the cells with shaking at 250 rpm in an orbital shaker at 30 °C for 2.5 h. An aliquot (100–200 μL) is plated onto a LB agar plate containing 100 μg/mL ampicillin. Incubate the plates in a 30 °C incubator for 1–2 days.

2. Pick colonies from LB + ampicillin plates and grow the individual colonies in 5 mL LB medium containing 100 μg/mL ampicillin.

3. Isolate recombinant plasmid DNA from individual colonies containing the DENV4 replicon in a mini scale following standard recombinant DNA methods.

4. Verify the sequence of the replicon cDNA. Prepare the replicon cDNA from 250 mL LB + ampicillin (100 μg/mL).

3.9 In Vitro Transcription

1. For the linearization of the plasmid, add the following: H_2O (4 μL), 40 μg of plasmid pRS424 DENV4-*Rluc*-Neo^r replicon, 10× buffer (10 μL), 10 mg/mL acetylated BSA (1 μL), and 50 U of Sac II (5 μL). Incubate overnight at 37 °C.

2. Purify linearized replicon DNA by extraction with phenol/chloroform/isoamyl alcohol and precipitation with ethanol. Resuspend the DNA in 10 μL of H_2O.

3. Mix the following components for the in vitro transcription reaction: H_2O (5 μL), AmpliCap™ 10× transcription buffer (2 μL), 5 mM each of NTPs (4 μL), 40 mM cap analog (2 μL; the final concentration is 4 mM), 100 mM dithiothreitol

(2 μL), 2 μL of the purified linear pRS424 DENV4-*Rluc*-Neo[r] replicon or DENV4-*Rluc* replicon, RNase inhibitor (1 μL), and SP6 DNA-dependent RNA polymerase enzyme (2 μL from the AmpliCap™ SP6 High Yield Message Maker Kit) in a final volume of 20 μL. Incubate for 2 h at 37 °C. If desired, treat the in vitro transcription reaction with 1 μL RNase-free DNase I and incubate for 15 min at 37 °C (*see* **Note 17**). Purify the RNA by phenol/chloroform/isoamyl alcohol extraction and ethanol precipitation. Resuspend the pellet in 12 μL of H_2O.

4. Verify the integrity of the DENV4-*Rluc*-Neo[r] replicon RNA and DENV4-*Rluc* replicon RNA produced in the in vitro transcription reaction by electrophoresis of a 0.5 μL aliquot on a 0.6 % agarose gel. Aliquot the in vitro transcript and store at −80 °C (*see* **Note 18**).

5. For expression of both replicon RNAs in mammalian cells and quantification of *Renilla* luciferase activity, transfect ~1×10^6 cells of a chosen mammalian cell type (Vero cells in this case) by electroporation with 1.5 μg of the replicon RNA. Immediately after electroporation, allow the cells to recover by addition of growth medium containing 10 % FBS and kept at 37 °C. Incubate for 5 min and plate the cells into 6-well tissue culture plates. Lyse cells using the *Renilla* luciferase cell lysis solution provided in the kit at different time points (*see* **Note 19**). Determine the luciferase activity in the cell lysates collected at various time points (*see* **Note 20**).

3.10 Establishment of Stable Cell Line Expressing DENV4-Rluc-Neo[r] Replicon

1. Estimation of cytotoxicity of G418 to Vero cells.
Before establishment of a stable cell line, it is important to assess the concentration of G418 that would be cytotoxic to normal Vero cells but not toxic to cells replicating pRS424 DENV4-*Rluc*-Neo[r] replicon RNA-encoding Neo[r] gene which would inactivate G418. This protocol is described below.

2. Grow Vero cells from a stock vial stored in liquid nitrogen. When the cells become confluent, remove the medium and wash once with PBS. Add 2 mL of 0.25 % trypsin-EDTA to the cells and incubate 2 min in a humidified 37 °C incubator with 5 % CO_2. Remove trypsin by aspiration and incubate again for 2 min in the 37 °C incubator. Resuspend cells in 10 mL of PBS and count the cells using a hemocytometer or a suitable alternative cell counter following standard cell culture techniques.

3. Plate ~1.25×10^5 cells in 2 mL of complete medium in each well of a 6-well plate; nine wells are needed.

4. After 24 h, remove medium by aspiration, and add 2 mL of complete medium with different concentrations of G418 ranging from 0 to 800 μg/mL (*see* **Note 21**).

5. For the cytotoxicity curve, add the following stock solution of G418 (50 mg/mL) to 2 mL each of medium: (1) 0 μL of G418; (2) 100 μg/mL (4 μL G418); (3) 200 μg/mL (8 μL of G418); (4) 300 μg/mL (12 μL of G418); (5) 400 μg/mL (16 μL of G418); (6) 500 μg/mL (20 μL of G418); (7) 600 μg/mL (24 μL of G418); (8) 700 μg/mL (28 μL of G418); and (9) 800 μg/mL (32 μL of G418).

6. Replenish medium every other day with fresh G418-containing medium.

7. After 2 weeks, select the lowest concentration of antibiotic that kills all the cells, and use this concentration to create the stable cell lines.

3.11 Electroporation

1. Plate ~2.5×10^6 Vero cells in a sterile tissue culture 100 mm dish. 48 h after seeding, use them for transfection of DENV4-*Rluc*-Neo^r replicon RNA by electroporation (*see* **Note 22**).

2. Trypsinize, take a cell count, and centrifuge the cells at $600 \times g$ for 8 min. Resuspend the cells in chilled PBS at 4 °C to have ~1×10^6 cells per 100 μL.

3. Add 100 μL of the cells and 1.5 μg of the DENV4 replicon RNA in a 1.5 mL Eppendorf tube. Mix and transfer to the electroporation cuvette.

4. Transfect by electroporation using the Bio-Rad Gene Pulser Xcell™ and the following settings: 450 V and 50 μF, resistance ∞, and one pulse.

5. After electroporation, immediately add 1 mL of pre-warmed complete medium to the cuvette containing the cells.

6. Transfer the cells to a 100 mm sterile tissue culture dish with medium without any G418.

7. 24 h post-transfection, change the medium, and add fresh medium containing G418 at the appropriate concentration (determined as described under Subheading 3.9).

8. Every 48 h incubation, remove the medium, and add 10 mL of fresh medium with 60 μL of G418 (50 mg/μL stock) at a final concentration of 300 μg/mL.

9. After a week when the cells become confluent, trypsinize the cells, and place half of the cells in a new 100 mm sterile tissue culture dish (*see* **Note 23**). Repeat the replenishment of medium every 48 h followed by trypsinization and replating half of the cells into a new 100 mm sterile tissue culture dish.

10. After 15 days, the majority of the cells that do not have replicating DENV4-*Rluc*-Neo^r replicon RNA would die off leaving several foci of stable clones.

3.12 Cell Cloning

1. Identify the cell foci with an inverted microscope and mark them using a permanent ink marker.

2. Place 4 mL of trypsin in a new 100 mm sterile tissue culture dish and let the cloning disks soak.

3. In parallel, remove the medium from the plates containing the cell foci, add 10 mL of PBS to wash the cells, and remove all traces of liquid by aspiration.

4. Place one presoaked cloning disk in each cell clone, and incubate for 4 min in a 37 °C CO_2 incubator.

5. Take the cloning disk with a sterile fine-point forceps and place each cloning disk in a well of a 24-well plate that has 500 μL of medium without G418.

6. After 24 h incubation, remove the medium and cloning disk from the wells. Replenish with 500 μL fresh complete medium and add 3 μL of G418 (50 mg/μL).

7. Change medium every 48 h and add G418.

8. When cells become confluent, trypsinize the cells, and plate each clone in an individual well of a 6-well plate in 1 mL of complete medium containing 300 μg/μL of G418.

9. Change medium containing G418 every 48 h.

10. When cells become confluent, trypsinize the cells, and plate each clone in an individual 100 mm sterile tissue culture dish in 10 mL of complete medium with 60 μL of G418 (300 μg/μL).

11. When cells become confluent, assay for luciferase activity (described under, **steps 12–14** below). Make stocks of stable Vero cell clones and store in liquid nitrogen.

12. Quantify the *Rluc* activity using an appropriate luminometer (in this case, Centro LB 960 Microplate Luminometer and MikroWin Version 4.0 Software from Berthold Technologies is used).

13. Wash cells in wells of a 6-well plate using PBS and lyse by scraping into 250 μL *Renilla* luciferase assay lysis buffer. The cells are collected and mixed vigorously. In some cases, the lysate is centrifuged in a microcentrifuge to remove any cellular debris. The cell lysates can be stored at −20 °C until all the samples are collected.

14. Luciferase assays are performed in accordance with the manufacturer's instructions. An aliquot of the lysate (50 μL) is placed in wells of an opaque 96-well plate. For each sample, 50 μL of *Renilla* luciferase assay buffer with substrate (Promega) is injected into each well and delayed for 2 s before reading luminescence over a period of 10 s. The *Rluc* activities of six cloned cell lines are shown in Fig. 4.

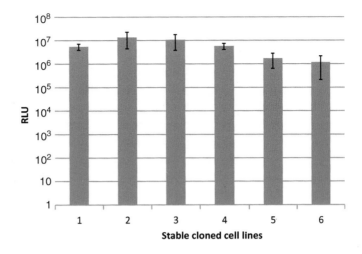

Fig. 4 *Rluc* expression levels of stable Vero cell clones. The details for isolation of stable Vero cell clones expressing *Rluc* reporter and *Rluc* activity assay are described in the text. The numbers in *Y* axis represent the levels of *Rluc* activity expressed as relative luciferase units (RLU) present in the same amounts of cell lysates from six independent clones, 1–6

4 Notes

1. The ligation buffer has ATP and dithiothreitol that are very labile; so freeze-thaw cycles are not recommended. Store aliquots of 5 μL each at –20 °C (for short-term storage of 1–2 months) or –80 °C (for long-term storage). Once an aliquot is thawed and partly used, the reminder is discarded. The dithiothreitol in the buffer has a strong smell. Prepare aliquots in a fume hood to avoid inhalation.

2. Ligase requires ATP as a substrate. dATP forms a more stable complex thus inhibiting the ligation reaction.

3. BSA is added to stabilize the restriction enzyme from denaturation due to dilution.

4. Thaw at room temperature. If any white precipitate is observed, incubate for 10 min at 37 °C or until it is completely dissolved.

5. Store the cap analog stock solution in 5 μL aliquots at –20 °C (~6 months) or –80 °C for long term.

6. The NTP mixture can be frozen and thawed once. Discard any unused NTP mixture after thawed once.

7. If purchased in a powder form, verify the amount of active ingredient, e.g., if the active ingredient is 775 μg/mg, take 645.2 mg and dissolve in 10 mL PBS to prepare a stock solution of 50 mg/mL. Filter-sterilize by passing through a

0.22 μm filter fitted onto a 50 mL syringe. The antibiotic is stable for several months at 4 °C.

8. Leave out the amino acid which is used for selection. In case of using another amino acid for selection other than tryptophan, add 0.2 g of tryptophan and leave out the amino acid of selection.

9. Thermal cycler machine can be programmed at the end of PCR cycles to store samples at a preset temperature. Setting the storage temperature at 16 °C for short periods of time is recommended by a technical representative of the manufacturer as (1) DNA samples are stable at that temperature and (2) it is good for optimal performance of the PCR machine.

10. The T-tailed pGEM-T Easy vector is useful for cloning PCR products which are A tailed as Taq polymerase catalyzes non-templated addition of usually A residue to the 3′-end of PCR products. The sequence GTTTCTT added to the 5′-end of reverse primer promotes very efficient non-templated A to the 3′-end of forward strand [21]. Thus, the cloning efficiency of products of PCR amplification depends on the design of PCR primers. In addition, the pGEM-T Easy vector is useful in verifying the sequence of the insert using primers annealing to either T7 or SP6 promoter in the vector.

11. For purification of DNA fragments from agarose gels, there are a number of protocols available; some are based on selective adsorption to column matrices and others on glass beads. We prefer protocols based on column matrices and elution into minimum volume from the column. With regard to protocols based on beads, it is important to ensure that no beads remain in the eluted and purified DNA which could interfere with solubility and inhibit the subsequent ligation reaction.

12. DENV4 cDNA clone is linearized by digestion with a suitable restriction enzyme digestion which cuts once near the site of insertion. The PCR fragment should have sufficient overlapping sequences at both sides of the restriction enzyme cleavage site for yeast homologous recombination to work efficiently. If there are two unique restriction sites in the vicinity of the insertion site, both restriction enzymes could be used if the PCR fragment selected for insertion contains sufficient length of homologous sequences outside of the two restriction sites chosen for cleavage. There is no need to separate the small fragment released from the double digestion. This latter strategy would reduce the background colonies arising due to intramolecular circularization in yeast cells without recombination with the PCR fragment, if only one restriction enzyme cleavage is used.

13. Store the DNA in precipitated form at –80 °C. The DNA is resuspended in a buffer in the subsequent step (Subheading 3.3, **step 5**).

14. The PCR insert is flanked by two EcoRI restriction recognition sites within the polylinker site of the pGEM-T Easy plasmid, which makes it easy to release the fragment.

15. Extension of the time is important for initial cell breakage and liberation of DNA into the solution and denaturation.

16. For multiple samples, according to the manufacturer's protocol, add 15 µL Zymolyase™ for each mL of Solution 1 and mix. Use 200 µL of this mixture to resuspend the pellet from each sample.

17. We have observed that it is not necessary to digest the DNA template by treatment with DNase I after the in vitro transcription.

18. RNA is very unstable to freeze-thaw cycles. Therefore, we recommend the in vitro-transcribed RNA to be stored as 1.5 µL aliquots at –80 °C.

19. If desired, the cell extracts can be stored in a 96-well plate at –20 °C until samples at all time points are collected and then the luciferase activity can be measured. This is recommended for optimal use of the buffers and luciferase substrate in the kit.

20. Based on published results reported for different flavivirus replicons including ours for transient expression of DENV4 replicon [19], a peak of *Renilla* luciferase activity is observed in the sample collected 2–4 h post-transfection. This *Rluc* activity represents translation of the transfected viral RNA. The luciferase activity declines gradually up to 20–24 h post-transfection. From 24 to 96 h post-transfection, a steady increase in luciferase signal is obtained due to increase in copy number and translation of the newly replicated RNA.

21. Cells are at ~25 % confluence at this stage.

22. For Vero cells a density of ~2.5×10^6 are needed to reach confluence in a 100 mm sterile tissue culture dish in 48 h. Cells passaged 2–3 days prior to electroporation are at 70–85 % confluence on the day of electroporation. Lower or higher cell densities may cause lower transfection efficiencies and increased cell death.

23. Antibiotics work best when cells are actively dividing. If the cells become too dense, the antibiotic efficiency will decrease. It is best to split cells such that they are not more than 25 % confluent.

Acknowledgments

The research was supported by NIH grants R01 AI-32078 and U01 AI 54776 (R.P.).

References

1. Khromykh AA, Meka H, Guyatt KJ, Westaway EG (2001) Essential role of cyclization sequences in flavivirus RNA replication. J Virol 75(14):6719–6728

2. Lo MK, Tilgner M, Bernard KA, Shi PY (2003) Functional analysis of mosquito-borne flavivirus conserved sequence elements within 3′ untranslated region of West Nile virus by use of a reporting replicon that differentiates between viral translation and RNA replication. J Virol 77(18):10004–10014

3. Alvarez DE, Lodeiro MF, Luduena SJ, Pietrasanta LI, Gamarnik AV (2005) Long-range RNA–RNA interactions circularize the dengue virus genome. J Virol 79(11): 6631–6643

4. Jones M, Davidson A, Hibbert L, Gruenwald P, Schlaak J, Ball S, Foster GR, Jacobs M (2005) Dengue virus inhibits alpha interferon signaling by reducing STAT2 expression. J Virol 79(9):5414–5420

5. Jones CT, Patkar CG, Kuhn RJ (2005) Construction and applications of yellow fever virus replicons. Virology 331(2):247–259

6. Ng CY, Gu F, Phong WY, Chen YL, Lim SP, Davidson A, Vasudevan SG (2007) Construction and characterization of a stable subgenomic dengue virus type 2 replicon system for antiviral compound and siRNA testing. Antivir Res 76(3):222–231. doi:10.1016/j.antiviral.2007.06.007

7. Puig-Basagoiti F, Tilgner M, Forshey BM, Philpott SM, Espina NG, Wentworth DE, Goebel SJ, Masters PS, Falgout B, Ren P, Ferguson DM, Shi PY (2006) Triaryl pyrazoline compound inhibits flavivirus RNA replication. Antimicrob Agents Chemother 50(4):1320–1329.doi:10.1128/AAC.50.4.1320-1329.2006, 50/4/1320 [pii]

8. Manzano M, Reichert ED, Polo S, Falgout B, Kasprzak W, Shapiro BA, Padmanabhan R (2011) Identification of cis-acting elements in the 3′-untranslated region of the dengue virus type 2 RNA that modulate translation and replication. J Biol Chem 286(25):22521–22534. doi:10.1074/jbc.M111.234302

9. Alcaraz-Estrada SL, Reichert ED, Padmanabhan R (2013) Construction of self-replicating subgenomic West Nile virus replicons for screening antiviral compounds. Methods Mol Biol 1030:283–299. doi:10.1007/978-1-62703-484-5_22

10. Spencer F, Ketner G, Connelly C, Hieter P (1993) Targeted recombination-based cloning and manipulation of large DNA segments in yeast. Meth Comp Meth Enzymol 5: 161–175

11. You S, Padmanabhan R (1999) A novel in vitro replication system for Dengue virus. Initiation of RNA synthesis at the 3′-end of exogenous viral RNA templates requires 5′- and 3′-terminal complementary sequence motifs of the viral RNA. J Biol Chem 274(47): 33714–33722

12. Filomatori CV, Lodeiro MF, Alvarez DE, Samsa MM, Pietrasanta L, Gamarnik AV (2006) A 5′ RNA element promotes dengue virus RNA synthesis on a circular genome. Gene Dev 20(16):2238–2249

13. Clyde K, Harris E (2006) RNA secondary structure in the coding region of dengue virus type 2 directs translation start codon selection and is required for viral replication. J Virol 80(5):2170–2182

14. Falgout B, Chanock R, Lai CJ (1989) Proper processing of dengue virus nonstructural glycoprotein NS1 requires the N-terminal hydrophobic signal sequence and the downstream nonstructural protein NS2a. J Virol 63(5): 1852–1860

15. Ruggli N, Rice CM (1999) Functional cDNA clones of the Flaviviridae: strategies and applications. Adv Virus Res 53:183–207

16. Rice CM, Grakoui A, Galler R, Chambers TJ (1989) Transcription of infectious yellow fever RNA from full-length cDNA templates produced by in vitro ligation. New Biol 1(3): 285–296

17. Kapoor M, Zhang L, Mohan PM, Padmanabhan R (1995) Synthesis and characterization of an infectious dengue virus type-2 RNA genome (New Guinea C strain). Gene 162(2):175–180

18. Polo S, Ketner G, Levis R, Falgout B (1997) Infectious RNA transcripts from full-length dengue virus type 2 cDNA clones made in yeast. J Virol 71(7):5366–5374

19. Alcaraz-Estrada SL, Manzano MI, Del Angel RM, Levis R, Padmanabhan R (2010) Construction of a dengue virus type 4 reporter replicon and analysis of temperature-sensitive mutations in non-structural proteins 3 and 5. J Gen Virol 91(Pt 11):2713–2718. doi:10.1099/vir.0.024083-0, doi:vir.0.024083-0 [pii]

20. Kouprina N, Eldarov M, Moyzis R, Resnick M, Larionov V (1994) A model system to assess the integrity of mammalian YACs during transformation and propagation in yeast. Genomics 21(1):7–17. doi:10.1006/geno.1994.1218

21. Khromykh AA, Kenney MT, Westaway EG (1998) trans-Complementation of flavivirus RNA polymerase gene NS5 by using Kunjin virus replicon-expressing BHK cells. J Virol 72(9):7270–7279

Chapter 10

Targeted Mutagenesis of Dengue Virus Type 2 Replicon RNA by Yeast In Vivo Recombination

Mark Manzano and Radhakrishnan Padmanabhan

Abstract

The use of cDNA infectious clones or subgenomic replicons is indispensable in studying flavivirus biology. Mutating nucleotides or amino acid residues gives important clues to their function in the viral life cycle. However, a major challenge to the establishment of a reverse genetics system for flaviviruses is the instability of their nucleotide sequences in *Escherichia coli*. Thus, direct cloning using conventional restriction enzyme-based procedures usually leads to unwanted rearrangements of the construct. In this chapter, we discuss a cloning strategy that bypasses traditional cloning procedures. We take advantage of the observations from previous studies that (1) unstable sequences in bacteria can be cloned in eukaryotic systems and (2) *Saccharomyces cerevisiae* has a well-studied genetics system to introduce sequences using homologous recombination. We describe a protocol to perform targeted mutagenesis in a subgenomic dengue virus 2 replicon. Our method makes use of homologous recombination in yeast using a linearized replicon and a PCR product containing the desired mutation. Constructs derived from this method can be propagated in *E. coli* with improved stability. Thus, yeast in vivo recombination provides an excellent strategy to genetically engineer flavivirus infectious clones or replicons because this system is compatible with inherently unstable sequences of flaviviruses and is not restricted by the limitations of traditional cloning procedures.

Key words Dengue, Yeast in vivo recombination, Targeted mutagenesis, 3'UTR, Shuttle vector

1 Introduction

Reverse genetics has proven to be a powerful tool in advancing our understanding of different processes in the flavivirus life cycle. Such systems have allowed the generation of targeted mutations in regions of interest in the context of infectious cDNA clones. The phenotypic effects of these mutant infectious clones could then be studied by transfecting them into their host cells. Several infectious clones have been established for different flaviviruses [1–6]. However, their construction has proven to be technically challenging as these viral sequences appear to be genetically unstable in *Escherichia coli* and resulted in rearranged clones after recovery from bacteria. This prompted earlier infectious clones to be

Radhakrishnan Padmanabhan and Subhash G. Vasudevan (eds.), *Dengue: Methods and Protocols*, Methods in Molecular Biology, vol. 1138, DOI 10.1007/978-1-4939-0348-1_10, © Springer Science+Business Media, LLC 2014

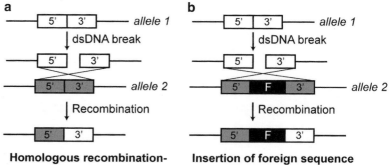

Fig. 1 Homologous recombination repair induced by double-stranded DNA breaks. (**a**) One of the pathways eukaryotic cells use to repair dsDNA breakages is homologous recombination. When a cell senses a dsDNA lesion (allele 1), the DNA repair complex finds the homologous allele (*gray*, allele 2) corresponding to the damaged region. Homologous recombination is induced that results in the replacement of the homologous region 5′ of the breakage with the fragment from allele 2. (**b**) Exploiting this mechanism, one can introduce sequences or mutations (deletions, insertions, or substitutions) by flanking this foreign sequence ("F," in *black*) with regions of homology

comprised of two plasmids each containing a segment of the viral genome [1, 4, 5]. These two cDNAs are digested from the plasmids, ligated together, and used directly as a template for in vitro transcription. An alternative approach described for Kunjin virus and dengue virus type 4 (DENV4) is to construct a full-length cDNA clone in the low-copy plasmid pBR322 at the cost of high DNA recovery [2, 6].

It is known that DNA sequences that are unstable in *E. coli* could be efficiently cloned in eukaryotic systems. Thus, to circumvent the problem of flavivirus cDNA instability in bacteria while not sacrificing DNA yield, we and others have exploited homologous recombination in *Saccharomyces cerevisiae* to construct infectious clones and subgenomic replicons of flaviviruses [3, 7, 8, 11–14]. In this method, we take advantage of the yeast homologous recombination repair pathway that is induced by double-stranded DNA (dsDNA) breaks. Once a breakage is sensed by the cell in a DNA, the DNA damage complex finds an intact and homologous allele (allele 2) with sequences identical to the damaged region. Recombination occurs between these two alleles at the regions of homology ultimately resulting in the exchange of DNA fragments (Fig. 1a). Although this event requires the sequences that are 5′ and 3′ of the dsDNA damage to be identical to sequences in allele 2, these corresponding sequences in allele 2 need not be adjacent to each other. Thus, artificial sequences could be introduced to the location of the DNA damage break for as long as the foreign DNA is flanked by sequences homologous to the damaged strand (Fig. 1b). The clear advantage of this method over traditional cloning procedures is that this removes the need for convenient restriction sites. In addition, small or large fragments can be inserted with similar ease and efficiency. Lastly, simultaneous insertion of multiple fragments in a desired region is possible provided that they have corresponding homologous flanking regions.

Another chapter in this book describes the assembly of DENV subgenomic replicons into the yeast shuttle vector pRS424 using yeast recombination (S. Alcaraz-Estrada, R. del Angel, and R. Padmanabhan, Chapter 9, in this series). In this section, we describe a recombination-based strategy to introduce targeted mutations into these constructs, particularly in the 3′UTR of the DENV2 New Guinea C replicon. This method has been successfully used with great ease to create several mutant replicons analyzed previously [12]. The protocol in this chapter can easily be adapted to the targeted mutagenesis of virtually any region in the DENV genome.

We first introduce a dsDNA break in the DENV2 replicon by linearization of the plasmid. Here, we use the unique restriction enzyme site *Bbv*CI which is found near the 3′ end of the 3′UTR. It is critical that the mutation site is upstream of this dsDNA breakage as this will be the region replaced by "allele 2." This "allele 2" is a 3′UTR PCR product containing the desired mutation (deletion, insertion, or substitution). Both the linearized replicon and mutant 3′UTR PCR fragment are transformed into *S. cerevisiae* and grown in synthetic dextrose (SD) medium without the amino acid tryptophan. The pRS424 backbone encodes the phosphoribosylanthranilate isomerase *TRP1* gene which is critical for tryptophan biosynthesis. Thus, only colonies that have successfully repaired the dsDNA breakage will grow on this SD-Trp dropout (SD-Trp⁻) medium. Resulting plasmids from these colonies are sequenced and then transformed into *E. coli* Stbl2, a strain that is often used to clone and produce plasmids with unstable or toxic sequences. Stbl2 cells have an *end*A1 mutation which allows for high plasmid recovery. It is important to note that although Stbl2 cells are designed for cloning unstable inserts, attempts to directly clone the DENV2 sequences into pRS424 in these *E. coli* cells have repeatedly failed (E.R. Reichert and R. Padmanabhan, unpublished observations and [3]). Thus, cloning in *S. cerevisiae* appears to be necessary in constructing DENV clones.

2   Materials

1. pRS424 DENV2 NGC replicon ([12].

2. Primers. The forward and reverse primers used to amplify the 3′UTR have the following sequences: 5′-AAGGCAAAAC TAACATGAAACAAGGCTAG-3′ (Primer 1) and 5′-AGAA CCTGTTGATTCAAC-3′ (Primer 2), respectively. Additional

oligos used for site-directed mutagenesis have the mutated/deleted nucleotides at their 5′ ends and are designed by the user.

3. PCR reagents. Any standard DNA polymerase would suffice, but high-fidelity enzymes are preferred to avoid unwanted mutations.

4. Agarose gel extraction kit.

5. *BbvCI* restriction enzyme.

6. Molecular biology-grade chemicals for organic DNA extraction: phenol, chloroform-isoamyl alcohol (24:1), ethanol, 3 M sodium acetate (pH 5.5).

7. *S. cerevisiae* YPH857 strain.

8. Frozen-EZ Yeast Transformation Kit (Zymo Research Cat. No. T2001).

9. YPD medium.

10. Amino acid dropout mix (Trp−): 1.5 g valine, 0.2 g arginine, 0.3 g isoleucine, 0.2 g methionine, 0.5 g phenylalanine, 0.3 g lysine, 1.0 g glutamic acid, 3.7 g serine, 0.3 g tyrosine, 0.3 g leucine, 1.0 g aspartic acid, 0.2 g histidine, 2.0 g threonine, 0.2 g uracil, and 0.2 g adenine sulfate.

11. Synthetic dextrose Trp− medium (SD Trp−): dissolve 6.7 g yeast nitrogen base (without amino acid), 20 g dextrose, 0.87 g amino acid dropout mix, and 20 g agar (if Petri dishes are made) in 1 L of water. Autoclave.

12. Zymoprep Yeast Plasmid Miniprep II (Zymo Research Cat. No. D2004).

13. MAX Efficiency Stbl2 Competent Cells (Life Technologies Cat. No. 10268-019).

14. Reagents for bacterial transformation: 17×100 mm culture tube, water bath, LB or Super Optimal broth with Catabolite repression (SOC) medium (2 % tryptone, 0.5 % yeast extract, 10 mM NaCl, 2.5 mM KCl, 10 mM $MgCl_2$, 10 mM $MgSO_4$, 20 mM glucose), LB-ampicillin plates (0.1 mg/mL).

15. Vacuum concentrator.

16. PCR purification kit.

17. Plasmid DNA purification kit.

3 Methods

3.1 Preparation of Yeast Competent Cells

1. Using a sterile pipette tip or inoculation loop, streak cells from a frozen glycerol stock of yeast onto a YPD agar plate and incubate overnight at 30 °C.

2. On the next day, culture an isolated colony in 10 mL YPD broth.

3. Incubate overnight at 30 °C.

4. The following morning, dilute an aliquot of the overnight culture to 10 mL YPD broth in a 50 mL tube at a final OD_{600} of 0.1–0.2. Loosen the cap of the tube and tape it on the tube to allow aeration.

5. Culture the tube with 250 rpm shaking at 30 °C until OD_{600} 0.6–0.7. Centrifuge at $800 \times g$ for 10 min.

6. Wash the cells with 10 mL EZ 1 solution.

7. Resuspend in 1 mL EZ 2 solution.

8. Dispense cells in 50 µL aliquots.

9. Place microfuge tubes in a Styrofoam box and slow-freeze at –80 °C. Competent cells are viable for up to 6 months.

3.2 Preparation of Linearized pRS424 DENV2 Replicon and Mutant 3′UTR PCR Product

Day 1

3.2.1 Linearization of pRS424 DENV2

Digest 42.5 µg of pRS424 DENV2 with 5 U of BbvCI in an appropriate buffer recommended by the supplier at 37 °C overnight (*see* **Note 1**).

3.2.2 Mutant 3′UTR

Mutations (insertions, deletions, or substitutions) are introduced by standard overlap-extension PCR technique. For each mutant clone, forward and reverse primers that contain the mutation at their 5′ ends are required (Primers 3 and 4, respectively, to be designed by the user). These primers should overlap at least 15 nucleotides at their 5′ ends to facilitate overlap extension (*see* **Note 2**).

1. Set up standard PCR reactions to amplify the 5′ and 3′ fragments of the mutated 3′UTR. This first round of PCR uses the primer pairs 1 and 4 and 2 and 3. A good starting point for annealing temperature is 55 °C. For this chapter, we have used the DyNAzyme II DNA polymerase from Finnzymes (Thermo Scientific), and any suitable alternative enzyme can be substituted. Each 30 µL PCR reaction contained 0.83 mM of each dNTP, 1× DyNAzyme Buffer, 1 ng of WT replicon DNA, 83 nM of each primer, and 0.01 U DyNAzyme II DNA polymerase. Amplification was done using the following conditions: initial denaturation at 94 °C for 2 min, 35 cycles of 94 °C for 30 s, 55 °C for 30 s, 72 °C for 15 s, and a final extension of 72 °C for 2 min. *Optional stop point.*

2. Run the PCR reactions in a 1–1.5 % agarose gel.

3. Excise the bands of expected size.

4. Purify PCR products using any commercial gel extraction kit.

5. Elute the DNA in 50 μL water or elution buffer. *Optional stop point.*

6. Perform four 50 μL PCR reactions each containing 1 mM each dNTP, 1× DyNAzyme buffer, 0.75 μL of each PCR product purified from **step 5**, Subheading 3.2.2, 0.1 μM each of Primers 1 and 2, and 0.02 U DyNAzyme II DNA polymerase. Amplification was done using the following protocol: initial denaturation at 94 °C for 2 min, 35 cycles of 94 °C for 30 s, 60 °C for 30 s, 72 °C for 30 s, and a final extension of 72 °C for 2 min. *Optional stop point.*

7. Run the PCR reactions in a 1–1.5 % agarose gel. The expected overlap product is ~450 bp.

8. Excise and purify final PCR products using gel extraction kit.

9. Elute the DNA in 30 μL water and combine in one tube.

Day 2

3.2.3 Purification of Digested pRS424 DENV2 Replicon

1. Run a 1 μL aliquot of the digested pRS424 DENV2 replicon in an agarose gel together with an uncut plasmid to verify that digestion is complete. It is critical that all copies of the plasmid are cut as undigested replicons will produce a significant number of wild type colonies.

2. After confirmation of complete linearization, bring the volume to 100 μL with H_2O.

3. Add an equal volume (100 μL) of freshly prepared phenol-chloroform-isoamyl alcohol mixture at a ratio of 25:24:1.

4. Mix the solution by inversion for about 10–15 s.

5. Centrifuge at maximum speed for 5 min.

6. The aqueous layer (top) contains the protein-free DNA. Transfer the top layer to a new tube, avoiding the interface or organic layer (*see* **Note 3**).

7. The total volume should now be ~100 μL. Add 1/10 volumes (10 μL) of 3 M NaOAc pH 5.5 and 2.5 volumes (250 μL) of ice cold ethanol to precipitate the DNA.

8. Place the tube on dry ice or –80 °C for at least 15 min.

9. Pellet the DNA at maximum speed for 15 min.

10. Add 500 μL 70 % ethanol and remove it manually.

11. Spin the tubes for a couple of seconds.

12. Remove residual ethanol and air-dry for ~10 min.

13. Dissolve the DNA in 50 μL H_2O.

3.3 Yeast In Vivo Recombination

1. Add 2 μL of linearized replicon DNA to the purified PCR product containing the mutation.

2. Vacuum dry the liquid until the final volume is ~5 μL.

3. Add 500 µL of EZ 3 solution from the Frozen-EZ Yeast Transformation II kit and mix by inversion.

4. Add a 50 µL aliquot of competent yeast cells. Mix by inversion.

5. Incubate the transformation reaction at 30 °C for 1 h, occasionally mixing every 10–15 min.

6. Plate the entire reaction on a SD-Trp⁻ agar plate.

7. Place the plate for 2 days at 30 °C until visible colonies form.

Day 4

8. Pick at least five colonies and streak individually on a new SD-Trp⁻ agar plate.

9. Incubate at 30 °C overnight.

Day 5

10. Amplify the 3′UTR using Primers 1 and 2 following **step 6**, Subheading 3.2.2. Dab a pipette tip to the colony streak and swirl in the PCR reaction tube.

11. To lyse the cells, set up the initial denaturation step for 5 min. Cycle the rest using the same conditions as in **step 1**, Subheading 3.2.2.

12. Purify the amplicons using a commercial PCR purification kit.

13. Send an aliquot for sequencing to confirm the presence of the mutation. *Optional stop point.*

Day 7

3.4 Plasmid Extraction from Yeast (See Note 4)

1. After confirming the presence of the mutation, pick the correct colony from the original master SD-Trp plate and inoculate into 3 mL SD-Trp⁻ medium at the end of the day at 30 °C overnight. Do not overgrow the yeast culture.

Day 8

2. Pellet 1.5 mL of cells in the morning.

3. Resuspend the yeast cells in 200 µL Solution 1 and 15 U of Zymolyase of the Zymoprep Yeast Plasmid Miniprep II kit.

4. Incubate for 1–1.5 h at 37 °C with occasional mixing every 10 min to digest the cell wall.

5. Add 200 µL Solution 2 and mix immediately.

6. Incubate for 5 min at room temperature.

7. Add 400 µL Solution 3.

8. Spin for 4 min at $12,000 \times g$.

9. Transfer supernatant to Zymo-Spin-I column and spin for 30 s.

10. Wash with 550 µL Wash Buffer for 30 s.

11. Elute DNA in 10 µL H_2O.

3.5 Transformation into Stbl2 Cells (See Note 5)

1. Add 1 µL of the purified DNA with chemically competent 20 µL MAX Efficiency Stbl2 *E. coli* cells in a prechilled 17 × 100 mm culture tube.

2. Incubate on ice for 30 min.

3. Heat shock in a 42 °C water bath for 25 s.

4. Immediately transfer to ice for 2 min.

5. Add 400 µL LB or SOC medium and incubate at 30 °C for 2 h (*see* **Note 6**).

6. Pellet the cells and resuspend in 100 µL medium.

7. Plate the entire transformation reaction on LB-ampicillin agar plate (0.1 mg/mL).

8. Incubate overnight at 30 °C (*see* **Note 7**).

Day 9

9. Pick three colonies for inoculation into 2 mL each LB-ampicillin broth overnight.

Day 10

10. Extract plasmid DNA using commercial DNA miniprep extraction kit.

11. Sequence 3′UTR region of DNA again to confirm the presence of mutation. Positive clones can now be propagated in larger cultures to produce greater amounts of DNA and treated as a standard bacterial plasmid with the exception of requirement for a 30 °C incubation. Make sure to freeze a glycerol stock as soon as possible prior to further handling.

4   Notes

1. It is better to digest with two unique restriction enzymes if such sites are available to prevent intramolecular ligation of the vector without the mutant fragment. If two restriction sites are chosen for cleavage, then the PCR fragment containing the desired mutation should have a sufficient overlap with the regions on either side of the two restriction sites for homologous recombination to be successful. This step would yield low background of wild type colonies.

2. In some cases, the overlap PCR step needs extensive optimization. This is most likely due to the failure of the two initial PCR products to anneal over a short overlap. To avoid this problem, the user might consider designing the mutant primers such that the one is a perfect reverse complement of the other.

3. It is critical that there is no phenol or chloroform carryover at the aqueous phase as this will inhibit transformation. If you accidentally touch the interphase or organic layer, transfer as much of the aqueous phase to a new tube, including some of the organic layer. Repeat the spin and retransfer the top layer to a new tube.

4. Our initial protocol to extract plasmid DNA from yeast involved the use of glass bead lysis and organic extraction. However, we found that this method produces crude and low-purity DNA that often results in an insoluble DNA pellet or failure to get Stbl2 colonies. Use of the Zymoprep Yeast Plasmid Miniprep II kit or equivalent yeast plasmid purification system gives purer DNA and successful Stbl2 transformations.

5. Because the plasmid DNA extracted from yeast using the Zymoprep Yeast Plasmid Miniprep II kit is pure, chemically competent Stbl2 cells that are prepared in the lab can be used.

6. While Stbl2 cells can be grown at 37 °C, we found that incubating cultures at 30 °C reduced the amount of colonies with rearranged plasmids.

7. Do not incubate plates for more than 24 h. Incubate until tiny colonies appear. Once colonies appear, longer incubation is not desirable.

Acknowledgments

The research was supported by NIH grants AI 57705 and AI 70791 (to R.P.).

References

1. Kapoor M, Zhang L, Mohan PM, Padmanabhan R (1995) Synthesis and characterization of an infectious dengue virus type-2 RNA genome (New Guinea C strain). Gene 162(2):175–180, doi:037811199500332Z [pii]

2. Khromykh AA, Westaway EG (1994) Completion of Kunjin virus RNA sequence and recovery of an infectious RNA transcribed from stably cloned full-length cDNA. J Virol 68(7):4580–4588

3. Polo S, Ketner G, Levis R, Falgout B (1997) Infectious RNA transcripts from full-length dengue virus type 2 cDNA clones made in yeast. J Virol 71(7):5366–5374

4. Sumiyoshi H, Hoke CH, Trent DW (1992) Infectious Japanese encephalitis virus RNA can be synthesized from in vitro-ligated cDNA templates. J Virol 66(9):5425–5431

5. Rice CM, Grakoui A, Galler R, Chambers TJ (1989) Transcription of infectious yellow fever RNA from full-length cDNA templates produced by in vitro ligation. New Biol 1(3):285–296

6. Lai CJ, Zhao BT, Hori H, Bray M (1991) Infectious RNA transcribed from stably cloned full-length cDNA of dengue type 4 virus. Proc Natl Acad Sci U S A 88(12):5139–5143

7. Puri B, Polo S, Hayes CG, Falgout B (2000) Construction of a full length infectious clone for dengue-1 virus Western Pacific,74 strain. Virus Gene 20(1):57–63

8. Kelly EP, Puri B, Sun W, Falgout B (2010) Identification of mutations in a candidate dengue 4 vaccine strain 341750 PDK20 and construction of a full-length cDNA clone of the PDK20 vaccine candidate. Vaccine 28(17):3030–3037. doi:10.1016/j.vaccine.2009.10.084, S0264-410X(09)01633-8 [pii]

9. Yang CC, Tsai MH, Hu HS, Pu SY, Wu RH, Wu SH, Lin HM, Song JS, Chao YS, Yueh A (2013) Characterization of an efficient dengue virus replicon for development of assays of discovery of small molecules against dengue virus. Antiviral Res 98(2):228–241. doi:10.1016/j.antiviral.2013.03.001, S0166-3542(13)00055-7 [pii]

10. Yang CC, Hsieh YC, Lee SJ, Wu SH, Liao CL, Tsao CH, Chao YS, Chern JH, Wu CP, Yueh A (2011) Novel dengue virus-specific NS2B/NS3 protease inhibitor, BP2109, discovered by a high-throughput screening assay. Antimicrob Agents Chemother 55(1):229–238. doi:10.1128/AAC.00855-10, AAC.00855-10 [pii]

11. Alcaraz-Estrada SL, Manzano MI, Del Angel RM, Levis R, Padmanabhan R (2010) Construction of a dengue virus type 4 reporter replicon and analysis of temperature-sensitive mutations in non-structural proteins 3 and 5.

J Gen Virol 91(Pt 11):2713–2718. doi:10.1099/vir.0.024083-0, vir.0.024083-0 [pii]

12. Manzano M, Reichert ED, Polo S, Falgout B, Kasprzak W, Shapiro BA, Padmanabhan R (2011) Identification of cis-acting elements in the 3′-untranslated region of the dengue virus type 2 RNA that modulate translation and replication. J Biol Chem 286(25):22521–22534. doi:10.1074/jbc.M111.234302, M111.234302 [pii]

13. Zeng L, Falgout B, Markoff L (1998) Identification of specific nucleotide sequences within the conserved 3′-SL in the dengue type 2 virus genome required for replication. J Virol 72(9):7510–7522

14. Teramoto T, Kohno Y, Mattoo P, Markoff L, Falgout B, Padmanabhan R (2008) Genome 3′-end repair in dengue virus type 2. RNA 14(12):2645–2656. doi:10.1261/rna.1051208, rna.1051208 [pii]

Chapter 11

Identification of Dengue-Specific Human Antibody Fragments Using Phage Display

Moon Y.F. Tay, Chin Chin Lee, Subhash G. Vasudevan, and Nicole J. Moreland

Abstract

High-affinity antibodies are valuable tools for dengue research. A method for the selection of dengue-specific, human antibody fragments using naïve repertoires displayed on M13 filamentous bacteriophage is described. Naïve repertoires are unbiased, thus enabling the identification of antibodies to dengue structural and nonstructural proteins from the same library. Dengue-specific clones are enriched by binding to an immobilized dengue antigen, followed by washing, elution, and amplification of phage for subsequent rounds of selection. Dengue virus has four antigenically related serotypes, and the serotype of the antigen can be kept constant or alternated during the selection process depending on whether serotype-specific or cross-reactive antibodies are required. After the selection process, clones are screened, and specific clones are identified by phage ELISA and Western blot.

Key words Phage display, Antibody, Biopanning, Dengue

1 Introduction

Dengue virus (DENV) is a positive-sense, single-stranded RNA virus with four antigenically related serotypes (DENV1–4). The 11 kb genome encodes a single polyprotein that is proteolytically cleaved into three structural proteins (capsid, premembrane, and envelope) and seven nonstructural proteins (NS1, NS2A, NS2B, NS3, NS4A, NS4B, and NS5) [1]. High-affinity monoclonal antibodies toward the DENV envelope protein (E) have therapeutic potential in passive immunotherapy [2, 3], and antibodies toward both structural and NS proteins are valuable tools for DENV research. Anti-E antibodies enable DENV-infected cells to be identified by microscopy and the study of viral packaging and assembly [4, 5]. Anti-NS antibodies enable studies of DENV protein–protein interactions and replication both in infected cells and in biochemical assays. For example, anti-NS3 antibodies have been used to demonstrate the interaction of NS3 with NS5 in DENV-infected cells [6, 7], and an antibody

Radhakrishnan Padmanabhan and Subhash G. Vasudevan (eds.), *Dengue: Methods and Protocols*, Methods in Molecular Biology, vol. 1138, DOI 10.1007/978-1-4939-0348-1_11, © Springer Science+Business Media, LLC 2014

specific for the hydrophilic cofactor region of NS2B enabled a biochemical assay to be developed for the NS3–NS5 interaction [6]. DENV-specific monoclonal antibodies and their fragments also have applications in cellular localization studies [8, 9] and as cocrystallization reagents in structural biology of DENV proteins [10, 11].

Monoclonal antibodies are traditionally generated by hybridoma technology where antibody-secreting B cells from mice are fused with an immortal myeloma cancer cell line following immunization, and the hybrid cells (hybridomas) are screened for antigen specificity. However this method is restricted by poor immunogenicity of some targets, the need to immunize animals for each new antigen, and unwanted immune responses elicited by murine antibodies in the therapeutic setting [12]. Phage display is a powerful method for the isolation of antibodies directed against almost any antigen that completely bypasses the use of animals. Antibody fragments such as single-chain variable fragments (scFv) and the antigen-binding fragment (Fab) are readily expressed in the periplasm of *Escherichia coli* [13]. This has enabled selection of these antibody fragments by phage display where *E. coli*-specific M13 bacteriophages are genetically engineered to produce a terminal gene III protein fused to an antibody fragment [14, 15].

Synthetic, immune, and naïve phage libraries have been developed, each offering its own advantages [16]. Immune repertoires are generated by cloning V genes from peripheral blood lymphocytes (PBMCs) of immune donors, for example, those with a recent DENV infection, and will be enriched for antigen-specific antibodies. However these libraries are biased toward the antigen to which the immune response was induced. In DENV, the envelope (E) protein is the major target of the humoral response, and although B-cell responses toward premembrane and NS1 have been reported [17–19], the level of B-cell response toward other NS proteins is low. This is because these proteins are primarily expressed in the cytoplasm of infected cells. Synthetic libraries are artificially built in vitro, often using a single framework region, with diversity introduced into complementary determining regions (CDRs) [20]. They are unbiased and can yield high-affinity antibodies to almost any target but require significant effort to generate. Naïve libraries are cloned from V genes of unimmunized human donors and when sufficiently large can also generate antibodies to almost any target, including those not normally visible in a natural humoral response, e.g., DENV NS proteins. Several naïve phage antibody libraries are now available to the research community [21] making them an ideal starting point for identifying new anti-DENV antibodies.

This protocol describes a general approach for the selection of DENV-specific human antibody fragments from a naïve Fab-phage repertoire, but the protocols could be adapted for other antigens and phage repertoires. An overview of the phage selection process,

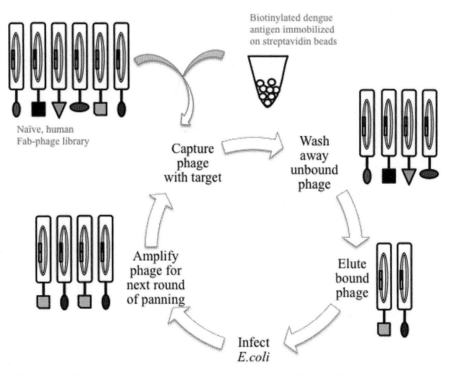

Fig. 1 Selection of specific Fab-phage clones by biopanning. A naïve library of Fab-phage clones is incubated with a biotinylated dengue antigen immobilized on streptavidin resin. Nonbinding phage is removed by washing, and eluted phage is used to infect *E. coli* for the next round of biopanning. Repeated rounds of panning lead to the enrichment of phage clones that are specific to the dengue antigen

called biopanning, is shown in Fig. 1. The Fab-phage library is incubated with a DENV antigen, specific clones are captured, and nonspecific clones are removed by a series of washes. Phages are amplified via an infection of *E. coli*, and repeated rounds of selection are performed. The biopanning process can be manipulated to enrich for clones with desired properties. In the case of DENV, for example, this enables the serotype of the antigen to be changed in subsequent selection rounds if cross-reactive antibodies that bind antigen from DENV1–4 are needed. Human antibody fragments identified by phage display can be re-cloned into IgG expression vectors for the generation of full-length human IgG as required [6, 22].

2 Materials

2.1 Biotinylation of Antigens and Preparation of Streptavidin Resin

1. Dengue proteins of interest at 1 mg/mL in PBS pH 7.4.

2. EZ-Link NHS-PEG$_4$-Biotin (Thermo Scientific, cat no. 21329).

3. 100 mM glycine solution.

4. No-Weigh™ HABA/Avidin Premix (Thermo Scientific, cat no. 1858746).

5. Dynabeads® M-280 Streptavidin (Life Technologies).

6. Milk blocking buffer: 4 % (wt/vol) low-fat dry milk powder in PBS (MB).

2.2 Biopanning and Phage Amplification

1. A naïve human Fab-phage library.

2. M13K07 Helper Phage (Life Technologies) (*see* **Note 1**).

3. *Escherichia coli* XL-1 Blue cells (Life Technologies). The F′ episome contains the tetracycline resistance gene. Growth of this strain in the presence of tetracycline ensures the F′ episome is maintained for the production of bacterial pili that are essential for M13 phage infection.

4. Filtered PBS pH 7.4.

5. 0.1 % Tween-20 in PBS (PBST).

6. Antibiotic stock solutions: 50 mg/mL kanamycin, 20 mg/mL tetracycline, and 100 mg/mL ampicillin.

7. A sterile 40 % glucose solution.

8. 2× TY medium: 16 g/L tryptone, 10 g/L yeast extract, and 5 g/L NaCl adjusted to pH 7.0 with 1 M NaOH prior to autoclaving.

9. 2× TY agar: 16 g/L tryptone, 10 g/L yeast extract, 5 g/L NaCl, and 15 g/L bacto-agar adjusted to pH 7.0 prior to autoclaving.

10. 2× TY medium and 2× TY agar supplemented with antibiotics and glucose as required. Final antibiotic concentrations are 100 μg/mL ampicillin (A), 10 μg/mL tetracycline (Tet), and 50 μg/mL kanamycin (K). The final concentration of glucose is 2 %.

11. Super Broth (SB) medium: 10 g MOPS, 30 g tryptone, and 20 g yeast extract adjusted to pH 7.0 with 1 M NaOH prior to autoclaving.

12. Milk blocking buffer as above.

13. 0.2 M glycine/HCl pH 2.2.

14. 2 M Tris–HCl pH 9.0.

15. Sterile 80 % glycerol.

16. A 2 % sodium azide solution.

17. PEG solution containing 20 % (wt/vol) PEG-6000 and 2.5 M NaCl.

2.3 Phage ELISA and Western Blot

1. 96-well deep well plates for growth cultures.

2. Breathable film for 96-well growth cultures.

3. Immunosorp flat-bottom microtiter plates (Nunc).

4. Filtered PBS pH 7.4.

5. Milk blocking buffer as above.

6. Anti-M13 horseradish peroxidase (HRP) conjugate (GE Healthcare).

7. TMB (3,3′,5,5′-Tetramethylbenzidine) solution (Sigma).

8. 3 M HCl.

9. 12 % SDS-PAGE Gel and gel running apparatus.

10. PVDF membrane and Western blotting transfer apparatus.

11. ECL Prime Western Blotting detection kit (GE Healthcare).

3   Methods

The biopanning process in this protocol utilizes antigen that has been immobilized on streptavidin resin. Alternatively, protein antigens can be coated directly onto a plastic surface by adsorption, but although practically straightforward, this can be problematic for antibody selection as up to 95 % of adsorbed proteins have been shown to be denatured or nonfunctional [23].

3.1 Biotinylation of Antigens and Preparation of Streptavidin Resin

This method describes chemical biotinylation, which is the most common way to obtain a biotinylated antigen. For other alternatives such as enzymatic biotinylation, refer to Chames et al. [24].

1. Use 500 μg of a dengue protein at 1 mg/mL in PBS pH 7.4. Calculate the amount of NHS-PEG$_4$-Biotin needed for a molar ratio of biotin/protein of 20:1. Add the biotin reagent and incubate on ice for 2 h (*see* **Note 2**).

2. Add 50 μL of 100 mM glycine solution to stop the reaction.

3. To remove excess biotin, dialyze overnight against PBS pH 7.4 or perform size exclusion chromatography with PBS.

4. Confirm the antigen is biotinylated using the biotin quantitation kit (follow manufacturer's instructions). This will also determine the moles of biotin added per mole of protein.

5. Wash 70 μL of streptavidin magnetic beads twice with 1 mL PBS. Place the Eppendorf tube containing resin in a magnetic rack for 1 min and carefully remove PBS once resin has collected on the side of the tube by the magnet.

6. Add 500 μL of biotinylated dengue protein at a final concentration of 1,400 nM in milk blocking buffer (MB).

7. Rock the mixture gently at room temperature for 1 h and then wash three times with 1 mL of MB (as in **step 5**).

3.2 Preparation of Phage Antibody Repertoire

These protocols were developed for use with a naïve human Fab-phage library, but can be readily adapted for other antibody domain libraries (*see* **Note 3**).

1. Libraries are generally stored in 1 mL aliquotes at –80 °C in 15 % glycerol. Thaw an aliquot of the Fab-phage library (10^{12} pfu).

2. Add one-fifth volume of PEG solution and incubate on ice for 1 h.

3. Centrifuge at $10,000 \times g$, 4 °C for 10 min to pellet phage.

4. Resuspend the white phage pellet in 200 μL PBS and centrifuge at $10,000 \times g$ for 2 min to pellet any remaining debris.

5. Transfer 200 μL of phage supernatant to a clean Eppendorf tube.

6. Add 800 μL of MB to phage supernatant and rock the mixture gently at room temperature for 1 h.

3.3 First Round of Biopanning

3.3.1 Binding and Elution of Phage

1. Mix pre-blocked phage (from subheading 3.2, **step 6**) with streptavidin beads (from subheading 3.1, **step 7**) coated with dengue protein (*see* **Note 4**).

2. Rock the mixture gently at room temperature for 2 h.

3. Wash the magnetic beads, using the magnetic rack, six times with 1 mL of PBST and two times with 1 mL of PBS.

4. Elute bound phage by adding 350 μL of 0.2 M glycine/HCl pH 2.2 to the washed beads. Incubate at room temperature for 10 min.

5. Remove supernatant to a clean Eppendorf tube containing 20 μL of 2 M Tris–HCl pH 9.0 and mix to neutralize.

3.3.2 Infection of XL-1 Blue Cells and Amplification of Eluted Phage

1. Inoculate 4 mL of SB with a single colony of XL-1 Blue and grow until an optical density 600 nm (OD_{600}) of 0.6–0.8 (log phase). Infect the culture with neutralized phage and stand at room temperature without shaking for 15 min.

2. Add 4 mL of pre-warmed (37 °C) SB medium supplemented with 20 μg/mL ampicillin and 10 μg/mL tetracycline.

3. Remove 2 μL of infected XL-1 Blue cells for *output* titering. Dilute this 2 μL in 200 μL of SB (1:100) and plate 10 and 100 μL of this on 2× TY/A-glucose plates. Incubate plates overnight at 37 °C and calculate the output titer by multiplying the number of colonies by the culture volume and dividing by the plating volume (*see* **Note 5**).

4. Shake the 8 mL culture from **step 2** at 37 °C (200 rpm) for 1 h.

5. Add further ampicillin for a final concentration of 50 μg/mL and continue shaking the culture at 37 °C (200 rpm) for another hour.

6. Add M13KO7 helper phage at a multiplicity of infection (MOI) of 20:1 (phage-to-cells ratio; *see* **Note 6**) and transfer the culture to 500 mL disposable flask.

7. Add pre-warmed (37 °C) SB medium supplemented with 100 μg/mL ampicillin and 10 μg/mL tetracycline for a final culture volume of 100 mL and shake at 37 °C (200 rpm) for 2 h.

8. Add kanamycin for a final concentration of 70 μg/mL and continue shaking the culture at 37 °C (200 rpm) overnight.

3.3.3 Preparation of Amplified Phage for Subsequent Rounds of Selection

1. Centrifuge the 100 mL overnight culture at $4{,}000 \times g$, 4 °C for 15 min.

2. Transfer the phage supernatant to a clean 250 mL centrifuge bottle and add one-fifth volume (20 mL) PEG solution.

3. Mix thoroughly and leave on ice for 1 h to precipitate phage.

4. Centrifuge at $10{,}000 \times g$, 4 °C for 15 min to pellet phage.

5. Discard supernatant and resuspend the white phage pellet in 2 mL PBS pH 7.4.

6. Centrifuge at $10{,}000 \times g$ for 2 min to pellet cell debris.

7. Transfer the phage supernatant into a clean tube and repeat the PEG precipitation (**steps 2–5** above).

8. Resuspend the pellet in 1–2 mL PBS, depending on viscosity. Remove 2 μL of the phage suspension for titering. The remaining phage can be stored at 4 °C or –80 °C for subsequent panning (*see* **Note 7**).

9. To titer the amplified phage, make serial dilutions of the phage preparation at 10^{-3}, 10^{-6}, 10^{-8}, and 10^{-9} in 2× TY.

10. Add 1 μL of 10^{-8} and 10^{-9} dilutions to 50 μL of log-phase XL-1 Blue cells and incubate at room temperature without shaking for 15 min.

11. Plate onto pre-warmed 2× TY/A-glucose plates and incubate at 37 °C overnight.

12. Count colonies and calculate the titer of amplified phage. This titer will determine the volume of phage used in the next round of biopanning.

3.4 Subsequent Rounds of Selection to Enrich for Dengue-Specific Clones

After the first round of biopanning and amplification, the pool of phage becomes slightly enriched for clones that bind the initial DENV antigen. Repeated rounds of panning (normally 3–6 rounds) are required to enrich and identify clones with the desired specificity. The serotype of the DENV protein can be kept constant to enrich for serotype-specific clones, or if cross-reactive clones are required that bind a DENV protein from serotypes 1–4, the serotype of the antigen may need to be alternated throughout the panning process. The need to alternate serotypes is driven by a

Table 1
Conditions used for subsequent rounds of biopanning to enrich for cross-reactive clones

	First round	Second round	Third round	Fourth round
Phage input (pfu)	Phage library stock (1×10^{12})	Amplified first round phage $(10^{11}–10^{12})$	Amplified second round phage $(10^{11}–10^{12})$	Amplified third round phage $(10^{11}–10^{12})$
Dengue serotype	DENV2	DENV3	DENV2	DENV3
Antigen concentration (nM) and volume (µL)	1,400; 500	300; 500	150; 500	150; 500
Magnetic bead volume (µL)	70	50	30	30
Number of washes	6× PBST 2× PBS	14× PBST 2× PBS	14× PBST 2× PBS	14× PBST 2× PBS

combination of sequence conservation and the location of immunogenic epitopes. A screen against NS3 from DENV4 in our laboratory yielded clones that cross-reacted with NS3 from DENV1–4, despite only panning against the DENV4 protein [25]. More recently, screens with other NS proteins have required the serotype of the antigen to be alternated to obtain cross-reactive clones as shown in Table 1.

3.5 Screening Clones by Monoclonal Phage ELISA

After 3–6 rounds of selection, individual colonies from the *output* titer plates (Subheading 3.3.2, **step 3**) can be tested for antigen binding by phage ELISA (enzyme-linked immunosorbent assay). Plates can be coated for antigens from two or more DENV serotypes to identify cross-reactivity clones (*see* **Note 8**).

3.5.1 Small-Scale Phage Rescue

XL-1 Blue phage colonies are picked from *output* titer plates of the last round of panning for small-scale phage rescue in a 96-well deep well plate.

1. Add 500 µl of 2× TY/A/Tet-glucose medium to each well of the master plate.

2. Seed each well with an XL-1 Blue phage colony using sterile pipette tips (*see* **Note 9**). Seal the plate with breathable film and shake overnight at 37 °C (800 rpm) in a thermomixer.

3. Centrifuge plate at $500 \times g$ for 3 min at room temperature to remove condensation from the breathable film prior to opening.

4. Transfer 5 µl of overnight culture to a new plate containing 500 µl 2× TY/A/Tet-glucose medium. Seal the new plate with breathable film and shake it at 37 °C (800 rpm) in a thermomixer for 3 h, OD ~0.5.

5. Add M13KO7 helper phage for a multiplicity of infection (MOI) of 20:1 (phage-to-cells ratio; *see* **Note 10**).

6. Reseal plate with breathable film and shake at 800 rpm for 30 s to mix. Stop the shaking and leave the plate undisturbed at 37 °C for 30 min.

7. Centrifuge the plate at 4,000×g for 5 min to pellet cells. Invert to remove supernatant and resuspend each cell pellet with 500 μL 2× TY/A/K/Tet.

8. Seal plate with breathable film and shake at 30 °C (800 rpm) overnight.

9. Spin the plate at 4,000×g for 5 min and remove 400 μL of the phage supernatant to a new 96-well deep well plate. Dilute 1:3 by adding 800 μL of MB and rest on ice until required.

3.5.2 Monoclonal Phage ELISA

1. Add 100 μL of 5 μg/mL unbiotinylated dengue protein in PBS to each well of a 96-well immunoplate and incubate overnight at 4 °C. Plates may be coated for dengue proteins from two or more serotypes.

2. Wash plates twice with PBS. To block add 200 μL of MB to each well and incubate at room temperature for 1 h or overnight at 4 °C.

3. Wash plate twice with PBST.

4. Add 100 μL of diluted phage supernatant to each well and incubate plates at room temperature for 1–2 h.

5. Wash plate 5× with PBST.

6. Add 100 μL of diluted HRP-anti-M13 conjugate (1:5,000 in MB; *see* **Note 11**) to each well and incubate plate at room temperature for 1 h.

7. Wash 3× with PBST and 1× with PBS.

8. Add 50 μL TMB substrate to each well.

9. Cover plate with aluminum foil and incubate at room temperature for 1–30 min until blue color develops (*see* **Note 12**).

10. Stop the reaction with 12.5 μL of 3 M HCl. The color of TMB substrate will turn yellow after quenching the reaction with HCl.

11. Read absorbance at 450 nm on an optical density plate reader.

3.5.3 Phage Western Blot

The cross-reactivity and specificity of the Fab-phage clones can be further investigated by phage Western blot.

1. Add 300 ng of unbiotinylated dengue protein from DENV1–4 to 15 μL of a soluble protein extract from XL-1 Blue cells (*see* **Note 13**). Resolve the mixture on a 12 % SDS-PAGE gel at 120 V for 1 h.

2. Transfer the proteins onto a PVDF membrane using standard Western blotting transfer protocols.

3. Block membrane at 4 °C overnight with MB.

4. Dilute 400 µL of phage supernatant (from Subheading 3.5.1, **step 9**) 1:10 with MB and incubate the membrane in this 4 mL phage at room temperature for 1 h.

5. Wash membrane four times, 15 min each time, with PBST.

6. Add the HRP-anti-M13 conjugate diluted 1:1,000 in MB and incubate at room temperature for 1 h.

7. Wash membrane four times, 15 min each time, with PBST.

8. Develop blot using ECL Prime Western Blotting detection kit. A specific Fab-phage clone will only detect the dengue antigen and not proteins in the *E. coli* soluble fraction. A cross-reactive clone will detect dengue proteins from more than one serotype (*see* **Note 14**).

4   Notes

1. Helper phage stock can be readily amplified as required. The protocols for this are well established. Refer to Lee et al. [21] and Coomber et al. [26]. Helper phage is stable at 4 °C for approximately 2 months. For long-term storage, add sterile 80 % glycerol to make 15 % glycerol stock and store at –80 °C.

2. Protein previously stored in a buffer containing amines, such as Tris–HCl, should be dialyzed overnight against PBS pH 7.4 to remove reactive amines prior to biotinylation. Proteins contain variable numbers of lysines, and it may be necessary to perform biotinylation reactions containing 10-, 20-, and 40-fold molar excess of biotin reagent to optimize the level of biotinylation.

3. Our laboratory has made use of a naïve human Fab-phage library named HX02 [25, 27] derived from the spleen and/or PMBCs of 14 nonimmune donors. Similar libraries are available from Source BioScience (domain antibody library and the Tomlinson I and J scFv libraries). Alternatively, a library can be generated in-house using published protocols [28, 29] or via a custom service company such as Creative Biolabs.

4. Neutravidin resin, rather than streptavidin resin, can be used in subsequent rounds to prevent the enrichment of streptavidin binders, though this is not absolutely necessary.

5. It is important to monitor the titers after each round of selection. With an input of 10^{12} phage, approximately 10^5–10^7 clones would be expected after the first round of selection. After several rounds of selection, a rise in titer can indicate the selection of binders. However, a lack in titer rise does not

necessarily mean the selection has not worked, and clones should always be checked by phage ELISA after the final round of panning.

6. Calculate the number of M13KO7 helper phage to add for an MOI 20 using the following formula: culture volume (mL) $\times$ OD600 $\times$ 3 $\times$ 10^8 $\times$ 20, where OD600 is the optical density (OD) of a suspension of cells at 600 nm (OD600) and 3 $\times$ 10^8 is the concentration of bacterial cells in 1 mL culture when OD600 = 1 and 20 is MOI.

7. PEG-precipitated phage supernatant can be stored for several weeks at 4 °C, with addition of sodium azide to final concentration of 0.02 %. For long-term storage, add sterile 80 % glycerol to make 15 % glycerol stock and store at –20 °C.

8. Clones can also be screened for specificity against an unrelated antigen (e.g., BSA).

9. Leave well A1 empty as a negative control. Master plate can be sealed with sealing tape and be stored at 4 °C for up to 2 weeks. If longer storage is required, sterile 80 % glycerol is added to a final concentration of 15 % and the plate is stored at –80 °C.

10. Calculate the number of M13KO7 helper phage to add at MOI 20 based on the formula in **Note 6**. Dilute helper phage in an appropriate amount of 2× TY and add 50 µL of helper phage dilution to each well.

11. Anti-M13 monoclonal antibody conjugate reacts with the bacteriophage major coat gene VIII protein.

12. 1–2 min incubation with TMB is usually adequate for sufficient color development.

13. To prepare a soluble protein extract from XL-1 Blue cells, pick a single colony and grow overnight at 37 °C in 10 mL of 2× TY/Tet. Lyse cells using a sonicator or chemical lysis kit and spin culture at 10,000 $\times g$ for 10 min to pellet cell debris. Use soluble material for Western blotting.

14. In some circumstances, a Fab-phage clone may give a positive signal in an ELISA but be Western blot negative. While this can raise concern over the antigen-binding potential of the clone, it might also suggest the clone binds a conformational epitope on the antigen that was destroyed by SDS-PAGE. These clones should not be discarded, but investigated further as soluble Fab. A dot blot will preserve antigen structure and can be used to confirm whether an antibody binds a conformational or a linear epitope. A Fab specific for a dengue NS protein recently identified in our laboratory was Western blot negative and dot blot positive and, as an IgG, has subsequently proved to be an excellent reagent for immunohistochemistry where native epitopes are preserved.

References

1. Chambers TJ, Hahn CS, Galler R, Rice CM (1990) Flavivirus genome organization, expression, and replication. Annu Rev Microbiol 44:649–688

2. Lai CJ, Goncalvez AP, Men R, Wernly C, Donau O, Engle RE, Purcell RH (2007) Epitope determinants of a chimpanzee dengue virus type 4 (DENV-4)-neutralizing antibody and protection against DENV-4 challenge in mice and rhesus monkeys by passively transferred humanized antibody. J Virol 81(23): 12766–12774

3. Beltramello M, Williams KL, Simmons CP, Macagno A, Simonelli L, Quyen NT, Sukupolvi-Petty S, Navarro-Sanchez E, Young PR, de Silva AM, Rey FA, Varani L, Whitehead SS, Diamond MS, Harris E, Lanzavecchia A, Sallusto F (2010) The human immune response to Dengue virus is dominated by highly cross-reactive antibodies endowed with neutralizing and enhancing activity. Cell Host Microbe 8(3):271–283

4. Lin SR, Zou G, Hsieh SC, Qing M, Tsai WY, Shi PY, Wang WK (2011) The helical domains of the stem region of dengue virus envelope protein are involved in both virus assembly and entry. J Virol 85(10):5159–5171

5. Zheng A, Umashankar M, Kielian M (2010) In vitro and in vivo studies identify important features of dengue virus pr-E protein interactions. PLoS Pathog 6(10):e1001157

6. Moreland NJ, Tay MY, Lim E, Rathore AP, Lim AP, Hanson BJ, Vasudevan SG (2012) Monoclonal antibodies against dengue NS2B and NS3 proteins for the study of protein interactions in the flaviviral replication complex. J Virol Methods 179(1):97–103

7. Kapoor M, Zhang L, Ramachandra M, Kusukawa J, Ebner KE, Padmanabhan R (1995) Association between NS3 and NS5 proteins of dengue virus type 2 in the putative RNA replicase is linked to differential phosphorylation of NS5. J Biol Chem 270(32): 19100–19106

8. Pryor MJ, Rawlinson SM, Butcher RE, Barton CL, Waterhouse TA, Vasudevan SG, Bardin PG, Wright PJ, Jans DA, Davidson AD (2007) Nuclear localization of dengue virus nonstructural protein 5 through its importin alpha/beta-recognized nuclear localization sequences is integral to viral infection. Traffic 8(7): 795–807

9. Westaway EG, Mackenzie JM, Kenney MT, Jones MK, Khromykh AA (1997) Ultrastructure of Kunjin virus-infected cells: colocalization of NS1 and NS3 with double-stranded RNA, and of NS2B with NS3, in virus-induced membrane structures. J Virol 71(9):6650–6661

10. Lok S-M, Kostyuchenko V, Nybakken GE, Holdaway HA, Battisti AJ, Sukupolvi-Petty S, Sedlak D, Fremont DH, Chipman PR, Roehrig JT, Diamond MS, Kuhn RJ, Rossmann MG (2008) Binding of a neutralizing antibody to dengue virus alters the arrangement of surface glycoproteins. Nat Struct Mol Biol 15(3): 312–317

11. Hunte C, Michel H (2002) Crystallisation of membrane proteins mediated by antibody fragments. Curr Opin Struct Biol 12(4):503–508

12. Hwang WY, Foote J (2005) Immunogenicity of engineered antibodies. Methods 36(1): 3–10

13. Kwong KY, Rader C (2009) E. coli expression and purification of Fab antibody fragments. Current protocols in protein science/editorial board, John E Coligan [et al] Chapter 6: Unit 6 10

14. Barbas CF, Kang AS, Lerner RA, Benkovic SJ (1991) Assembly of combinatorial antibody libraries on phage surfaces: the gene III site. Proc Natl Acad Sci U S A 88(18):7978–7982

15. de Haard HJ, van Neer N, Reurs A, Hufton SE, Roovers RC, Henderikx P, de Bruïne AP, Arends JW, Hoogenboom HR (1999) A large non-immunized human Fab fragment phage library that permits rapid isolation and kinetic analysis of high affinity antibodies. J Biol Chem 274(26):18218–18230

16. Hoogenboom HR (2002) Overview of antibody phage-display technology and its applications. Methods Mol Biol 178:1–37

17. Falconar AK (1999) Identification of an epitope on the dengue virus membrane (M) protein defined by cross-protective monoclonal antibodies: design of an improved epitope sequence based on common determinants present in both envelope (E and M) proteins. Arch Virol 144(12):2313–2330

18. Falconar AK (2007) Antibody responses are generated to immunodominant ELK/KLE-type motifs on the nonstructural-1 glycoprotein during live dengue virus infections in mice and humans: implications for diagnosis, pathogenesis, and vaccine design. Clin Vaccine Immunol 14(5):493–504

19. Huang KJ, Yang YC, Lin YS, Huang JH, Liu HS, Yeh TM, Chen SH, Liu CC, Lei HY (2006) The dual-specific binding of dengue virus and target cells for the antibody-dependent enhancement of dengue virus infection. J Immunol 176(5):2825–2832

20. Lee CV, Liang W-C, Dennis MS, Eigenbrot C, Sidhu SS, Fuh G (2004) High-affinity human antibodies from phage-displayed synthetic Fab libraries with a single framework scaffold. J Mol Biol 340(5):1073–1093

21. Lee CMY, Iorno N, Sierro F, Christ D (2007) Selection of human antibody fragments by phage display. Nat Protoc 2(11):3001–3008

22. Rader C, Popkov M, Neves JA, Barbas CF (2002) Integrin alpha(v)beta3 targeted therapy for Kaposi's sarcoma with an in vitro evolved antibody. FASEB J 16(14):2000–2002

23. Davies J, Dawkes AC, Haymes AG, Roberts CJ, Sunderland RF, Wilkins MJ, Davies MC, Tendler SJ, Jackson DE, Edwards JC (1994) A scanning tunnelling microscopy comparison of passive antibody adsorption and biotinylated antibody linkage to streptavidin on microtiter wells. J Immunol Methods 167(1–2):263–269

24. Chames P, Hoogenboom HR, Henderikx P (2002) Selection of antibodies against biotinylated antigens. Methods Mol Biol 178:147–157

25. Moreland NJ, Tay MY, Lim E, Paradkar PN, Doan DN, Yau YH, Geifman Shochat S, Vasudevan SG (2010) High affinity human antibody fragments to dengue virus non-structural protein 3. PLoS Negl Trop Dis 4(11):e881

26. Coomber DW (2002) Panning of antibody phage-display libraries. Standard protocols. Methods Mol Biol 178:133–145

27. Lim APC, Chan CEZ, Wong SKK, Chan AHY, Ooi EE, Hanson BJ (2008) Neutralizing human monoclonal antibody against H5N1 influenza HA selected from a Fab-phage display library. Virol J 5:130

28. Rader C (2012) Generation of human Fab libraries for phage display. Methods Mol Biol 901:53–79

29. Barbas CF, Burton DR, Scott JK, Silverman GJ (2001) Phage display, a laboratory manual. Cold Spring Harbor Press, , Cold Spring Harbor, New York

Chapter 12

Next-Generation Whole Genome Sequencing of Dengue Virus

Pauline Poh Kim Aw, Paola Florez de Sessions, Andreas Wilm,
Long Truong Hoang, Niranjan Nagarajan, October M. Sessions,
and Martin Lloyd Hibberd

Abstract

RNA viruses are notorious for their ability to quickly adapt to selective pressure from the host immune system and/or antivirals. This adaptability is likely due to the error-prone characteristics of their RNA-dependent, RNA polymerase [1, 2]. Dengue virus, a member of the Flaviviridae family of positive-strand RNA viruses, is also known to share these error-prone characteristics [3]. Utilizing high-throughput, massively parallel sequencing methodologies, or next-generation sequencing (NGS), we can now accurately quantify these populations of viruses and track the changes to these populations over the course of a single infection. The aim of this chapter is twofold: to describe the methodologies required for sample preparation prior to sequencing and to describe the bioinformatics analyses required for the resulting data.

Key words Dengue, Intra-host genetic diversity, Next-generation sequencing, Dengue serotypes

Abbreviations

TBE   Tris–Borate–EDTA
EtBr   Ethidium bromide
PCR   Polymerase chain reaction
RT-PCR   Reverse transcriptase polymerase chain reaction
cDNA   Complementary DNA
DNA   Deoxyribonucleic acid
RNA   Ribonucleic acid
dNTP   Deoxyribonucleotide triphosphate
kb   Kilobase
bp   Base pairs

Radhakrishnan Padmanabhan and Subhash G. Vasudevan (eds.), *Dengue: Methods and Protocols*, Methods in Molecular Biology,
vol. 1138, DOI 10.1007/978-1-4939-0348-1_12, © Springer Science+Business Media, LLC 2014

1 Introduction

Sequencing technology has evolved over the last 6 years, moving from Sanger sequencing or capillary sequencing to NGS. Dengue NGS publications, reporting on various aspects of the virus and its hosts started to emerge in 2010. Skalsky et al. [4] used NGS to identify conserved and novel micro RNAs in *Culex* and *Aedes* mosquitoes. The investigators found that female Culex mosquitoes infected with West Nile Virus, a close flavivirus relative to Dengue, showed significant expression level changes in two micro RNAs upon flavivirus infection. Both of these mosquito species, *Culex* and *Aedes*, are important flavivirus vectors in tropical and subtropical areas worldwide. The host transcriptome response to dengue infection has also been analyzed by NGS in both the mosquito and human hosts. David et al. [5] explored aspects of pollutants and insecticides on the *Aedes* mosquito transcriptome. In this study, they were attempting to raise awareness of how mosquito vector control using pollutants can have greater effects on our ecosystem than just on the target organism. In 2013, Sessions et al. explored the human transcriptome response to infection with a wild-type strain of dengue virus and an attenuated derivative vaccine candidate [6]. The authors postulated that the large extent of previously uncharacterized transcriptional regulation might constitute a novel human innate immune response, which was more successfully evaded by a wild-type strain compared to an attenuated strain. In addition to vector control and host immunology, NGS has also been used as a tool for clinical identification of Dengue virus. Yozwiak et al. characterized a gamut of viruses in acute serum from over 100 Nicaraguan patients enrolled in a prospective dengue study using NGS and Virochip microarray technologies [7]. These studies serve to elucidate the wide variety of facets that can be studied using NGS as a tool for both clinical and academic purposes.

When using the Sanger sequencing platform, consensus sequences can be inferred quite readily from the output chromatograms. However, non-consensus variants are obscured by the nucleotides of higher frequencies and cannot be easily quantified. In order to more accurately quantify the viral population and determine how the heterogeneity and subpopulations therein may be contributing to pathogenesis and transmission, an in-depth analysis of the viral population is essential. NGS technologies permit rapid and cost-effective acquisition of millions of short DNA sequences, which can then be used to reconstruct a full-length genome. Due to the relatively small size of viral genomes, it is possible to combine them into a single sequencing reaction and thus significantly reduce the cost of sequencing. These millions of reads, if distributed with relative uniformity across the genome, can be used to calculate single nucleotide variations (SNV) in a viral

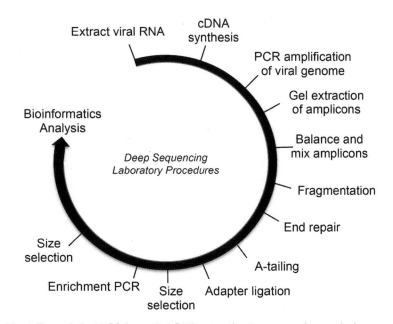

Fig. 1 Flow of viral NGS from viral RNA extraction to sequencing analysis

population with digital precision. The frequencies of these variants are detectable long before the SNV is fixed in the population [8]. Indeed, with sufficient coverage, variants present at far less than 1 % frequency can be reliably calculated [9]. These variant positions within the viral genome can either be of benefit to the replicative fitness, deleterious or neutral, depending on the viral host environment [10]. This type of analysis on the virus population structure presented here is broadly applicable to any virus [11–17]. Although we will focus on the protocols and analysis pipelines used to create and interpret sequencing data on the Illumina platform, this kind of analysis is in theory agnostic and can be performed using other sequencing platforms as well [18, 19].

An overview of preparing samples for sequencing on an Illumina machine is presented schematically in Fig. 1 and is described below. A single RT-primer is designed to bind specifically to the 3′-end of dengue virus genomes. Reverse transcriptase is used to make complementary DNA (cDNA). Next, the primer binding locations are designed by identifying regions of high conservation among strains. Briefly, representative dengue virus sequences for each serotype were downloaded from NCBI (www.ncbi.nlm.nih.gov/nuccore/). It is important to include highly related strains with the same geographical and genetic background as the virus intended for NGS since the areas of conservation can vary significantly between viruses from different regions and genotypes [20]. Next, the MEGA software was used to align the sequences and conserved regions of each serotype were identified.

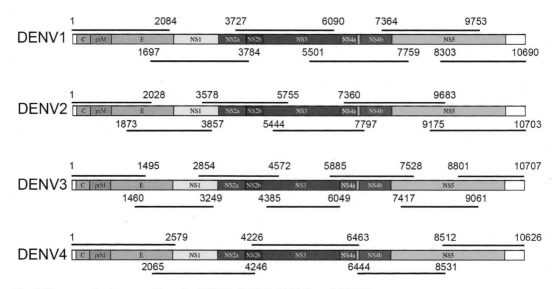

Fig. 2 Diagram of primer positions for DENV1, DENV2, DENV3, and DENV4

Primers targeting these conserved regions were then generated so that each amplicon produced is approximately 2 kb in length. These primer pairs are then used with a high fidelity DNA polymerase to amplify the viral genome in 5–7 fragments (Fig. 2 and Table 1). These fragments, once generated, are then separated and extracted from a 1 % agarose gel and their concentrations measured using the Agilent Bioanalyzer. Equal concentrations of the DNA products are pooled and are fragmented to ~300 bp. After fragmentation, the samples are purified and optionally checked for quality using the Agilent Bioanalyzer. The ends of the DNA samples, which were damaged during the fragmentation step, are repaired with Klenow polymerase to form uniform blunt ends. To allow for efficient ligation of the sequencing adapters in the next step, the blunt ends of the DNA are then adenylated with polynucleotide kinase (PNK). Ligation of the sequencing adapters adds common primer binding sites to the end of each fragmented piece of DNA and is the step that ultimately allows for unbiased amplification and sequencing. If desired, one can check again for quality of the ligated product using an Agilent Bioanalyzer. After ligation, ~300 bp fragments are excised from 2 % agarose gel. Samples are subjected to 14 PCR cycles of enrichment PCR to incorporate indices and are size selected on a 2 % agarose gel one final time to ensure a tight distribution of library size. Following a final quality check, these samples are then ready to be sequenced.

Although the bioinformatics analysis of deep sequencing can initially seem overwhelming, it can in actuality be accomplished with a cursory understanding of Linux and a handful of useful programs. While it is true that there are programs to analyze deep sequencing data on all three major computing platforms (Windows,

Table 1
Primer sequences for DENV1, DENV2, DENV3, and DENV4

DENV 1	Primer sequence (5′–3′)
D1f1F	GTTAGTCTACGTGGACCGAC
D1f1R	CATCGTGATAGGAGCAGGTG
D1f3F4	TCACAAGAAGGAGCAATGCACA
D1f2R2	AAGAAGAACTTCTCTGGATGTTA
D1f5F	ACCAATGTTTGCTGTAGGGC
D1f5R	TATTCCCCGTCTATTGCTGC
D1f7F	CAGAGCAACGCAGTTATCCA
D1f7R	CAATTTAGCGGTTCCTCTCG
D1f9F	TCACAGATCCTCTTGATGCG
D1f9R	CATGGCACCACTATTTCCCT
D1f10F	ATGGCTCACAGGAAACCAAC
D1f10R	TGCCTGGAATGATGCTGTAG
DENV2	Primer sequence (5′–3′)
D2f1F	AGTWGTTAGTCTACGTGGAC
D2f1R	TGGGCTGTCTTTTTCTGTGA
D2f2F	GCAGAAACACAACATGGAACA
D2f2R	AACGCGTCAGTCAGTTCAAG
D2f3F	GAAAGCTGACCTCCAAGGAA
D2f3R	CTGAAATGTCTGTCGTGACCA
D2f4F3	AGGCAGCTGGGATTTTCATGA
D2f4R3	TTTCCCTTCTGGTGTGACCA
D2f5F3	ACTCAAGTATTGATGATGAGGA
D2f5R3	TGTGTCCAATCGTTCCATCCT
D2f12F	GCAGGATGGGACACAAGAAT
D2f12R	AGAACCTGTTGATTCAACAG
DENV 3	Primer sequence (5′–3′)
D3f1F	AGTTGTTAGTCTACGTGGAC
D3f1R	TGTCCGTGGTGAGCATTCTA
D3f2F2	TATGGAACCCTTGGGCTAGAA
D3f2R2	TAGTTGAAGTCCAGCTCCAATT

(continued)

Table 1
(continued)

DENV 1	Primer sequence (5′–3′)
D3f3F	AAGAGCATGGAATGTGTGGGA
D3f3R	TCTTCCAGTTCTGCTTTCTGTGT
D3f4F3	GATGATGAGACTGAGAAYATC
D3f4R	TGGTTCAAAGAGAGCTGGTAT
D3f5F	AGGAGAGGGAGAGTTGGCA
D3f5R	AGCAAGCCCAGCTCCTGCTA
D3f6F	ACCAATAACAACACTCTGGGA
D3f6R	ACGCGAGAACCARTGGTCTT
D3f7F	ACAGAGGAGAACCAATGGGA
D3f7R	AGAACCTGTTGATTCAACAGCA
DENV 4	Primer sequence (5′–3′)
D4-1F	GATGAGGGAAGATGGGGAGTTGTTAGTCTGTGT GGACCGAC
D4-2579R	GGGCATTYAATATTGCAGACGCTA
D4-2065F	CATAGTGATAGGTGTTGGAG
D4-4246R	GCAAGCCTCCTGCCACCATTG
D4-4226F	CAATGGTGGCAGGAGGCTTGC
D4-6463R	CTATGTTGTCAAGGGCCAGC
D4-6444F	GCTGGCCCTTGACAACATAG
D4-8531R	CCATTCACCATGGAGGATGC
D4-8512F	GCATCCTCCATGGTGAATGG
D4-10626R	CCATCTCGCGGCGCTCTGTGCC

Mac OS, and Linux), these programs are generally offered at a premium price and tend to lag behind open source software development. For this reason, most bioinformatics analyses are done on the Linux platform. The learning curve for the Linux operating system is approximately the same as switching from Windows to Mac OS and vice versa. There are many different "flavors" of the Linux operating system such as Ubuntu, Fedora, SUSE, RedHat, etc. For the purposes of this tutorial we will be using Ubuntu as it is freely available and the interface will seem more familiar to those used to working with Windows and/or Mac OS. Installation of Ubuntu can be done in parallel to an existing operating system and will cause no interference with the normal operation of the existing installation.

Once a Linux environment is successfully installed on your system, there are a handful of programs needed for the analysis. Several approaches exist that are tailored towards the analysis of low frequency variants and haplotype estimations for small viral and bacterial genomes, such as V-Phaser [21], VICUNA [22], SNVer [23], Breseq [24], and LoFreq [9]. Most use an existing mapping of the sequencing reads as a starting point. To create such a mapping a short read mapping software and usually also a format conversion tool is needed. Here we chose the Burrows-Wheeler Aligner (BWA) [25] for short read mapping and Samtools as a swiss-army knife for format conversion. Although there are many other mapping programs that could be used as a drop-in replacement for the efficient mapping of next-generation sequencing reads to a reference genome, we chose BWA since it is widely used and able to handle both single and paired end reads generated from multiple platforms such as Illumina, 454, Ion Torrent, etc. In this chapter we focus on the utilization of reads from the most widely used platform, Illumina. The file format generated by an Illumina sequencing run is FASTQ, which includes both the raw sequence information and the quality measurement for each base in the read. BWA uses this information during the mapping process. Although BWA works well with high performance, multi-node computer clusters, alignments can also be accomplished on typical laptop configurations. Naturally, more computing power will result in shorter alignment times. It is critical to emphasize here the importance of the best possible reference genome. Reads should ideally be mapped to a reference as close as possible to the actual sample to achieve highest possible coverage and mapping quality. If a reference is not available, one can in theory be created using a de novo assembler such as SOAPdenovo2 [26]. One should note, however, that de novo assembly of viral sequences is often very difficult due to the extremely high and variable coverage and the presence of haplotypes, which tend to violate assumptions made by assemblers: high coverage is usually interpreted as repeats and the presence of haplotypes will be interpreted as sequencing errors. Consequently, most de novo assemblers will produce only short contigs, instead of the desired full-length genome. One viable alternative is based on iterative mapping: it starts by mapping the reads to an initial user-provided reference, then constructs a consensus sequence from that mapping, remaps the reads against that consensus and repeats this process until no more additional reads map [27]. The output of a BWA alignment of deep sequencing reads to a reference genome is a Sequence Alignment Map (SAM) [28] file. This type of file is a non-compressed, human readable file that contains all of the information from the FASTQ file, the location that each individual read maps to on the reference and the associated quality of the mapping (if supported by the mapping program) for each sequence read. In order to save space and improve computation efficiency when dealing with these large files, SAM files are typically compressed into a binary file

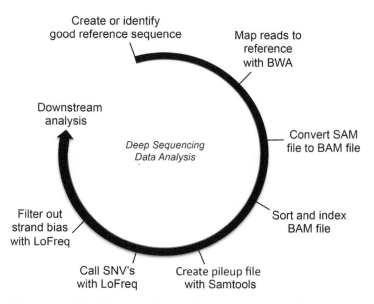

Fig. 3 Overview of the bioinformatics analysis of deep sequencing data

format that contains all of the same information at a fraction of the original size. This format is called binary alignment map (BAM) and will serve as the input for all subsequent steps. Conversion from SAM to BAM can be performed using a freely available software suite called Samtools. Once a BAM file is created one can readily start to predict low frequency variants from it. Optionally one might want to recalibrate base-call quality scores first with a program such as GATK [29], which usually results in fewer spurious low frequency variant calls in the downstream analysis. For our downstream analysis, we utilize the LoFreq program [9]. LoFreq is a fast and sensitive variant-caller and is designed to robustly call low-frequency variants by exploiting base-call and read-mapping quality values. Variant calls made by LoFreq down to frequencies as low as 0.5 % have been laboratory validated. LoFreq outputs a file containing only variants that have significant p-values after multiple testing correction. These should be filtered to remove positions with strand-bias, which is known to be associated with mapping issues and false positive SNV calls [30]. Ideally, primer regions are ignored in the final analysis, as they are known to contain artifactual low-frequency SNVs. A schematic of this workflow is shown in Fig. 3.

The analyses steps described above allow for great flexibility: at each step alternative programs can be used as replacement or quality checks can be performed before proceeding. Alternatively, a pipeline can be used that automates the execution of all steps. One such pipeline is called Vipr (available at https://github.com/CSB5/vipr). Basically it performs all steps described, i.e., it computes a consensus sequence in an iterative fashion starting from a given reference sequence, maps reads against the consensus, masks primer regions, performs base-call quality recalibration and finally calls low frequency SNVs by means of LoFreq.

Further analyses steps from this point onwards will depend largely on the experimental design of each investigators and individual studies.

2 Materials

2.1 Dengue RNA Extraction Using QIAamp Viral RNA Mini Kit (Qiagen)

1. Ethanol (100 %).
2. 1.5 ml microcentrifuge tubes.
3. Sterile, RNase-free pipet tips.
4. Microcentrifuge.

   Reagents supplied in the kit

5. QIAamp Mini Spin Columns.
6. Collection Tubes (2 ml).
7. Buffer AVL.
8. Buffer AW1.
9. Buffer AW2.
10. Buffer AVE.
11. Carrier RNA.

2.2 cDNA Preparation Using Maxima H Minus First Strand cDNA Synthesis Kit (Thermo Scientific)

1. Template RNA.
2. Primers.
3. 10 mM dNTP Mix.
4. Nuclease-free water.
5. 5× RT Buffer.
6. Maxima H Minus Enzyme Mix.
7. 0.2 ml nuclease-free tubes.
8. Microcentrifuge.
9. Thermocycler.

2.3 PCR Amplification Using PfuUltra II Fusion HS DNA Polymerase (Agilent)

1. Distilled water (dH_2O).
2. 10× PfuUltra II reaction buffer.
3. dNTP mix (25 mM each dNTP).
4. DNA template.
5. Forward primer (10 µM).
6. Reverse primer (10 µM).
7. PfuUltra II fusion HS DNA polymerase.
8. 0.2 ml nuclease-free tubes.
9. Microcentrifuge.
10. Thermocycler.

2.4 Running Gel	1. Agarose gel.
	2. Loading dye.
	3. 1× TBE.
	4. 1× TAE.
	5. Ethidium bromide (EtBr).
	6. Ladder.
	7. Items required for running gel: Power pack, casting tray, weighing machine, microwave, comb.
	8. X-tracta Gel Extraction Tool.
	9. UV transilluminator.
2.5 Isolate PCR Product from Gel Using QIAGEN Gel Extraction kit	1. Absolute Ethanol.
	2. Buffer QG.
	3. Buffer PE (diluted with 100 % Ethanol).
	4. Buffer EB.
	5. Microcentrifuge.
	6. 1.5 or 2 ml microcentrifuge tubes.
	7. QIAquick column (2 ml).
	8. Isopropanol (100 %).
	9. Heating block or water bath set at 37 °C.
	10. NanoDrop to quantify individual amplicons for amplicon balancing and mixing.
2.6 Fragmentation	1. Covaris S2 sonicator (Covaris, Woburn, MA, USA).
	2. microTUBE AFA Fiber Pre-Slit Snap-Cap 6 × 16 mm (Covaris).
2.7 DNA Purification Using QIAquick PCR Purification Kit	1. QIAquick Spin Columns.
	2. Buffer PB.
	3. Buffer PE (diluted with 100 % Ethanol).
	4. Buffer EB.
	5. Collection Tubes (2 ml).
2.8 DNA Quality Check (Agilent Bioanalyzer)	1. Chip priming station (reorder number 5065-4401).
	2. IKA vortex mixer.
	3. Microcentrifuge.
	4. Bioanalyzer.
	5. DNA1000 chip.
2.9 Library Preparation Using KAPA Library Preparation Kit (KAPA Biosystems)	1. End Repair Enzyme Mix.
	2. 10× End Repair Buffer with dNTPs.
	3. A-Tailing Enzyme.
	4. 10× A-Tailing Buffer.

5. DNA Ligase.

6. 5× Ligation Buffer.

7. PfuUltra II Fusion HS DNA Polymerase.

8. Microcentrifuge.

9. Thermocycler.

10. 0.2 ml tubes.

2.10 Linux Tutorial

Many excellent tutorials are freely available for familiarization with the Linux operating system. To get started, we recommend: http://www.linux.org/forums/beginner-tutorials.53/

2.11 Ubuntu Linux

The Ubuntu operating system and detailed instructions for its installation and use can be found here: http://www.ubuntu.com

2.12 SOAPdenovo2

The SOAPdenovo2 program and detailed instructions for its installation and use can be found here: http://soap.genomics.org.cn/soapdenovo.html

2.13 BWA

The BWA program and detailed instructions for its installation and use can be found here: http://bio-bwa.sourceforge.net

2.14 Samtools

The Samtools program and detailed instructions for its installation and use can be downloaded here: http://samtools.sourceforge.net

2.15 LoFreq

The LoFreq program and detailed instructions for its installation and use can be downloaded here: http://sourceforge.net/projects/lofreq/

3 Methods

3.1 Dengue RNA Extraction with the QIAamp-Viral-RNA-Mini Kit (See Note 1)

1. Add 310 µl Buffer AVE to the tube containing 310 µg lyophilized carrier RNA to obtain a solution of 1 µg/µl.

2. Follow **steps 1–10** of the QIAamp-Viral-RNA-Mini-handbook—June-2012-EN(1).pdf, pages 22–24.

3. Elute samples in 1×40 µl of Buffer AVE.

3.2 cDNA Preparation with the Maxima H Minus First Strand cDNA Synthesis Kit (See Note 2)

Prepare the component 1 and 2 as listed below:

Component 1 (for one reaction)	
Nuclease-free water	8 µl
10 mM dNTPmix	1 µl
Primer antisense (*see* **Note 3**)	1 µl
RNA	5 µl
Total	15 µl

1. Incubate at 65 °C for 5 min, then place on ice for at least 1 min.

Component 2 (for one reaction)	
5× RT buffer	4 µl
Maxima H Minus Enzyme Mix	1 µl
Total	5 µl

Add component 2 (5 µl) into the mix in component 1, mix gently, and centrifuge to bring liquid droplets down to bottom of the tube.

Incubate the mix in the thermocycler at 50 °C for 30 min followed by 85 °C for 5 min. Cool the mix to 4 °C before starting the next step (see **Note 4**).

3.3 PCR Amplification (See Notes 4 and 5)

Prepare the following components in a 0.2 ml tube as listed below:

Nuclease-free water	39 µl
10 mM dNTPmix	1 µl
10× Pfu Ultra II rxnbuffer	5 µl
PfuUltraII fusion HS DNA Polymerase	1 µl
cDNA	2 µl
Primer sense (see **Note 6**)	1 µl
Primer antisense (see **Note 6**)	1 µl
Total	50 µl

Briefly vortex the reagents and spin down the reaction mix prior to PCR amplification according to these conditions:

Activation	2 min at 92 °C
40 cycles	10 s at 92 °C 20 s at 55 °C 1.5 min at 68 °C (30 s per kb)
Final extension	5 min at 68 °C
Hold	4 °C

3.4 Running Gel

To select the desired PCR product size, run the entire PCR product on a 1 % agarose gel.

1. Cast a 1 % agarose gel in 1× TBE buffer. Add 1 µl of EtBr for every 50 ml of 1× TBE buffer used.

2. Add 20 µl of water to 10 µl of 1 kb ladder and 6 µl of 6× loading dye.

3. Add 8 µl of 6× loading dye to the PCR product.

4. Load the samples and ladder into the gel.

5. Run the gel according to the time and voltage as listed below:
   - 90 V for 40 min for smallest gel (70 ml)
   - 100 V for 40 min for medium gel (250 ml)
   - 120 V for 40 min for large gel (400 ml)

3.5 DNA Product Isolation from Gel Using QIAquick Gel Extraction Kit (Qiagen) (See Note 7)

Excise the correct DNA fragment from the agarose gel using the x-tracta gel extraction tool. Place the gel slice into a colorless tube and weigh the gel slice (*see* **Note 8**). Add 3 volumes of Buffer QG to 1 volume of gel (100 mg–100 μl). Incubate at 37 °C for 15 min or until the gel slice has completely dissolved. To help dissolve gel, vortex the tube every 2–3 min during the incubation. Follow **steps 5–12** EN-QIAquick-Spin Handbook.pdf, pages 25–26. To elute DNA, add 32 μl of Buffer EB (10 mM Tris–Cl, pH 8.5) to the center of the QIAquick membrane and centrifuge the column for 1 min.

3.6 Measure the Concentration of the PCR Amplicons Using an Agilent Bioanalyzer DNA1000 Chip (See Note 9)

Prepare the gel–dye mix according to the manufacturer's instructions and load onto the chip (*see* **Note 10**). Load the marker, followed by the ladder and finally the samples into the designated wells. Place the chip into the machine and run the pre-programmed DNA1000 chip analysis (*see* **Note 11**).

3.7 Balance the PCR Amplicons

Using the precise concentration measurement from the Agilent Bioanalyzer, balance the total amount (e.g., in nanograms) of each PCR amplicon to the fragment with the lowest concentration and aliquot into a 1.5 ml tube. To insure that the subsequent library creation steps are successful, we recommend that once the amplicons are mixed together, the total quantity should be at least 400 ng (*see* **Note 4**).

3.8 Fragmentation Using the Covaris Machine (See Note 12)

1. Place the pooled PCR amplicons into a Covaris microTUBE. Add EB buffer to obtain a total volume of 130–140 μl.

2. Shear the pooled PCR amplicons using the following settings:

Duty cycle	10 %
Intensity	5.0
Cycles per burst	200
Duration	110 s
Mode	Frequency sweeping
Power	Covaris S2-23W
Temperature	4–5 °C

Agilent Bioanalyzer DNA1000 chip can be run for an optional quality check following fragmentation.

3.9 Perform End Repair on Your Sheared Products (See Notes 13 and 14)

1. Prepare the reaction mix:

DNA sample	30 µl
Water	55 µl
10× End repair buffer	10 µl
End Repair enzyme mix	5 µl
Total	100 µl

2. Incubate in thermal cycler for 30 min at 20 °C.

3.10 DNA Purification Using QIAquick PCR Purification Kit

1. Follow **steps 1–8** of the QIAquick-Spin Handbook.pdf (Qiagen), pages 19–20.
2. To elute DNA, add 30 µl Buffer EB (10 mM Tris–Cl, pH 8.5) to the center of the QIAquick membrane and centrifuge the column for 1 min.

3.11 "A" Tailing (See Notes 13 and 15)

1. Prepare the reaction mix:

DNA sample:	30 µl
Water	12 µl
10× A-Tailing Buffer:	5 µl
A-Tailing Enzyme	3 µl
Total	50 µl

2. Incubate for 30 min at 30 °C.
3. Purify the samples using QIAquick PCR Purification Kit (*see* Subheading 3.10).

3.12 Ligate Adapters to DNA Fragments

1. Prepare the reaction mix:

DNA sample	30 µl
5× ligation buffer	10 µl
DNA ligase	5 µl
DNA adaptor (15 µM)	5 µl
Total	50 µl

2. Incubate in a thermal cycler for 30 min at 20 °C.
3. Agilent Bioanalyzer DNA1000 chip can be run (*see* Subheading 3.6.) for an optional quality check following adaptor ligation.

3.13 Size Selection of the Desired 300bpproduct from Gel Using QIAquick Gel Extraction Kit (Qiagen) (See Notes 7and 16)

Prepare 2 % agarose gel in 1× TAE buffer and EtBr.

1. Load 50 or 100 bp ladder.

2. Load 32 µl sample together with 4 µl of 6× loading dye.

3. Run the gel at 120 V for 60 min.

4. View and capture the gel on a UV transilluminator.

5. Excise DNA fragments in the region of 300 bp (*see* **Note 17**).

6. Do a gel extraction using the QIAquick Gel Extraction Kit (*see* Subheading 3.5).

7. An Agilent Bioanalyzer DNA1000 chip can be run (*see* Subheading 3.6) for an optional quality check following gel clean up.

3.14 PCR Amplification of the Library (See Notes 4, 13 and 18)

1. Prepare the reaction mix:

Nuclease-free water	Balance to 50 µl
10× PfuUltra™ II reaction buffer	5 µl
Pfu UltraII HS polymerase	1 µl
dNTPs	1 µl
PCR Primer InPE1.0	1 µl
PCR Primer InPE2.0	1 µl
PCR Primer Index (*see* **Note 19**)	1 µl
Sample	Variable (*see* **Note 18**)
Total	50 µl

2. Amplify using the following PCR protocol:

    2 min at 92 °C

    14 cycles of:

        10 s at 92 °C

        20 s at 55 °C

        30 s at 68 °C

    5 min at 68 °C

    Hold at 4 °C

3. Purify the PCR product using MinElute PCR Purification Kit (*see* Subheading 3.10); however, use the PCR MinElute PCR Purification Kit columns and elute in 22 µl of EB buffer.

4. Run Agilent Bioanalyzer DNA1000 chip to check for the concentration and fragment size (*see* Subheading 3.6).

5. Dilute samples to 10 nM.

6. Consolidate equal volumes of diluted samples into a single 1.5 mL tube. Final volume will depend on the sequencing facility requirements (*see* **Note 20**).

Fig. 4 An example of a command entered into a terminal window

3.15 Submit
Your Samples to
a Sequencing Facility
and Get Your Reads
Back in FASTQ Format
(See Note 21)

Copy the following command into a terminal (*see* **Notes 22** and **23**):

    bwa index /home/October/Desktop/Sample1-Reference.fa

An example of what a command should look like in a terminal window is shown in Fig. 4.

3.16 Index Reference
with BWA

3.17 Align Reads to
Reference with BWA

Copy the following command into a terminal (*see* **Notes 22** and **23**):

bwa aln -R 2 /home/October/Desktop/Sample1-Reference /home/October/Desktop/Sample1_read1.fastq > /home/October/Desktop/Sample1_read1_sa.sai && bwa aln -R 2 /home/October/Desktop/Sample1-Reference /home/October/Desktop/Sample1_read2.fastq > /home/October/Desktop/Sample1_read2_sa.sai

3.18 Map Reads to
Reference with BWA

Copy the following command into a terminal (*see* **Notes 22** and **23**):

bwa sampe -r '@RG\tID:120113_SN513_0262_AC0C6RACXX-Sample1\tLB:OMS\tPL:ILLUMINA\tCN:Your-Sequencing-Facility\tPU:1\tSM:Sample1' /home/October/Desktop/Sample1-Reference /home/October/Desktop/Sample1_read1_sa.sai /home/October/Desktop/Sample1_read2_sa.sai /home/October/Desktop/Sample1_read1.fastq /home/October/Desktop/Sample1_read2.fastq > /home/October/Desktop/Sample1.sam

3.19 Convert SAM
File to a Compressed
BAM File

Copy the following command into a terminal (*see* **Notes 22** and **23**):

samtools view -bS /home/October/Desktop/Sample1.sam > /home/October/Desktop/Sample1.bam

3.20 Sort Your BAM File	Copy the following command into a terminal (*see* **Notes 22** and **23**):

samtools sort /home/October/Desktop/Sample1.bam /home/October/Desktop/Sample1-SORTED

3.21 Index Your Sorted BAM File	Copy the following command into a terminal (*see* **Notes 22** and **23**):

samtools index /home/October/Desktop/Sample1-SORTED.bam

3.22 Create a Pileup File with Samtools mpileup	Copy the following command into a terminal (*see* **Notes 22** and **23**):

samtools mpileup -B -d500000 -f /home/October/Desktop/Sample1-Reference.fa /home/October/Desktop/Sample1-SORTED.bam > /home/October/Desktop/Sample1_pileup

3.23 Call SNVs, Including Low Frequency SNVs with LoFreq	Copy the following command into a terminal (*see* **Notes 22** and **23**):

python /home/October/bin/LoFreq-0.2/scripts/lofreq_snpcaller.py -Q 3 -i /home/October/Desktop/Sample1_pileup -o /home/October/Desktop/Sample1.snp

3.24 Filter SNPs with Strand Bias Out of Your LoFreq Output	Copy the following command into a terminal (*see* **Notes 22** and **23**):

lofreq_snpcaller.py -i /home/October/Desktop/Sample1.snp -o /home/October/Desktop/Sample1_filtered.snp --strandbias-holmbonf

---

4 Notes

1. The QIAamp-Viral-RNA-Mini Kit (Qiagen) is a non-phenol–chloroform extraction method for viral RNA. The column-based technology uses a silica membrane to bind viral RNA from patient plasma, patient serum, or cell culture supernatant. Below is the RNA Mini Kit protocol with minor modifications. Described below is the method we chose to implement but any other method that gives you a clean RNA extraction product would also be suitable.

2. The Maxima H Minus First Strand cDNA Synthesis Kit from Thermo Scientific is used to generate cDNA from the viral RNA. We have achieved reliable and reproducible results with this kit, but it is by no means the only available kit for cDNA preparation. This is the case for all kits used in subsequent sections of the methods seconds. If you are generating more than ten fragments to amplify your viral genome, you will need to make multiple cDNA preparations at this step.

3. The primer for making cDNA is the reverse primer at the 3′ end of the viral genome.

4. The protocol can be paused here if necessary and stored over night at −80 °C

5. PFU Ultra II Fusion HS DNA Polymerase from Agilent serves to amplify viral amplicons. In the case of whole genome sequencing, it is ideal to have the PCR amplicons be approximately equal in size as this makes balancing the molarity of your amplicons more straightforward as well as more homogenous for fragmentation.

6. Ideally, the fewest number of amplicons required to cover the entire genome should be used so as to reduce the amount of primer masking that occurs in the downstream analysis. See Appendix for the primers used to amplify DENV1-4.

7. This step is essentially following the QIAquick-Spin Handbook. pdf (Qiagen), pages 25–26.

8. Weigh the empty tube first and then weigh again with the gel slice in the tube. Subtract the weight of the empty tube from the weight of the tube with the gel slice to get the weight of the gel slice alone.

9. This step is performed to check the size, concentration, and quality of the DNA material collected after the gel extraction in **step 3.5 prior to pooling the amplicons.

10. It is important to ensure that you have allowed sufficient time for your reagents to equilibrate to room temperature. Failure to do so can be a source of error.

11. After filling all the wells, ensure that the run is started within 5 min for best results. It is important to minimize vibrations around the machine while the analysis is running, as this can be a cause of errors.

12. This step shears your gel extracted amplicons to a range ~200 base pairs. Covaris implements adaptive focus acoustics in a temperature-controlled environment to mechanically process samples to a specific size range of interest by using bursts of ultrasonic acoustic energy. The settings below have been optimized for 1,500–2,500 base pair amplicons. Other amplicon sizes would need further optimization.

13. Library preparation of your cleaned fragmented cDNA is necessary to ensure proper binding of your library onto the flow-cell of the sequencer. KAPA biosystems makes the kit we will describe in subsequent steps, but as the field evolves there will inevitably be more choices for library preparation.

14. This step converts overhangs generated during the fragmentation step into blunt ends using the end repair mix.

15. This step adds an "A" to the 3′ end of the blunt DNA fragments, which prevents self-ligation of fragments and allows for adapter ligation in the following step.

16. This step ensures that the fragments entering the PCR step are relatively homogenous in size. 300 bp libraries are typically preferred when doing 2×76bp Illumina sequencing. If your application requires longer reads, than the excised product will need to be increased accordingly to ensure that your reads do not overlap on the final product.

17. Depending on what your starting concentration was, it is possible that you may not see a band at this position. Don't panic. Cut as closely as you can to where the 300 bp product should be and proceed to amplification step.

18. Primers binding to the ligated adapters are used along with Kapa high-fidelity polymerase to amplify adapter-ligated products, which are at this point present in very low concentrations. The optimal amount of adapter-ligated product to be used as starting material for the enrichment PCR is 40 ng. If your sample is less concentrated, you can add up to the entire volume of the purified adapter-ligated product, 32 μl, to begin the enrichment PCR. It is critical that during the enrichment step, the number of cycles used to enrich the library be kept to an absolute minimum to avoid introducing PCR-derived error.

19. Each sample will be assigned a unique barcode, consisting of six random nucleotides.

20. For example, if there are 12 samples (12 indexes), take 2 μl of each sample.

21. The amount of time it takes to sequence your sample will depend on the platform you are using. An Illumina MiSeq can generate data in as little as 4 h while the Illumina HiSeq can take as long as 2 weeks.

22. For simplicity, all files are assumed to be located in the "home" directory of the user "October" on the "Desktop." This location of the files is not a requirement and will change depending on the name of the user and where the actual files are located. For a better understanding of Paths in a Linux environment, please see the link provided in Subheading 2.10.

23. It is important to type this command all on one line without pressing, "return" until the entire command has been entered.

References

1. Coffey LL, Beeharry Y, Borderia AV, Blanc H, Vignuzzi M (2011) Arbovirus high fidelity variant loses fitness in mosquitoes and mice. Proc Natl Acad Sci U S A 108(38):16038–16043. doi:10.1073/pnas.1111650108

2. Jenkins GM, Rambaut A, Pybus OG, Holmes EC (2002) Rates of molecular evolution in RNA viruses: a quantitative phylogenetic analysis. J Mol Evol 54(2):156–165. doi:10.1007/s00239-001-0064-3

3. Holmes EC (2003) Molecular clocks and the puzzle of RNA virus origins. J Virol 77(7):3893–3897

4. Skalsky RL, Vanlandingham DL, Scholle F, Higgs S, Cullen BR (2010) Identification of microRNAs expressed in two mosquito vectors, *Aedes albopictus* and *Culex quinquefasciatus*. BMC Genomics 11:119. doi:10.1186/1471-2164-11-119

5. David JP, Coissac E, Melodelima C, Poupardin R, Riaz MA, Chandor-Proust A, Reynaud S (2010) Transcriptome response to pollutants and insecticides in the dengue vector Aedes aegypti using next-generation sequencing technology. BMC Genomics 11:216. doi:10.1186/1471-2164-11-216

6. Sessions OM, Tan Y, Goh KC, Liu Y, Tan P, Rozen S, Ooi EE (2013) Host cell transcriptome profile during wild-type and attenuated dengue virus infection. PLoS Negl Trop Dis 7(3):e2107. doi:10.1371/journal.pntd.0002107

7. Yozwiak NL, Skewes-Cox P, Stenglein MD, Balmaseda A, Harris E, DeRisi JL (2012) Virus identification in unknown tropical febrile illness cases using deep sequencing. PLoS Negl Trop Dis 6(2):e1485. doi:10.1371/journal.pntd.0001485

8. Borderia AV, Stapleford KA, Vignuzzi M (2011) RNA virus population diversity: implications for inter-species transmission. Curr Opin Virol 1(6):643–648. doi:10.1016/j.coviro.2011.09.012

9. Wilm A, Aw PP, Bertrand D, Yeo GH, Ong SH, Wong CH, Khor CC, Petric R, Hibberd ML, Nagarajan N (2012) LoFreq: a sequence-quality aware, ultra-sensitive variant caller for uncovering cell-population heterogeneity from high-throughput sequencing datasets. Nucleic Acids Res 40(22):11189–11201. doi:10.1093/nar/gks918

10. Domingo E, Holland JJ (1997) RNA virus mutations and fitness for survival. Annu Rev Microbiol 51:151–178. doi:10.1146/annurev.micro.51.1.151

11. Cordey S, Junier T, Gerlach D, Gobbini F, Farinelli L, Zdobnov EM, Winther B, Tapparel C, Kaiser L (2010) Rhinovirus genome evolution during experimental human infection. PLoS One 5(5):e10588. doi:10.1371/journal.pone.0010588

12. Eckerle LD, Becker MM, Halpin RA, Li K, Venter E, Lu X, Scherbakova S, Graham RL, Baric RS, Stockwell TB, Spiro DJ, Denison MR (2010) Infidelity of SARS-CoV Nsp14-exonuclease mutant virus replication is revealed by complete genome sequencing. PLoS Pathog 6(5):e1000896. doi:10.1371/journal.ppat.1000896

13. Henn MR, Boutwell CL, Charlebois P, Lennon NJ, Power KA, Macalalad AR, Berlin AM, Malboeuf CM, Ryan EM, Gnerre S, Zody MC, Erlich RL, Green LM, Berical A, Wang Y, Casali M, Streeck H, Bloom AK, Dudek T, Tully D, Newman R, Axten KL, Gladden AD, Battis L, Kemper M, Zeng Q, Shea TP, Gujja S, Zedlack C, Gasser O, Brander C, Hess C, Gunthard HF, Brumme ZL, Brumme CJ, Bazner S, Rychert J, Tinsley JP, Mayer KH, Rosenberg E, Pereyra F, Levin JZ, Young SK, Jessen H, Altfeld M, Birren BW, Walker BD, Allen TM (2012) Whole genome deep sequencing of HIV-1 reveals the impact of early minor variants upon immune recognition during acute infection. PLoS Pathog 8(3):e1002529. doi:10.1371/journal.ppat.1002529

14. Nasu A, Marusawa H, Ueda Y, Nishijima N, Takahashi K, Osaki Y, Yamashita Y, Inokuma T, Tamada T, Fujiwara T, Sato F, Shimizu K, Chiba T (2011) Genetic heterogeneity of hepatitis C virus in association with antiviral therapy determined by ultra-deep sequencing. PLoS One 6(9):e24907. doi:10.1371/journal.pone.0024907

15. Neverov A, Chumakov K (2010) Massively parallel sequencing for monitoring genetic consistency and quality control of live viral vaccines. Proc Natl Acad Sci U S A 107(46):20063–20068. doi:10.1073/pnas.1012537107

16. Parameswaran P, Charlebois P, Tellez Y, Nunez A, Ryan EM, Malboeuf CM, Levin JZ, Lennon NJ, Balmaseda A, Harris E, Henn MR (2012) Genome-wide patterns of intrahuman dengue virus diversity reveal associations with viral phylogenetic clade and interhost diversity. J Virol 86(16):8546–8558. doi:10.1128/JVI.00736-12

17. Wright CF, Morelli MJ, Thebaud G, Knowles NJ, Herzyk P, Paton DJ, Haydon DT, King DP (2011) Beyond the consensus: dissecting within-host viral population diversity of foot-and-mouth disease virus by using next-generation genome sequencing. J Virol 85(5):2266–2275. doi:10.1128/JVI.01396-10

18. Chin-inmanu K, Suttitheptumrong A, Sangsrakru D, Tangphatsornruang S, Tragoonrung S, Malasit P, Tungpradabkul S, Suriyaphol P (2012) Feasibility of using 454 pyrosequencing for studying quasispecies of the whole dengue viral genome. BMC Genomics 13(Suppl 7):S7. doi:doi:10.1186/1471-2164-13-S7-S7

19. Makhluf H, Buck MD, King K, Perry ST, Henn MR, Shresta S (2013) Tracking the evolution of dengue virus strains D2S10 and D2S20 by 454 pyrosequencing. PLoS One 8(1):e54220. doi:10.1371/journal.pone.0054220

20. Hoang LT, Lynn DJ, Henn M, Birren BW, Lennon NJ, Le PT, Duong KT, Nguyen TT, Mai LN, Farrar JJ, Hibberd ML, Simmons CP (2010) The early whole-blood transcriptional signature of dengue virus and features associated with progression to dengue shock syndrome in Vietnamese children and young adults. J Virol 84(24):12982–12994. doi:10.1128/JVI.01224-10

21. Macalalad AR, Zody MC, Charlebois P, Lennon NJ, Newman RM, Malboeuf CM, Ryan EM, Boutwell CL, Power KA, Brackney DE, Pesko KN, Levin JZ, Ebel GD, Allen TM, Birren BW, Henn MR (2012) Highly sensitive and specific detection of rare variants in mixed viral populations from massively parallel sequence data. PLoS Comput Biol 8(3):e1002417. doi:10.1371/journal.pcbi.1002417

22. Yang X, Charlebois P, Gnerre S, Coole MG, Lennon NJ, Levin JZ, Qu J, Ryan EM, Zody MC, Henn MR (2012) De novo assembly of highly diverse viral populations. BMC Genomics 13:475. doi:10.1186/1471-2164-13-475

23. Wei Z, Wang W, Hu P, Lyon GJ, Hakonarson H (2011) SNVer: a statistical tool for variant calling in analysis of pooled or individual next-generation sequencing data. Nucleic Acids Res 39(19):e132. doi:10.1093/nar/gkr599

24. Barrick JE, Yu DS, Yoon SH, Jeong H, Oh TK, Schneider D, Lenski RE, Kim JF (2009) Genome evolution and adaptation in a long-term experiment with *Escherichia coli*. Nature 461(7268):1243–1247. doi:10.1038/nature08480

25. Li H, Durbin R (2009) Fast and accurate short read alignment with Burrows-Wheeler transform. Bioinformatics 25(14):1754–1760. doi:10.1093/bioinformatics/btp324

26. Luo R, Liu B, Xie Y, Li Z, Huang W, Yuan J, He G, Chen Y, Pan Q, Liu Y, Tang J, Wu G, Zhang H, Shi Y, Liu Y, Yu C, Wang B, Lu Y, Han C, Cheung DW, Yiu SM, Peng S, Xiaoqian Z, Liu G, Liao X, Li Y, Yang H, Wang J, Lam TW, Wang J (2012) SOAPdenovo2: an empirically improved memory-efficient short-read de novo assembler. Gigascience 1(1):18. doi:10.1186/2047-217X-1-18

27. Otto TD, Sanders M, Berriman M, Newbold C (2010) Iterative Correction of Reference Nucleotides (iCORN) using second generation sequencing technology. Bioinformatics 26(14):1704–1707. doi:10.1093/bioinformatics/btq269

28. Li H, Handsaker B, Wysoker A, Fennell T, Ruan J, Homer N, Marth G, Abecasis G, Durbin R, Genome Project Data Processing S (2009) The sequence alignment/Map format and SAMtools. Bioinformatics 25(16):2078–2079. doi:10.1093/bioinformatics/btp352

29. McKenna A, Hanna M, Banks E, Sivachenko A, Cibulskis K, Kernytsky A, Garimella K, Altshuler D, Gabriel S, Daly M, DePristo MA (2010) The Genome Analysis Toolkit: a MapReduce framework for analyzing next-generation DNA sequencing data. Genome Res 20(9):1297–1303. doi:10.1101/gr.107524.110

30. Guo Y, Li J, Li CI, Long J, Samuels DC, Shyr Y (2012) The effect of strand bias in Illumina short-read sequencing data. BMC Genomics 13:666. doi:10.1186/1471-2164-13-666

Part III

Modern Methods to Study Conformation of RNA Proteins and Their Molecular Interactions

Chapter 13

MPGAfold in Dengue Secondary Structure Prediction

Wojciech K. Kasprzak and Bruce A. Shapiro

Abstract

This chapter presents the computational prediction of the secondary structures within the 5′ and 3′ untranslated regions of the dengue virus serotype 2 (DENV2), with the focus on the conformational prediction of the two dumbbell-like structures, 5′ DB and 3′ DB, found in the core region of the 3′ untranslated region of DENV2. For secondary structure prediction purposes we used a 719 nt-long subgenomic RNA construct from DENV2, which we refer to as the minigenome. The construct combines the 5′-most 226 nt from the 5′ UTR and a fragment of the capsid coding region with the last 42 nt from the non-structural protein NS5 coding region and the 451 nt of the 3′ UTR. This minigenome has been shown to contain the elements needed for translation, as well as negative strand RNA synthesis. We present the Massively Parallel Genetic Algorithm MPGAfold, a non-deterministic algorithm, that was used to predict the secondary structures of the DENV2 719 nt long minigenome construct, as well as our computational workbench called StructureLab that was used to interactively explore the solution spaces produced by MPGAfold. The MPGAfold algorithm is first introduced at the conceptual level. Then specific parameters guiding its performance are discussed and illustrated with a representative selection of the results from the study. Plots of the solution spaces generated by MPGAfold illustrate the algorithm, while selected secondary structures focus on variable formation of the dumbbell structures and other identified structural motifs. They also serve as illustrations of some of the capabilities of the StructureLab workbench. Results of the computational structure determination calculations are discussed and compared to the experimental data.

Key words Dengue virus, Massively Parallel Genetic Algorithm, MPGAfold, RNA secondary structure prediction, 3′ untranslated core region, Dumbbell structures, StructureLab workbench

1 Introduction

The dengue virus (DENV) is a mosquito-borne flavivirus (MBFV) from the family *Flaviviridae* [1]. In humans it causes dengue fever, and in approximately 10 % of the most severe cases dengue hemorrhagic fever and dengue shock syndrome, which are life-threatening [2–4]. Flaviviruses include three subgroups: DENV (dengue serotypes DENV1 through 4), Yellow fever virus, and Japanese encephalitis virus (JEV) [5, 6]. The DENV viral genome is a single positive-stranded RNA, approximately 11,000 nucleotides long (10,723 nt for the DENV2 New Guinea strain C used in this study,

Radhakrishnan Padmanabhan and Subhash G. Vasudevan (eds.), *Dengue: Methods and Protocols*, Methods in Molecular Biology, vol. 1138, DOI 10.1007/978-1-4939-0348-1_13, © Springer Science+Business Media, LLC 2014

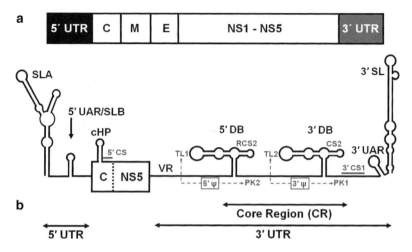

Fig. 1 (**a**) A schematic view of the DENV2 genomic RNA and the minigenome construct used for computational structure prediction. (**a**) Full genome regions are labeled as follows: UTRs—untranslated regions; C—capsid, M—membrane, and E—envelope structural protein coding regions; NS1–NS5 non-structural proteins coding regions. DENV2 New Guinea strain C (GenBank# M29095) is 10,723nt long. (**b**) The schematic representation of the 719nt long minigenome construct of DENV2, which was used in the computational structure predictions performed with the aid of the MPGAfold program. The minigenome includes the 5′ UTR, the C region fragment, the end of the NS5 and the 3′ UTR (VR indicates the variable region). *Small red labels* indicate sequence motifs of interest discussed in the text and involved in the predicted secondary structures and inferred pseudoknots (5′ψ and 3′ψ). Refer to the text for more details (Color figure online)

GenBank# M29095 [7]). It is depicted schematically in Fig. 1a. Its 5′ end is capped (type I cap), while the 3′ end is not polyadenylated [8]. The short 5′ and 3′ untranslated regions (UTRs) contain conserved *cis*-acting secondary structure elements needed for translation and replication [9–18]. The UTRs precede and follow one long open reading frame coding for a polyprotein, ultimately processed into three structural proteins (C—capsid, M—membrane, and E—envelope) and seven non-structural proteins (NS1, NS2a, NS2b, NS3, NS4a, NS4b, NS5). For more information *see* the reviews [8, 19, 20] and references therein. In the current replication model, the viral replicase complex assembles in cytoplasmic membrane compartments. Inside it the negative RNA strand is synthesized starting from the 3′ end of the viral genome RNA (positive strand). The negative strand is then used as a template for transcription into the positive genomic RNA strand [21] (also *see* reviews in refs. 1, 22).

Approximately 100nt long, the 5′ UTRs of flaviviruses form stable secondary structures [13, 23, 24] that include two stem-loop structures, SLA and SLB (Fig. 1b). This un-translated region is important for binding the NS5 protein, which includes the viral

RNA-dependent RNA polymerase (RdRp) in its C terminal region [13]. Another identified functional secondary structure called capsid-coding region hairpin (cHP) is located downstream of the 5′ UTR (Fig. 1b) and plays a role in the initiation of translation [25, 26], as well as DENV and WNV replication [25, 26]. Another motif known as the 5′–3′ DAR ("downstream of AUG region") is also required for viral RNA replication [27–29]. While the 3′ UTRs of flaviviruses show a lot of heterogeneity in their sequences and sizes, they also contain conserved sequences that are required for replication [9–11, 13, 15–18, 30]. The 3′ UTRs have a variable region (VR), the core region (CR), and the terminal stem-loop (3′ SL) region (Fig. 1b). The terminal stem-loop is formed within approximately the last 100 nucleotides [31, 32], the region that plays a significant role in replication [30, 33–36]. The core region, CR, includes the 3′ cyclization sequence (3′ CS1), which is complementary to the 5′ cyclization sequence (5′ CS) located within the capsid coding region (Fig. 1b). A long-range interaction between these two complementary RNA sequences results in circularization of the genome [11, 37] that has been shown to be essential for RNA synthesis in vitro [18, 38] and for replication of subgenomic replicons or infectious clones [9, 15, 16, 39, 40]. Two other complementary sequence fragments referred to as 5′ and 3′ UARs (upstream AUG) are found in the terminal regions of the genomic RNA, and they can form a long-distance UAR interaction. These are also required for cyclization and replication [10, 11, 41]. Relative to the CS/CS1 sequences, the 5′ and 3′end UAR sequences are located closer to both ends of the genomic RNA; the 5′ UAR upstream of the 5′ CS, and the 3′ UAR downstream of the 3′ CS1 (Fig. 1b). The core region maintains a high degree of sequence conservation among the mosquito-borne flaviviruses, and it is believed to fold into well-defined secondary structures local to the region [42, 43]. Also within the CR and upstream of the 3′ CS1, another conserved sequence, called CS2, is found in all three subgroups in the mosquito-borne flaviviruses. The repeat conserved sequence 2 (RCS2), on the other hand, is found only in DENV and JEV subgroups [37] (Fig. 1b). The CS2 and RCS2 sequences are part of two nearly identical secondary structures, referred to as the dumbbell-like structures (DBs) that can form in DENV (Fig. 1b). The hairpin loops at the end of the longer arms of the 5′ and 3′ DBs contain identical sequences (5′-GAAGCUGUA-3′) [42]. Different DENV serotypes conserve the inner five nucleotides in these loops (5′-GCUGU-3′), referred to as TL1 in the 5′ DB and TL2 in the 3′ DB (Fig. 1b). Also conserved are two complementary sequences, PK2 (5′-GCAGC-3′) and PK1 (5′-ACAGC-3′), downstream of the 5′ and 3′ dumbells, respectively. This conservation of sequence motifs suggests potential base pair interactions resulting in pseudoknot structures 5′ψ and 3′ψ, involving the 5′DB (sequences TL1/PK2) and the 3′ DB (sequences TL2/PK1) [42].

This chapter focuses on the computational tools we have developed and applied to the study of the core region within the 3′-untranslated region (UTR) of the dengue virus by focusing on the functional roles of the two dumbbell structures in the CR region and the TL and PK sequences [29]. For secondary structure prediction and analysis, we used our programs, the Massively Parallel Genetic Algorithm, MPGAfold, and StructureLab [44–52] to fold a 719 nt subgenomic RNA construct from the DENV2 (New Guinea C strain). The construct contained the 5′-most 226 nt (5′ UTR and a fragment of the capsid coding region C), the last 42 nt from the non-structural protein NS5 coding region including the translation termination codon, and the 451nt of the 3′ UTR [29]. It has been shown to contain the elements needed for translation [12, 25, 26, 53] and negative strand RNA synthesis in vitro [18, 38, 54, 55]. We refer to this construct as the minigenome. Combination of the experimental and computational data in our study showed that all four sequence motifs (TL1, PK2, TL2, PK1) played a crucial role in replication. On the other hand, the experimental results suggested that TL1/PK2 and TL2/PK1 played differential roles in translation. The structure prediction results generated by MPGAfold indicated differences in the stabilities of the two DBs and frequency of their occurrence in the sets of predicted structures [29]. These results offer a potential explanation of the fact that the TL1 and TL2 are not functionally identical despite having identical sequences within the two DBs.

2   Materials: Software Packages

Two of several software packages developed by our group were used to elucidate the secondary structure of the dengue minigenome. These programs, MPGAfold and StructureLab, are presented below with specific examples of how they were used in the dengue study. Together with many other downloadable programs and Web servers, the tools mentioned in this chapter are publicly available on our Web site: www.ccrnp.ncifcrf.gov/users/bshapiro/software.html and webserver_index.html.

3   Methods for RNA Secondary Structure Prediction and Analysis

3.1   RNA Secondary Structure Representations

The primary structure of an RNA molecule is its sequence that can be represented as a string built from an alphabet of just four letters: A, G, C, U. These letters correspond to the nucleic acid bases, Adenine, Guanine, Cytosine and Uracil. The bases tend to interact with each other preferentially via hydrogen bonds to form Watson-Crick base pairs (G-C and A-U) and a wobble base pair G-U. These interactions are additionally stabilized by base stacking interactions.

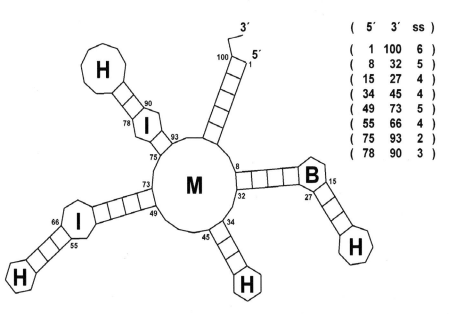

Fig. 2 An RNA secondary structure represented as a 2D drawing and a corresponding region table. Commonly found morphological features are labeled as follows: M—multi-branch loop, I—internal loop, B—bulge loop, and H—hairpin loop. Base-paired regions, listed explicitly in the region table, are represented are *rectangles* with cross-links corresponding to the number of base pairs in each region

Many other base interactions that occur less frequently in nature are possible and were classified by Westhof and Leontis [56]. However, to keep the introduction brief, they will not be discussed here.

Figure 2 illustrates how a generic RNA secondary structure with its different types of motifs, can be represented. In the drawing, the crosshatched lines correspond to base pairs forming stems, which are 2D equivalents of three-dimensional helices. The unpaired positions in the drawing form loop regions and are labeled according to their motif type. Labels M, I, B, and H indicate multibranch, internal, bulge, and hairpin loops respectively. The same topology can be represented (encoded) in the form of a region table, also shown in Fig. 2. The region table explicitly lists all the secondary structure stems, sorted based on their 5′ positions. For example, the first stem, represented by the triple (1 100 6) is a stem whose 5′ position is 1, its 3′ position is 100 and its size is 6. In this compact representation the unpaired regions of the structure are not listed explicitly and have to be derived as a complement of the paired regions for a given sequence. Other representations can also be used that are more explicit (the CT file format or the parenthesis notation employed by the mfold and RNAfold servers [57, 58]).

Regardless of its representation the relative stability of the secondary structure can be characterized by its free energy, the more negative the free energy the more stable the structure. Base-paired and additionally stabilized by base pair stacking, stem regions tend to add to the overall stability of the structure, while the loops generally decrease it. Additional stabilizing contributions can be made by stem stackings and higher order interactions, which, however, are poorly characterized in terms of their free energy contributions. Several free energy rule sets, with different degrees of context sensitivity, have been adopted for use in free energy calculations of an RNA secondary structure, the latest allowing for stem stacking calculations (efn2) [59] (also *see* reviews [60, 61]).

3.2 Massively Parallel Genetic Algorithm (MPGAfold)

Many algorithms have been developed for predicting RNA secondary structure. Broadly speaking, these fall into two main categories which may be defined as deterministic and or stochastic, e.g., genetic algorithm-based. The algorithm that was applied in this case was MPGAfold, a massively parallel genetic algorithm for secondary structure prediction that was developed by our group [45, 48–52].

3.2.1 Genetic Algorithm Fundamentals

Originally presented by John Holland [62], genetic algorithms (GAs) are optimization procedures that can be used to search large solution spaces for "best" or near optimal results that are achievable within the limits of the parameters describing the problem and the objective function. MPGAfold is based on the principles of GAs. In the process of RNA secondary structure determination, the objective function MPGAfold uses is the free energy of the RNA secondary structure. Thus the algorithm searches for an RNA fold such that its free energy is optimal or near the optimal for the given input sequence and a set of energy rules. The term "fitness" is used alternatively to free energy, with the better of higher fitness corresponding to lower free energy of a secondary structure. A key feature and strength of the algorithm is its Boltzman-like preference for the most probable solutions, rather than exclusively the lowest energy ones. This is a non-deterministic algorithm that must be run repeatedly in order to determine the consensus solution of multiple runs. This feature also implies that one or more frequent alternative secondary structures can be revealed in multiple runs, a feature that we used in the dengue minigenome study. The alphabet of MPGAfold, i.e., the set of fundamental building blocks used in the optimization procedure is comprised of all theoretically possible, contiguous, fully base-paired stems (i.e., equivalents of perfectly base-paired helical fragments). This set of stems, called the stem pool, is pre-computed from the given input sequence in a procedure preceding the optimization phase. As it is explained later, in the actual algorithm runs these stems may be shortened by a peel-back operator that (optionally) resolves conflicts between overlapping stems. An expansion of the basic alphabet adds to the stem pool motifs that consist of two neighboring stems separated by small bulge or internal loops of sizes 1×1, 1×2, and 2×2 (i.e., loops with combinations of one to two nucleotides in their 5' and/or 3' sides). In the optimization phase of the algorithm the GA applies operations of mutation, recombination, and selection to the maturing structures from the current (parent) population in order to generate new secondary structures that constitute the next population generation (children). These are discussed in more detail in the next paragraph. The process is repeated until the improvements to the fitness (free energy) of the population of the structures maturing in parallel fall below a selected criterion. Secondary structures, one in each virtual processing element reside on a rectangular or square grid representing a population (*see* Fig. 3). The most often employed population sizes range from as

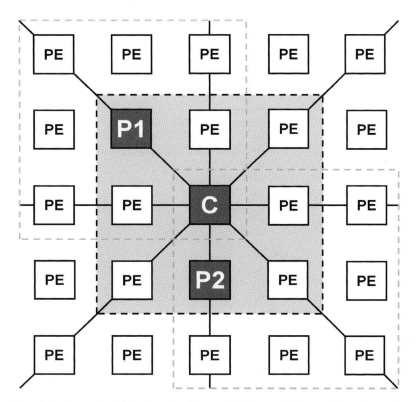

Fig. 3 A schematic illustration of a fragment of the grid of population elements (PEs) employed by the Massively Parallel Genetic Algorithm (MPGAfold). Only the full 8-way connectivity of the central element (C) in the central nine element neighborhood (3-by-3 neighborhood with the *light-gray* background) is shown. Labels P1, P2 and C in the central neighborhood indicate two current generation parent structures and the next generation child structure. New child structures are generated in parallel for all the central elements of all the overlapping 3-by-3 neighborhoods, three of which are delineated with the *dashed lines.* The full population grid is connected toroidally, i.e., the top PEs connect with the bottom, and the left with the right. At every generation MPGAfold replaces all the current structures (in all the PEs) with the child structures, calculated for all the overlapping 3-by-3 neighborhoods

low as $2K$ ($1K = 1,024$) to as high as $128K$ with a power of 2 increments. The population size is chosen as a function of the input sequence length, with the longer sequences usually subjected to MPGAfold runs over a broader range of sizes (i.e., runs are performed at several population levels, which is discussed below). Each population element in the grid, which consists of an RNA secondary structure, is connected vertically, horizontally and diagonally with its eight neighbors (i.e., N, S, E, W, NE, NW, SE, and SW, to employ the direction of the compass). From the functional point of view, the grid can be viewed as a toroid. In other words, its top and bottom as well as left and right boundaries are wrapped so that all population elements have eight neighbors and there are no boundaries. For example, the Left/West neighbor of a population element on the conceptual left edge of the grid will be an element on the right edge of the grid. Figure 3 illustrates a 5-by-5 square section of a larger grid of the population elements. MPGAfold has been implemented to run on a parallel cluster computer, distributing the population across a set of power of 2 CPUs, and its speed scales up almost linearly with the doubling of processors (or halving of the population size) [51].

3.2.2 MPGAfold Operators

MPGAfold fundamental operations are mutation, recombination, and selection. Mutation randomly draws stems or the small stem motifs (i.e., the fundamental building block, as was explained in the previous section) from the stem pool and adds them to existing structures, replacing or partly opening (via the so-called peel-back) any conflicting stems already part of the structure. Mutations are a precursor step to each recombination operation. In the process of recombination each population element and its eight nearest neighbors from a 3-by-3 neighborhood (the central box in Fig. 3) are placed in an array, sorted in the decreasing fitness order (i.e., from the lowest free energy to the highest). A sampling biased toward selection of the best fit structures chooses two parent structures, and their stems are distributed to two child structures, which have been initialized with stems from mutations. Both mutations and recombinations employ a probabilistic stem peel-back scheme that resolves structural conflicts (overlapping stem fragments) between the stems being added to a structure and the ones already a part it. Peel-back shortens stems (effectively removing some of their base pairs) in order to fit them into the rest of the structure. This mechanism has two advantages. First, it effectively increases the resolution of the algorithm by allowing stems to be peeled-back to a single base pair. Second, it keeps the stem pool smaller and stem pool sampling more effective by including in it only the stems of maximum size. Finally, in the selection operation the free energies of the two child structures are computed, and the lower free energy structure (better fit) is selected to replace the center element in the 3-by-3 neighborhood.

3.2.3 MPGAfold Output

The three GA operators described above are applied iteratively in parallel across the entire population grid. Initially, the entire population may contain all new secondary structures after calculations of the new generation of structures. MPGAfold employs an annealing mutation scheme that gradually lowers the mutation rate. As a consequence of this and the natural maturation, i.e., fitness improvements in individual structures and the whole population on average, the population stabilizes. This may mean convergence by a vast majority of the population elements to the same secondary structure or a situation in which multiple secondary structure populations remain at equilibrium with respect to each other. Once this population stabilization is detected, the algorithm run is terminated. MPGAfold can output either the population-wide consensus (population histogram peak) structure or the best fit (lowest free energy) structure. In some runs these two may be the same, but most often they are not. A histogram of the free energy values is calculated for an evolving population at every generation, and the structures corresponding to the peak value or the lowest energy value can be output for every generation of a run as well. Thus, MPGAfold provides a wealth of information on the intermediate as well as the final structures in multiple runs. In the dengue study presented here we collected both kinds of data.

3.2.4 Transient or Intermediate Structures

The significant transient structures (dominant for a significant fraction of generations of the whole run) that the algorithm outputs during each of many runs at the constant population level (grid size) can also be captured as the final population consensus structures in the runs with lower population sizes. The algorithm has been demonstrated to capture RNA secondary structures that are representative of significant intermediates in runs with varying population sizes [45, 63–66]. A folding RNA sequence may form intermediate conformations that are themselves functional alternatives to the better fit final states or play an important role in the folding pathway of the RNA helping it to reach its final state. At lower population levels, the algorithm has a tendency to converge to less fit population consensus structures, which are indicative of these significant intermediates. In the runs at higher population levels, the maturing structures will transition through potentially several intermediates on their path to the best fit, low free energy final states.

3.2.5 H-Type Pseudoknots

By using stems as the fundamental building blocks, MPGAfold can relatively easily generate structures including general pseudoknots. The key limitation on taking full advantage of this ability to generate general secondary structure topologies is the limited knowledge of the energies of complex pseudoknots, which makes it impossible to evaluate fitness of such structures using the same objective function applied to secondary structures without pseudoknots. For this reason the algorithm currently considers only the

H-type pseudoknots for which the free energies can be calculated. H-type pseudoknots allow base pairs to form between the nucleotides of simple hairpin loops and those of single stranded regions immediately upstream or downstream of them. It is worth noting that a scheme of 3D geometric constraints guiding prediction of general pseudoknotted structures has been successfully implemented in our CyloFold algorithm [67]. Potentially, it could be included in MPGAfold, by adopting a kind of a hybrid objective function. (See the CyloFold Web server: http://cylofold.abcc. ncifcrf.gov).

3.2.6 Other MPGAfold Capabilities

The MPGAfold program is also capable of co-transcriptional folding and can be monitored and queried during execution with the aid of a satellite visualization module [47, 51].

RNA is transcribed by the sequential addition of nucleotides to its 3′ end. RNA folding can be influenced by this elongation because of different folding pathways (intermediate structures) resulting from the changing sequence length. Final co-transcriptional folding conformations may differ from the full-length sequence folding, such as in quench-cooling experiments. MPGAfold offers an option of folding with sequence elongation. This is implemented by gradually increasing the size of the sampled stem pool space in correspondence to the increments in the input sequence's 3′ position. In other words, the rate of elongation determines how fast stems with the increasing 3′ positions can be drawn from the stem pool by the mutation and recombination operators. The rate of elongation is parameterized (e.g., increments by n nucleotides for every m algorithmic generations) to fine-tune its impact on the maturation of a predicted structure [45, 66]. However, the elongation rate adjustments were not designed to reflect the experimentally known transcription rates. The standard operators of the algorithm determine how quickly the newly available stems propagate throughout the evolving population or become parts of the population-wide consensus structures. An opposite feature of the algorithm allows for sequence shortening (from the 3′ end), which may be useful in the cases of post-transcriptional processing leading to refolding. This feature proved to be critically important in our study of the host-killing/suppression-of-killing (hok/sok) mechanism of *E. coli* plasmid R1 [45]. Employment of such an option requires multi-phase runs, which can be performed by MPGAfold. In the first phase the regular full length or the elongated sequence is folded. In the subsequent phase(s) the initial GA population is randomly seeded with selected secondary structures and the algorithm is run with other options, such as the 3′ end truncation from the full (starting point) sequence length until the foreshortened sequence length is reached. Then the algorithm is allowed to terminate normally. Other options available in MPGAfold allow for the imposition of constraints in cases where outside knowledge is available (e.g., structure probing data). Selected stems can be "forced" to be part of solutions or be designated as "sticky," in which case they are designated to be passed to the next generation from parent structures to their progeny. However, the resulting child structures are subject to the standard selection process, which means that the "sticky" stems persist only as natural parts of best fit structures.

MPGAfold can be run in batch mode or in conjunction with an interactive Visualizer program, which acts as a Graphical User Interface (GUI). Two-dimensional interactive maps of the population grid are displayed and can be manipulated while the algorithm is running. In this way runs can be paused and structural information queried at any stage. Color-coded maps allow for the monitoring of population energy distributions, formation of pseudoknots, and the presence and persistence of user-defined stems. Finally, MPGAfold can be run in synchronization with the Visualizer and StructureLab (see the next section), in which case StructureLab can draw individual RNA secondary structures selected by the user from the population displays [47].

3.2.7 Non-deterministic Nature of MPGAfold and Its Benefits

The non-deterministic nature of MPGAfold means that for the same input sequence no two runs will follow exactly the same folding intermediate states and are not guaranteed to converge to the same final structure. For the same reason the performance of the algorithm cannot be described as a function of the input sequence. Multiple runs for the same input sequence or for several sequences of the same length may result in highly variable numbers of generations required by the algorithm to converge to the final answers. Careful analysis of such differences in the MPGAfold output can help indicate—in a coarse way—multiple folding pathways and stable intermediates of potential biological significance. In the studies referenced here biologically significant conformations were identified as population consensus structures, both in the case of the intermediate and the final states [45, 65, 66].

3.3 StructureLab

StructureLab is a graphical data mining program developed in our lab and employed in the dengue minigenome study. The program helps to interactively explore large numbers of RNA conformations produced by secondary structure prediction programs [44, 46, 47]. These may include MPGAfold described above, or any of the DPA-based programs, such as the Vienna package RNAfold [57, 68, 69], Mfold [58, 59], or RNAstructure [59, 60, 70]. Also, *see* the reviews [60, 61]. The design and functional features of StructureLab have been presented in-depth before [44, 46, 47]. In this chapter we are going to review briefly one of the StructureLab tools, called StemTrace that was employed extensively in this dengue minigenome folding study.

3.3.1 StemTrace

StemTrace presents a set of secondary structures (MPGAfold or DPA output data) in the form of an interactive, two-dimensional plot of all the stems found in the input secondary structures. Each entry along the horizontal axis (X axis) represents an RNA secondary structure, while entries along the vertical axis (Υ axis) represent all stems found in at least one of the plotted structures. A stem is defined as a unique triplet of values corresponding to the 5′-start position of a stem, its 3′-stop position, and the number of base pairs (i.e., the stem size). Conceptually, a vertical line intersecting all the Υ-axis entries for a given x-position corresponds to a stem table describing one RNA secondary structure. Stem entries on the Υ-axis is can be ordered either by their first appearance in the solution space (i.e., order in which they are generated by an algorithm), which is the default option, or by a user-selected sort criterion (e.g., increasing 5′-start positions of the stems, stem size, or the 5′–3′ distance). Stems appearing repeatedly in the predicted secondary structures, each plotted as a pixel will form horizontal bands appearing at specific y-positions. The bands are color-coded on a 10 color bin scale, to help identify frequency of occurrence of individual stems within the displayed (input) set of structures. The user can interact with StemTrace plots by "pointing and clicking" over any graphed positions. Depending on which mouse button is pressed, one stem or information about a full structure can be retrieved, and, optionally, it can be drawn as a secondary structure (thanks to the integration of multiple tools within StructureLab).

The flexibility with which StemTrace can represent the secondary structure solution space is well suited to the many types of output that can be generated by MPGAfold. A StemTrace plot can depict an ensemble of RNA secondary structures predicted for a single input sequence in multiple runs (Fig. 4) thus giving insight into the consensus structure or consensus structural motifs. Results for several sequences in a family that can be reasonably aligned (with interactive corrections informed by structural considerations) [44, 47] can be plotted together to aid in searching for common structural features. StemTrace can also depict the process of conformation maturation in one MPGAfold run (*see* Fig. 5). Each of these application examples is briefly discussed below.

3.3.2 Plotting the Final Structures from Multiple MPGAfold Runs

In this case a StemTrace plot could represent, for example, 100 final consensus structures generated in 100 independent folding runs of the same input sequence (Fig. 4). Results can be thresholded based on the frequency of occurrence of individual stems. Such filtering, combined with the 2D drawing option provided by other integrated StructureLab tools, allows the user to quickly depict a consensus structure.

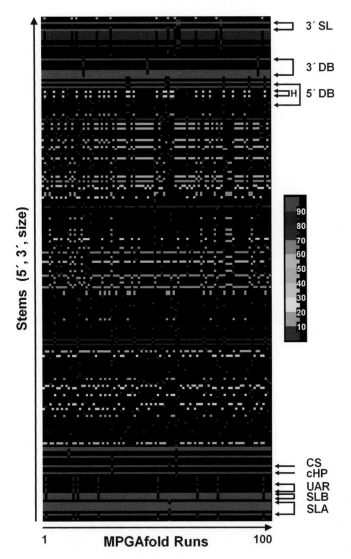

Fig. 4 A Stem Trace graph of the final results of 100 runs of MPGAfold with a population of 16K (16,384 structures maturing in parallel). All the stems found in the final structures, defined by unique triples (5′ pos., 3′ pos., stem size), are plotted on the vertical axis and sorted in the increasing order of their 5′ positions. Entries along the horizontal axis correspond to the final results of each MPGAfold run. The color-coding of individual stems in the structures shown is based on their frequencies of occurrence in the ensemble of the final predicted (final results of 100 runs). The 10 bin color scale is shown to the *right* of the plot. The stems from the structural motifs drawn and labeled in the same way in Figs. 6 and 7 are indicated with *arrows* and *labels*. The Stem Trace plot shows clearly the difference in the frequency of the full 3′ DB and the 5′ DB structures (refer to Fig. 6b), as well as the higher frequency of the 5′ DB head than the full 5′ DB (labeled as H above and 5′ DBH in Fig. 7b). Another result worth noting is the 8 % frequency of the UAR structures among the predicted conformations

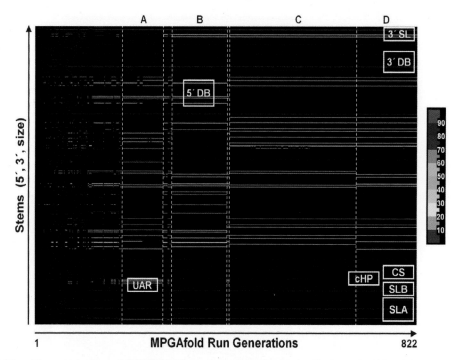

Fig. 5 A Stem Trace plot of one full MPGAfold run corresponding to the final result of the 86th run at the 16*K* population level, shown in Fig. 4. The color-coding of individual stems is based on their frequencies of occurrence in the 822 generations of this run. The 10 bin color scale is shown to the *right* of the plot. Letters A through D above the plot indicate key conformations that became dominant (population histogram peak structures) after the initial phase of rapid maturation and frequent conformation changes. A: Two secondary structure variants (−210.7 and −212.5 kcal/mol) with the long distance UAR structures, 5′ DBH and 3′ DB, overall similar to that shown in Fig. 7b. B: A secondary structure with both full dumbbells, but without the UAR interactions (−216.7 kcal mol). C and D: Variants of the secondary structures without the 5′ DB, but with the 3′ DB (−222.0 kcal/mol for C and −225.5 kcal/mol for D), both similar to the structure shown in Fig. 6a. In addition to the 3′ DB, all other key structural motifs (SLA, SLB, cHP, CS, and 3′ SL) form early in this MPGAfold run and are present in the above described transitional conformations. The short transient states between conformers A and B and conformers B and C contain the UAR and 5′ DBH motifs, with the second transient conformer reaching the free energy level of −218.1 kcal/mol

3.3.3 Plotting the Final Structures for a Family of Related Sequences

Solutions for each input sequence are plotted in separate blocks of runs along the *X*-axis (e.g., 25 at a time) and visually outlined with thin vertical separators. The *Y*-axis position for stems from different sequences of the family can be adjusted based on the sequence alignment to account for the difference (insertions and deletions), yielding one common *Y*-axis range. The most often used and simpler variant of this representation is applied to a series of multiple solution sets for the same input sequence, for which the secondary structures were run at different MPGAfold population (grid size) levels. In this case, no sequence alignment corrections are necessary.

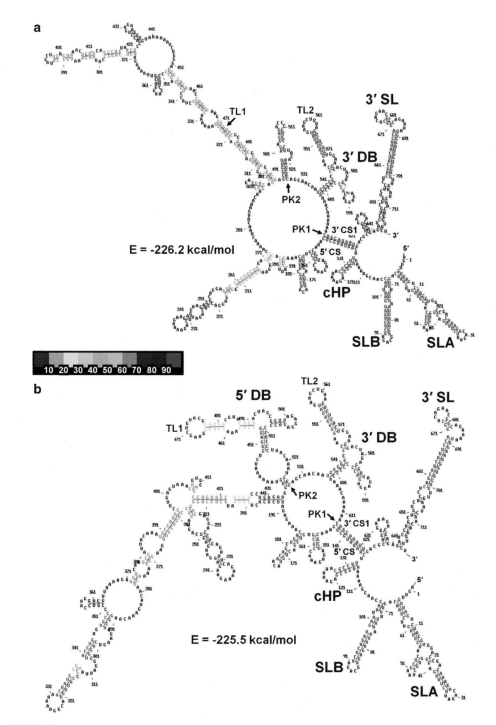

Fig. 6 MPGAfold-predicted secondary structures of the DENV2 minigenome construct (719nt). (**a**) A best fit and most frequent structure from the 16K population level runs (−226.2 kcal/mol, run 15). It contains only the 3′ DB structural motif. (**b**) The 5′ DB motif was found in lower fitness (−225.5 kcal/mol, run 55), metastable structures. The color-coding of individual stems in the structures shown is based on their frequencies of occurrence in the ensemble of the final predicted (final results of 100 runs). The 10 bin color scale is shown *in-between* the panels, increasing from left to right in 10 % increments. As the two colors of the 5′ DB stems in (**b**) indicate, the TL1 arm of this DB, referred to as the 5′ DB head motif (5′ DBH) was predicted to be a part of more final structures (26 %, *yellow*) than the full 5′ DB (6 %, *red*) in the 16K population level runs

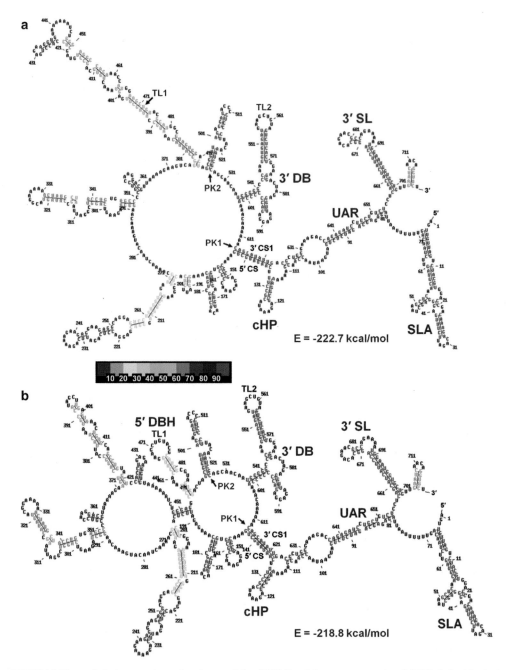

Fig. 7 MPGAfold-predicted secondary structures of the DENV2 minigenome construct (719nt) that include the long-distance 5′–3′ UAR interactions predicted in 16*K* population level runs. (**a**) The best fit structure containing the long-distance UAR interactions (−222.7 kcal/mol, run 14) contains only the 3′ full DB. (**b**) The best-fit structure combining the UAR and the elements of the 5′ DB (head motif, 5′ DBH) has a lower fitness (−218.86 kcal/mol, run 68). The color-coding of individual stems in the structures shown is based on their frequencies of occurrence in the ensemble of the final predicted (final results of 100 runs). The 10 bin color scale is shown *in-between* the panels

3.3.4 Plotting the Evolution of an RNA Secondary Structure Generated in a Single MPGAfold Run

This helps to gain more insight into the folding states (rough pathway) (Fig. 5). The early structure entries (low *x*-values) correspond to RNA secondary structures in the early stages of development. As one examines the RNA structures captured along the *X*-axis, they become more mature with more stems and decreasing free energies. Middle of the run results begin to display longer persistence of specific population histogram peak structures, measured in the number of generations. Usually in the final phase of an MPGAfold run StemTrace plots depict one persistent set of stems, indicating that the population of solutions has converged to one stable structure.

3.4 MPGAfold Parameter Settings and Simulation Run Protocols

MPGAfold requires the Local Area Multicomputer/Message Passing Interface (LAM/MPI-2) environment. It can be run as a Unix shell process or a batch job handled by a queuing system. In either case the key command invoking the program is the same. An example of a parallel run on 8 CPUs would look as follows:

prompt> mpirun –np 8 /path/GA.1.2.2/x86_64/GA.dir/bin/ga_mpi –f ga_param_file.com

The sample path may be installation-specific and in the example shown here and in the selected fragments of the parameters file ("ga_param_file.com"), shown below, is based on our setup.

The MPGAfold distribution package, which is available from our laboratory upon request (see www.ccrnp.ncifcrf.gov/users/bshapiro/software.html; contact shapirbr@mail.nih.gov), provides sample shell scripts that make the above call and perform some housekeeping operations. A sample parameters file is also provided. MPGAfold recognizes nearly 100 parameters, most of which have either been used to tune its performance and should be changed very carefully, or specify less often used options. A subset of the parameters is listed below in the form of an input file, with the hash marks indicating comments (explanations and examples).

```
# SELECTED INPUT PARAMETERS most often modified by the user -----------
#
#  PROCESSORS and GA POPULATION (Note: PhysicalPEs must not exceed the
#  number of processors (NUM) specified in the call: mpirun -np NUM …
PhysicalPEs = 8
totalVirtualPEs = 16384
#
#  NUMBER of runs, offset for output file names, and limit of generations per run
startRun = 1
stopRun = 100
outputStart =1
numGen = 1000
#
#  INPUT and OUTPUT file specifications.
sequenceFile = /home/username/DEN719/dengue-719.seq
OutputPath = /home/username/DEN719/gaout. dir/
FilePrefix = DEN719-16K
#  Solution output type (ex: output population histogram peak structures)
```

```
dumpSolFlag = true
dumpSolType = peak
#
#  ENERGY rule set to be used (binary format tables provided with MPGAfold,
#  including the H-type pseudoknot energies and stem-stacking energies - efn2)
energyRule = 4
energyTableFile = /bin/GA.1.2.2/x86_64/GA.dir/ene/rule_4/energyRule_4.ene
pkenergyTableFile = /bin/GA.1.2.2/x86_64/GA.dir/ene/rule_pk/pkenergy.ene
#  Rule set 4 option
efn2Flag = true
efn2energyTableFile = /bin/GA.1.2.2/x86_64/GA.dir/ene/rule_efn2/energyRule_efn2.ene
#
# SELECTED ALGORITHM PARAMETERS --------------------------------
#
#  RNGseed = 0 - nondeterministic runs; other seed number - repeatable runs (keeping
#  all other parameters unchanged)
RNGseed = 0
#  Algorithm termination criterion (Z score) and window size (interval of generations)
#  over which the score is calculated.
convergenceCriterion = 0.0001
windowWidth = 50
#
#  CORRELATED STEMS parameters - expand the stem pool sampling process so that
#  motifs consisting of stems separated by the loops specified below are selected as a unit
motifFlag = true
loop1by1 = true
loop1by2and2by1 = true
loop2by2 = true
#  Bulge motif Flag & Max bulge size in a motif
loopbulge = true
bulgemotifsize = 2
#
#  STEM FORCING or BIASING - allows for lists of stems to be included
#    unconditionally (forced) or conditionally (inherited, if yielding best fit
child)
# forceStemFile = /home/username/DEN719/UAR.stems
# stickyStemFile = /home/username/DENV719/DB5.sstem
#
#  TRACING - track presence of user-specified stems in population
#  (can also be visualized in the interactive runs).
# traceFile = /home/username/project/stems-to-trace-file.trace
# dumpTraceHits = true
#
#  FOLDING with SEQUENCE ELONGATION parameters
#  (ex: starting with a 30 nt-long fragment, add 2 nts at every GA generation)
sequentialFoldFlag = false
# startSeqMax = 30
# ntsPerExtend = 2
# gensPerNuc = 1
#
#  CROSSOVER, MUTATION and POPULATION SEEDING
#  (parameters for probabilistic structure conflict resolution by peelback)
peelToFitProb = 0.70
doCrossPeelToFit = true
doMutPeelToFit = true
doNumfillPeelToFIt = false
#
#  DISPLAY parameters for interactive runs (to be used with GUI front end)
displayFlag = false
pkDisplayFlag = false
```

MPGAfold outputs a list of parameters specified by the user and defaulted to by the program into a file name based on the FilePrefix parameter value and with extension "params," which documents a run. A typical set of output files is enumerated below for a batch job generating a set of data from 100 individual MPGAfold runs. Other output may depend on the type and amount of data requested by the user via specific parameters, such as stem tracing, based on parameters traceFile and dumpTraceHits (i.e., keeping track of requested stems in the population throughout the run).

1. DEN719-16K8P-100sol.reg—output file with (concatenated) region tables of the final structures (histogram peak in this case). This kind of file was used to plot the solution space shown in Fig. 4.

2. DEN719-16K8P-100sol.ene—output file listing the free energies of the structures from the DEN719-16K8P-100sol.reg file.

3. DEN719-16K8P-99.reg—one of the 100 output files (corresponding to the 99th run in this example), each containing region tables representing MPGAfold output at every generation on one run (*see* Fig. 5).

4. DEN719-16K.ene.hist_99—one of the 100 output files listing the population histogram at the end of every run; numbers of structures at all free energy levels observed in the population at the end of each run. This information may be helpful in deciding whether to run the algorithm at a higher population level, which in general tends to increase fitness of the dominant structures.

5. DEN719-16K8P-best-100sol.reg/ene—two files (region tables and energy data) analogous to the output files described in points 1 and 2, but with output reflecting the best fit structures encountered in the genetic algorithm populations that might not have reached the dominant status by the end of a run. Similarly useful as the population histogram information.

6. DEN719-16K.min/max/avg_99—one of the 100 output files listing minimum/maximum/average free energy in each generation of each run. This is auxiliary information useful in monitoring the energy landscape explored in the folds of a sequence.

7. DEN719-16K.convg_99—file tracking convergence of the population to a dominant solution at every generation of a run (99th run in this example). Output lists percentage of structures with the dominant free energy value and the Z-score (ratio of standard deviation to average population energy) within the convergence window (specified by the user as the windowWidth parameter).

8. DEN719-16K.efn2sol_99—an output file with stem stacking information for the final structures in runs utilizing efn2 calculations. The same information can be alternatively obtained from the secondary structure drawing and energy calculating applications within the StructureLab package.

To see a brief description of all MPGAfold parameters, one can issue the following help command:

prompt> mpirun –np 8 /path/GA.1.2.2/x86_64/GA.dir/bin/ ga_mpi -h

StemTrace was used to explore the generated solution spaces, as shown in Figs. 4 and 5, draw sample secondary structures presented in Figs. 6 and 7, and collect statistics on the motifs of interest present in final structures. Simple Unix shell scripts were used to collect the transient structure statistics from the individual runs (considering the observed minor variants, such as the CS long distance interactions 10 or 11 base pairs in length).

3.5 MPGAfold-Predicted Secondary Structures of the Dengue Minigenome

MPGAfold was used to predict the secondary structure of the 719 nt subgenomic RNA from the DENV2 (minigenome). We employed energy calculations with and without the efn2 coaxial stem-stacking energy calculations (parameter energyRule = 4) [59]. The results illustrated in the figures included in this chapter are based on the efn2 results, and the differences in the prediction runs performed without the efn2 calculations are briefly discussed below. Of particular interest was the core region of the 3′UTR of the dengue virus, with the two dumbbell structures (5′ DB and 3′ DB), and the impact of experimentally tested mutations on dumbbell formation. Given MPGAfold's ability to identify energetically suboptimal metastable conformations, the analysis of the results included results of runs at multiple population levels and intermediate conformations in individual runs (see Fig. 5). The results presented here are derived from 100 independent MPGAfold runs at a 16,384 population level. For more information refer to our 2011 publication (and its supplemental information) in the Journal of Biological Chemistry [29].

3.6 Results of MPGAfold Runs with efn2 Calculations

MPGAfold was run with 100 repeats for each sequence/MPGAfold population combination, in order not to miss any potentially low frequency motifs. Theoretically, structural motifs that may occur in 5–10 % of the runs could be found only once or be missed completely in the solution spaces with only 20 repeats, for example. Also, the secondary structures predicted by MPGAfold for the dengue minigenome construct exhibit a fairly flat energy landscape, and the higher number of runs could help in a more thorough exploration of the potential conformations.

The best-fit final state structures in the range of −225.6 to −226.2 kcal/mol (and up to −227.1 kcal/mol in the 32K population runs, results not shown) contain the full 3′ DB, and the CS long-distance interaction. However, the 5′ DB is replaced in them by alternative long-distance interactions, and the long distance and UAR interactions are missing (Fig. 6a). The best fit structure including the full 5′ DB motif together with the 3′ DB reached the free energy level of –225.5 kcal/mol (Fig. 6b). Finally, structures combining the 3′ DB, CS and UAR reach the best fitness of –222.7 kcal/mol (Fig. 7a, run 14 in Fig. 4), while a structure adding the head motif of the 5′ DB to the above reaches a fitness of –218.8 kcal/mol (run 68 in Fig. 4). Structures combining the full 5′ DB and UAR were found only as intermediates in individual MPGAfold runs, the best-fit of them reaching –216.3 kcal/mol (run 61 in Fig. 4). Among the final structures predicted in 100 MPGAfold runs at a $16K$ population level, the 3′ DB is found in 97 % of runs. The frequency of occurrence of the 5′ DB among the final structures is, on the other hand, only 6 %, while the 5′ DBH submotif is found in 26 % of the solutions. However, searching all the individual $16K$ population level runs for additional transient structures containing the 5′ DB brings the frequency to 68 % of the runs and 94 % for the 5′ DBH. These numbers increase to 73 and 99 % for the 5′ DB and 5′ DBH, respectively, in the $32K$ population level runs confirming a very strong, if transient, presence of the 5′ DB and its sub-motifs. It should also be considered that the best fit secondary structure with both full DBs is less than 1 % less fit than the best fit structures without the 5′ DB. Because both dumbbells can potentially form pseudoknot interactions between their TL and downstream PK sequences, these additional stabilizing interactions could change the energy landscape and allow the 5′ DB to "survive" until the end of the MPGAfold runs. At this stage we cannot calculate these pseudoknot energy contributions, because the TL1/PK2 and TL2/PK1 interactions are not of the H-type.

Full long distance UAR motifs (three individual stems) were predicted in 8 out of the 100 MPGAfold final solutions at a $16K$ population level (Fig. 4). Because the structures containing the UAR motif are more suboptimal than structures with the other monitored motifs, the UAR disappeared from the final solutions of the $32K$ population level runs, but it remained present in the intermediate structures at a relatively steady rate (35 % in $16K$ runs and 33 % in $32K$ runs).

3.7 Results of MPGAfold Co-transcriptional Runs Without the efn2 Calculations

In our earlier studies of the Hepatitis delta virus (HDV) we showed potential benefits of running MPGAfold in the co-transcriptional folding mode (folding with elongation) and without the efn2 calculations [65, 66]. Therefore we also tested these options with the DENV2 minigenome.

The minigenome construct was run 100 times without the efn2 component of the free energy calculations, both in full length folds and with sequence elongation (with 2nt added to the sequence per one GA generation).

Full length fold final structures contained a much higher percentage (54 %) of the full UAR, mostly because this motif was predicted to be part of the energetically best-fit structures. However, the frequency of the full 5′ DB (3 %) and 5′ DBH (11 %) went down among the final structures and the transient structures.

Despite the different fitness (free energy) rankings of the conformer types (structures with different combinations of the monitored motifs), the results of the co-transcriptional folding runs without the efn2 calculation at the $16K$ population level were very similar to the full length folding runs with efn2 calculations. The frequency of the full 5′ DBs was slightly increased (14 %), but the long-distance CYC motif frequency decreased (60 %) most probably due to the algorithmically delayed selection of the stems containing the 3′ CS1 sequence from the stem pool (as explained earlier in the "Other MPGAfold capabilities" section). Full UAR motifs were predicted with a comparable frequency in the final solutions (10 %). Overall, despite the changed free energy landscape, the difference in the frequencies of the key motifs was not altered significantly, and the unequal frequency of the occurrence of the difference in the 5′ and 3′ dumbbells was indicated by both folding options.

4 Summary

Our MPGAfold was used to predict the secondary structure of the 719 nt-long subgenomic RNA from the DENV2. The results agree with the experimental data with respect to the 5′ UTR and 3′ UTR structures SLA, SLB, 3′ SL and the cyclization long distance motif (CS/CS1). The other known long-distance interactions forming the UAR motif (and DAR) are predicted with a lower frequency in the lower fit (higher free energy) conformations, but are a constant feature in the intermediate structures in all the solution spaces. The analysis of MPGAfold results focused on the structures forming within the core region of the 3′ UTR. The results indicate that the minigenome RNA can fold into two dumbbell structures (5′ and 3′ DBs), but with different frequencies of occurrence for each dumbbell. Among the final structures predicted in multiple MPGAfold runs the 3′ DB is found in nearly all the runs at $16K$ (and $32K$) population levels. In contrast, the frequency of occurrence of the 5′ DB and the 5′ DBH submotif is dramatically lower among the final structures. However, among the transient structures the 5′ DB, and even more so the 5′ DBH, are present in the majority of runs. The predicted variable frequency of occurrence of the two dumbbell structures offers a potential explanation of the fact that the TL1 and TL2 sequence motifs in the heads of the two dumbbells are not functionally identical despite having identical sequences within the two DBs. One has to keep in mind, however, that these structures have the propensity to form two potential pseudoknots between identical terminal loop sequences TL1 and TL2 and their complementary pseudoknot motifs, PK2 and PK1. It may be that the 5′ DB is stabilized by the TL1/PK2 pseudoknot interaction, the energy of which we cannot evaluate now in MPGAfold because it is not an H-type pseudoknot, and the additionally stabilized 5′ DB is an integral structural feature of the dengue virus structure, which may open under certain conditions. On the other hand, due to an overlap of the PK1 sequence with the strong cyclization motif (CYC), the other potential pseudoknot TL2/PK1 may not form, leaving the more stable 3′ DB with the TL2 sequence free to interact. It is also conceivable that one or both of the dumbbell structures act as function-dependent structural switches. This idea would also be consistent with the experimental results indicating that the *cis-acting* RNA elements in the core region of DENV2 RNA, including the two DBs, are required for both RNA replication, as well as optimal translation. Recent SHAPE-based results [71] provide specific indications of the base pair interactions present in the minigenome, verifying the existence of the dumbbells and shedding further light on structural implications involved in the 3′ UTR.

Acknowledgements

This publication has been funded in part with Federal funds from the Frederick National Laboratory for Cancer Research, National Institutes of Health, under Contract No. HHSN261200800001E. This research was supported in part by the Intramural Research Program of the NIH, National Cancer Institute, Center for Cancer Research. The content of this publication does not necessarily reflect the views or policies of the Department of Health and Human Services, nor does mention of trade names, commercial products, or organizations imply endorsement by the US Government.

References

1. Westaway EG, Mackenzie JM, Khromykh AA (2003) Kunjin RNA replication and applications of Kunjin replicons. Adv Virus Res 59:99–140

2. Gubler DJ (1998) Dengue and dengue hemorrhagic fever. Clin Microbiol Rev 11: 480–496

3. Halstead SB, Lan NT, Myint TT, Shwe TN, Nisalak A, Kalyanarooj S, Nimmannitya S, Soegijanto S, Vaughn DW, Endy TP (2002) Dengue hemorrhagic fever in infants: research opportunities ignored. Emerg Infect Dis 8: 1474–1479

4. Monath TP (1994) Dengue: the risk to developed and developing countries. Proc Natl Acad Sci U S A 91:2395–2400

5. Gould EA, de Lamballerie X, Zanotto PM, Holmes EC (2001) Evolution, epidemiology, and dispersal of flaviviruses revealed by molecular phylogenies. Adv Virus Res 57:71–103

6. Heinz FX, Allison SL (2000) Structures and mechanisms in flavivirus fusion. Adv Virus Res 55:231–269

7. Irie K, Mohan PM, Sasaguri Y, Putnak R, Padmanabhan R (1989) Sequence analysis of cloned dengue virus type 2 genome (New Guinea-C strain). Gene 75:197–211

8. Lindenbach BD, Rice CM (2003) Molecular biology of flaviviruses. Adv Virus Res 59: 23–61

9. Alvarez DE, De Lella Ezcurra AL, Fucito S, Gamarnik AV (2005) Role of RNA structures

present at the 3'UTR of dengue virus on translation, RNA synthesis, and viral replication. Virology 339:200–212

10. Alvarez DE, Filomatori CV, Gamarnik AV (2008) Functional analysis of dengue virus cyclization sequences located at the 5' and 3'UTRs. Virology 375:223–235

11. Alvarez DE, Lodeiro MF, Luduena SJ, Pietrasanta LI, Gamarnik AV (2005) Long-range RNA-RNA interactions circularize the dengue virus genome. J Virol 79:6631–6643

12. Chiu WW, Kinney RM, Dreher TW (2005) Control of translation by the 5'- and 3'-terminal regions of the dengue virus genome. J Virol 79:8303–8315

13. Filomatori CV, Lodeiro MF, Alvarez DE, Samsa MM, Pietrasanta L, Gamarnik AV (2006) A 5' RNA element promotes dengue virus RNA synthesis on a circular genome. Genes Dev 20:2238–2249

14. Holden KL, Harris E (2004) Enhancement of dengue virus translation: role of the 3' untranslated region and the terminal 3' stem-loop domain. Virology 329:119–133

15. Khromykh AA, Meka H, Guyatt KJ, Westaway EG (2001) Essential role of cyclization sequences in flavivirus RNA replication. J Virol 75:6719–6728

16. Lo MK, Tilgner M, Bernard KA, Shi PY (2003) Functional analysis of mosquito-borne flavivirus conserved sequence elements within 3' untranslated region of West Nile virus by use of a reporting replicon that differentiates between viral translation and RNA replication. J Virol 77:10004–10014

17. Men R, Bray M, Clark D, Chanock RM, Lai CJ (1996) Dengue type 4 virus mutants containing deletions in the 3' noncoding region of the RNA genome: analysis of growth restriction in cell culture and altered viremia pattern and immunogenicity in rhesus monkeys. J Virol 70:3930–3937

18. You S, Padmanabhan R (1999) A novel in vitro replication system for Dengue virus. Initiation of RNA synthesis at the 3'-end of exogenous viral RNA templates requires 5'- and 3'-terminal complementary sequence motifs of the viral RNA. J Biol Chem 274:33714–33722

19. Mukhopadhyay S, Kuhn RJ, Rossmann MG (2005) A structural perspective of the flavivirus life cycle. Nat Rev Microbiol 3:13–22

20. Perera R, Kuhn RJ (2008) Structural proteomics of dengue virus. Curr Opin Microbiol 11:369–377

21. Westaway EG, Mackenzie JM, Khromykh AA (2002) Replication and gene function in Kunjin virus. Curr Top Microbiol Immunol 267:323–351

22. Bartenschlager R, Miller S (2008) Molecular aspects of Dengue virus replication. Future Microbiol 3:155–165

23. Brinton MA, Dispoto JH (1988) Sequence and secondary structure analysis of the 5'-terminal region of flavivirus genome RNA. Virology 162:290–299

24. Cahour A, Pletnev A, Vazielle-Falcoz M, Rosen L, Lai CJ (1995) Growth-restricted dengue virus mutants containing deletions in the 5' noncoding region of the RNA genome. Virology 207:68–76

25. Clyde K, Barrera J, Harris E (2008) The capsid-coding region hairpin element (cHP) is a critical determinant of dengue virus and West Nile virus RNA synthesis. Virology 379:314–323

26. Clyde K, Harris E (2006) RNA secondary structure in the coding region of dengue virus type 2 directs translation start codon selection and is required for viral replication. J Virol 80:2170–2182

27. Friebe P, Harris E (2010) Interplay of RNA elements in the dengue virus 5' and 3' ends required for viral RNA replication. J Virol 84:6103–6118

28. Friebe P, Shi PY, Harris E (2011) The 5' and 3' downstream AUG region elements are required for mosquito-borne flavivirus RNA replication. J Virol 85:1900–1905

29. Manzano M, Reichert ED, Polo S, Falgout B, Kasprzak W, Shapiro BA, Padmanabhan R (2011) Identification of cis-acting elements in the 3'-untranslated region of the dengue virus type 2 RNA that modulate translation and replication. J Biol Chem 286:22521–22534

30. Zeng L, Falgout B, Markoff L (1998) Identification of specific nucleotide sequences within the conserved 3'-SL in the dengue type 2 virus genome required for replication. J Virol 72:7510–7522

31. Brinton MA, Fernandez AV, Dispoto JH (1986) The 3'-nucleotides of flavivirus genomic RNA form a conserved secondary structure. Virology 153:113–121

32. Mohan PM, Padmanabhan R (1991) Detection of stable secondary structure at the 3' terminus of dengue virus type 2 RNA. Gene 108:185–191

33. Elghonemy S, Davis WG, Brinton MA (2005) The majority of the nucleotides in the top loop of the genomic 3' terminal stem loop structure are cis-acting in a West Nile virus infectious clone. Virology 331:238–246

34. Holden KL, Stein DA, Pierson TC, Ahmed AA, Clyde K, Iversen PL, Harris E (2006)

Inhibition of dengue virus translation and RNA synthesis by a morpholino oligomer targeted to the top of the terminal 3' stem-loop structure. Virology 344:439–452

35. Markoff L (2003) 5'- and 3'-noncoding regions in flavivirus RNA. Adv Virus Res 59:177–228

36. Yu L, Markoff L (2005) The topology of bulges in the long stem of the flavivirus 3' stem-loop is a major determinant of RNA replication competence. J Virol 79:2309–2324

37. Hahn CS, Hahn YS, Rice CM, Lee E, Dalgarno L, Strauss EG, Strauss JH (1987) Conserved elements in the 3' untranslated region of flavivirus RNAs and potential cyclization sequences. J Mol Biol 198:33–41

38. You S, Falgout B, Markoff L, Padmanabhan R (2001) In vitro RNA synthesis from exogenous dengue viral RNA templates requires long range interactions between 5'- and 3'-terminal regions that influence RNA structure. J Biol Chem 276:15581–15591

39. Corver J, Lenches E, Smith K, Robison RA, Sando T, Strauss EG, Strauss JH (2003) Fine mapping of a cis-acting sequence element in yellow fever virus RNA that is required for RNA replication and cyclization. J Virol 77:2265–2270

40. Suzuki R, Fayzulin R, Frolov I, Mason PW (2008) Identification of mutated cyclization sequences that permit efficient replication of West Nile virus genomes: use in safer propagation of a novel vaccine candidate. J Virol 82:6942–6951

41. Zhang B, Dong H, Stein DA, Iversen PL, Shi PY (2008) West Nile virus genome cyclization and RNA replication require two pairs of long-distance RNA interactions. Virology 373:1–13

42. Olsthoorn RC, Bol JF (2001) Sequence comparison and secondary structure analysis of the 3' noncoding region of flavivirus genomes reveals multiple pseudoknots. RNA 7:1370–1377

43. Proutski V, Gould EA, Holmes EC (1997) Secondary structure of the 3' untranslated region of flaviviruses: similarities and differences. Nucleic Acids Res 25:1194–1202

44. Kasprzak W, Shapiro B (1999) Stem Trace: an interactive visual tool for comparative RNA structure analysis. Bioinformatics 15:16–31

45. Shapiro BA, Bengali D, Kasprzak W, Wu JC (2001) RNA folding pathway functional intermediates: their prediction and analysis. J Mol Biol 312:27–44

46. Shapiro BA, Kasprzak W (1996) STRUCTURELAB: a heterogeneous bioinformatics system for RNA structure analysis. J Mol Graph 14:194–205

47. Shapiro BA, Kasprzak W, Grunewald C, Aman J (2006) Graphical exploratory data analysis of RNA secondary structure dynamics predicted by the massively parallel genetic algorithm. J Mol Graph Model 25:514–531

48. Shapiro BA, Navetta J (1994) A massively parallel genetic algorithm for RNA secondary structure prediction. J Supercomputing 8:195–207

49. Shapiro BA, Wu JC (1996) An annealing mutation operator in the genetic algorithms for RNA folding. Comput Appl Biosci 12:171–180

50. Shapiro BA, Wu JC (1997) Predicting RNA H-type pseudoknots with the massively parallel genetic algorithm. Comput Appl Biosci 13:459–471

51. Shapiro BA, Wu JC, Bengali D, Potts MJ (2001) The massively parallel genetic algorithm for RNA folding: MIMD implementation and population variation. Bioinformatics 17:137–148

52. Wu JC, Shapiro BA (1999) A Boltzmann filter improves the prediction of RNA folding pathways in a massively parallel genetic algorithm. J Biomol Struct Dyn 17:581–595

53. Edgil D, Polacek C, Harris E (2006) Dengue virus utilizes a novel strategy for translation initiation when cap-dependent translation is inhibited. J Virol 80:2976–2986

54. Ackermann M, Padmanabhan R (2001) De novo synthesis of RNA by the dengue virus RNA-dependent RNA polymerase exhibits temperature dependence at the initiation but not elongation phase. J Biol Chem 276: 39926–39937

55. Nomaguchi M, Ackermann M, Yon C, You S, Padmanabhan R (2003) De novo synthesis of negative-strand RNA by Dengue virus RNA-dependent RNA polymerase in vitro: nucleotide, primer, and template parameters. J Virol 77:8831–8842

56. Leontis NB, Westhof E (2001) Geometric nomenclature and classification of RNA base pairs. RNA 7:499–512

57. Hofacker IL (2003) Vienna RNA secondary structure server. Nucleic Acids Res 31: 3429–3431

58. Zuker M (2003) Mfold web server for nucleic acid folding and hybridization prediction. Nucleic Acids Res 31:3406–3415

59. Mathews DH, Sabina J, Zuker M, Turner DH (1999) Expanded sequence dependence of thermodynamic parameters improves prediction of RNA secondary structure. J Mol Biol 288:911–940

60. Mathews DH, Turner DH (2006) Prediction of RNA secondary structure by free energy minimization. Curr Opin Struct Biol 16:270–278

61. Shapiro BA, Yingling YG, Kasprzak W, Bindewald E (2007) Bridging the gap in RNA structure prediction. Curr Opin Struct Biol 17:157–165

62. Holland JH (1992) Adaptation in natural and artificial systems: An introductory analysis with applications in biology, control, and artificial intelligence. MIT Press, Cambridge, MA

63. Gee AH, Kasprzak W, Shapiro BA (2006) Structural differentiation of the HIV-1 polyA signals. J Biomol Struct Dyn 23:417–428

64. Kasprzak W, Bindewald E, Shapiro BA (2005) Structural polymorphism of the HIV-1 leader region explored by computational methods. Nucleic Acids Res 33:7151–7163

65. Linnstaedt SD, Kasprzak WK, Shapiro BA, Casey JL (2006) The role of a metastable RNA secondary structure in hepatitis delta virus genotype III RNA editing. RNA 12:1521–1533

66. Linnstaedt SD, Kasprzak WK, Shapiro BA, Casey JL (2009) The fraction of RNA that folds into the correct branched secondary structure determines hepatitis delta virus type 3 RNA editing levels. RNA 15:1177–1187

67. Bindewald E, Kluth T, Shapiro BA (2010) CyloFold: secondary structure prediction including pseudoknots. Nucleic Acids Res 38:368–372

68. Hofacker IL, Fontana W, Stadler PF, Bonhoeffer M, Tacker M, Schuster P (1994) Fast folding and comparison of RNA secondary structures. Monat Chem 125:167–188

69. Hofacker IL, Stadler PF (2006) Memory efficient folding algorithms for circular RNA secondary structures. Bioinformatics 22:1172–1176

70. Mathews DH, Disney MD, Childs JL, Schroeder SJ, Zuker M, Turner DH (2004) Incorporating chemical modification constraints into a dynamic programming algorithm for prediction of RNA secondary structure. Proc Natl Acad Sci U S A 101:7287–7292

71. Sztuba-Solinska J, Teramoto T, Rausch JW, Shapiro BA, Padmanabhan R, Le Grice SF (2013) Structural complexity of Dengue virus untranslated regions: cis-acting RNA motifs and pseudoknot interactions modulating functionality of the viral genome. Nucleic Acids Res 41:5075–5089

Chapter 14

Insights into Secondary and Tertiary Interactions of Dengue Virus RNA by SHAPE

Joanna Sztuba-Solinska and Stuart F.J. Le Grice

Abstract

Dengue virus (DENV) is a single-stranded positive-sense RNA virus belonging to the *Flaviviridae* family. The DENV RNA genome contains multiple *cis*-acting elements that continue to unravel their essential role in managing viral molecular processes. Attempts have been made to predict the secondary structure of DENV RNA using a variety of computational tools. Nevertheless, a greater degree of accuracy is achieved when these methods are complemented with structure probing experimentation. This chapter outlines detailed methodology for the structural study of DENV subgenomic minigenome RNA by applying high-throughput selective 2′-hydroxyl acylation analyzed by primer extension (SHAPE). High-throughput SHAPE combines a novel chemical probing technology with reverse transcription, capillary electrophoresis, and secondary structure prediction software to rapidly and reproducibly determine the structure of RNAs from several hundred to several thousand nucleotides at single-nucleotide resolution. This methodology investigates local structure for all positions in a sequence-independent manner and as such it is particularly useful in predicting RNA secondary and tertiary interactions.

Key words Dengue virus, RNA–RNA interactions, *cis*-Acting RNA motifs, SHAPE

1 Introduction

Dengue virus (DENV) is a single-stranded, positive (+)-sense RNA virus belonging to the *Flaviviridae* family together with other important human pathogens [1]. The DENV genome encodes a single open reading frame, flanked by highly structured untranslated regions (UTRs). DENV UTRs have been documented to contain conserved secondary structures involved in regulation of viral molecular processes by modulating the conformation of genomic RNA [2–4]. The intra-genomic long-distance interactions between the 3′-localized motifs and their complementary elements positioned at the 5′ terminus result in long-range RNA–RNA-mediated circularization of the viral genome [5, 6]. Hybridization of the 5′–3′ termini has been suggested to involve extensive rearrangements of both UTRs, resulting in formation of

Radhakrishnan Padmanabhan and Subhash G. Vasudevan (eds.), *Dengue: Methods and Protocols*, Methods in Molecular Biology, vol. 1138, DOI 10.1007/978-1-4939-0348-1_14, © Springer Science+Business Media, LLC 2014

a panhandle structure, a prerequisite to launching (–) strand RNA synthesis [3].

As the secondary structure of DENV genomic RNA continues to unravel its compelling role in supporting viral molecular processes, here we describe a detailed methodology for studying *cis*-acting RNA motifs in the context of subgenomic minigenome RNA (DENV-MINI). A DENV-MINI construct containing 226 nucleotides (nt) from the 5′-end and 493 nt from the 3′-end of DENV2 RNA [7] has been shown previously to harbor several essential *cis*-acting elements for efficient translation [8] as well as (–) strand RNA synthesis by RNA-dependent RNA polymerase in vitro [9, 10]. As such, it represents a valuable construct for studying the long-range interactions of viral genome that might occur in host cells. DENV-MINI RNA has been also studied using Massively Parallel Genetic Algorithm (MPGAfold) and the computational workbench called StructureLab, to predict RNA folding free energies and frequencies of occurrence of RNA folding intermediates ([3]; *see* also Chapter 13).

We applied high-throughput selective 2′-hydroxyl acylation analyzed by primer extension (SHAPE) approach to gain insight into DENV-MINI RNA conformation (Fig. 1) [11–13]. This methodology employs an electrophilic reagent that reacts selectively with the 2′-hydroxyl ribose group of single-stranded nucleotides to create a covalent 2′-*O*-ribose adduct, while the 2′-hydroxyl group of structurally constrained residues shows reduced nucleophilic reactivity [12]. In contrast to conventional chemical and enzymatic probing techniques, SHAPE is insensitive to nucleotide identity, and as such interrogates the global RNA fold, providing structural information at every nucleotide position. Sites of RNA modification are detected by reverse transcriptase-mediated primer extension employing fluorescently labeled primers, which result in a pool of cDNA fragments. A control reaction that omits the electrophilic reagent is performed in parallel to establish positions of natural reverse transcriptase pausing or nonspecific RNA degradation. Also, two dideoxy sequencing reactions are performed in parallel to assign the nucleotide reactivity to the RNA sequence. Subsequently, products of four reactions, the (+) and (–) cDNA reactions plus two dideoxy sequencing reactions, are combined and fractionated via automated capillary electrophoresis (CE) [14]. Data generated in the CE genetic analyzer is presented in the form of an electropherogram, which contains four overlapping traces corresponding to the combined reactions, and where each trace comprises peaks corresponding to individual DNA products. Electropherogram data is subsequently imported into ShapeFinder software, which performs a whole trace Gaussian integration to quantify the intensity of every peak in the (+) and (–) reaction trace and align these with the primary RNA sequence [15]. The "reactivity profile" for the analyzed fragment of RNA is

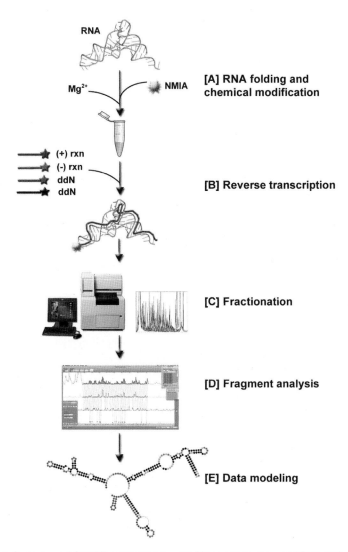

Fig. 1 Overview of SHAPE methodology. (**a**) Ribose 2′ hydroxyl of RNA at flexible or unpaired nucleotides are selectively modified by NMIA. (**b**) Positions of adduct formation are detected by reverse transcriptase-mediated primer extension. (**c**) Fluorescently labeled primer extension products are resolved by automated capillary electrophoresis. (**d**) Electropherograms are computationally deconvoluted to obtain normalized NMIA reactivities, used as pseudo-energy constraints for secondary structure prediction (**e**)

obtained by subtracting the (−) control values from the (+) values associated with each RNA nucleotide and normalizing the data. These reactivity values are then applied as pseudo-free energy constraints to the structure prediction algorithm of RNAstructure (v5.3) software, which generates the possible 2D models of the studied RNA [16, 17].

The combined SHAPE experimental data with conventional thermodynamic parameters facilitated generation of DENV-MINI

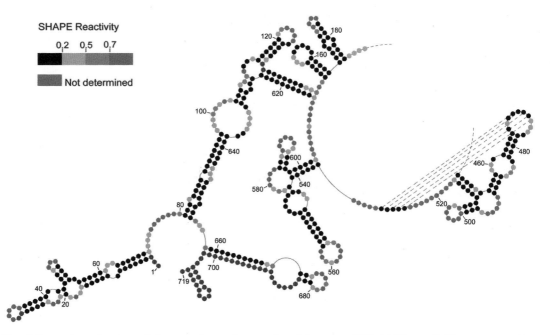

Fig. 2 Secondary structure of dengue virus subgenomic minigenome RNA. NMIA reactivities were used as pseudo-free energy constraints in a minimal free energy folding algorithm (RNAstructure). Nucleotide reactivities are color-coded as shown in the key and *numbered with a dash* every 20 nt

RNA 2D model with a higher degree of accuracy than is possible with either method alone. SHAPE verified the existence of previously proposed *cis*-acting RNA motifs, identified their potential binding partners, and supported the existence of a protein-independent communication between 5′ and 3′ terminal regions in DENV-MINI RNA (Fig. 2). We gained insight into putative long-range tertiary interactions existing within DENV-MINI RNA motifs in the 3′ terminal region leading to the formation of a pseudoknot, which could be further addressed using an antisense-interfered modification of SHAPE (aiSHAPE) [18]. In this method, hybridization of an antisense oligonucleotide, e.g., LNA/DNA hybrid or 2′-*O*-methyl-containing RNA oligonucleotides, to one partner of the proposed RNA duplex would displace its complementary strand (Fig. 3). As a result of disrupted long-distance interactions, an increased NMIA sensitivity of the displaced counterpart could be observed. Overall, high-throughput SHAPE methodology supports further understanding of viral molecular interactions within regulatory regions and as a result will endorse design of live attenuated vaccines and/or antiviral therapeutics.

2   Materials

Take precautions working with RNA material to prevent nonspecific degradation. Use molecular biology grade chemicals and reagents and Milli-Q water for preparation of all solutions.

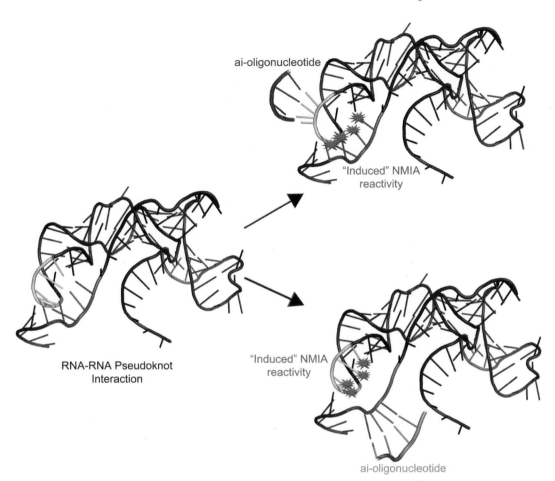

Fig. 3 Examining dengue virus RNA tertiary interactions by aiSHAPE. Hybridization of an ai-oligonucleotide to one partner of the proposed RNA duplex increases acylation sensitivity of its base-paired counterpart

1. Plasmid clone: plasmid encoding the dengue virus type 2 subgenomic minigenome (DENV-MINI) is provided from the Padmanabhan Laboratory [7].

2. PCR: for amplification of the DENV-MINI template, use ~50 ng of purified plasmid, 10 μM oligonucleotide primers 1 and 2 (Table 1), 10× High Fidelity PCR Buffer for Platinum *Taq*® High Fidelity polymerase (Invitrogen), and 10 mM dNTP Mix, PCR grade (Invitrogen). Purify the PCR product using PureLink™ Quick PCR Purification Kit (Life Technologies™) according to the manufacturer's protocol.

3. In vitro transcription: for the synthesis of DENV-MINI RNA, use 0.5 μg of purified PCR template with the MEGAscript Kit (Ambion) and follow the manufacturer's protocol. Purify the RNA using MEGAclear™ Kit (Ambion).

4. RNA folding: prepare 10× RNA renaturation buffer (100 mM Tris–HCl pH 8.0, 1 M KCl, 1 mM EDTA (Ambion)) and 5×

Table 1
Oligonucleotide sequences used for the study of DENV-MINI RNA structure

Name	Sequence	Function
Primer 1	5′-<u>GCTAATACGACTCACTATAG</u>AG TTGTTAGTCTACG-3′	T7 promoter (underlined) containing primer used for amplification of DENV-MINI DNA template
Primer 2	5′-AGAACCTGTTGATTCAACAGCA CCATTC-3′	Primer used for amplification of DENV-MINI DNA template
Primer 3	5′-AGAACCTGTTGATTCAACAGCAC-3′	Reverse transcription 5′-labeled primer that binds to the very 3′ end of the RNA
Primer 4	5′-CAAGCTATGGCATTTGTAATGGC-3′	Reverse transcription 5′-labeled primer that binds DENV-MINI RNA at the nt position 378–400

RNA folding buffer (200 mM Tris–HCl pH 8.0, 25 mM $MgCl_2$, 2.5 mM EDTA, 650 mM KCl (Ambion).

5. Chemical modification of DENV-MINI RNA: prepare fresh 10× *N*-methylisatoic anhydride (NMIA) stock solution (Invitrogen) in dimethyl sulfoxide (DMSO) (Sigma) (*see* **Note 1**). For RNA precipitation, use 100 % ethanol (Sigma), 3 M NaOAc (Ambion), and glycogen (10 mg/ml, Roche).

6. Reverse transcription: use SuperScript™ III Reverse Transcriptase (Invitrogen), 10 mM dNTP Mix, PCR grade (Invitrogen), and two differently 5′-labeled primers, e.g., with Cy5 for (+) RT reaction and Cy5.5 for (−) RT reaction (*see* **Note 2**). 5′ fluorescently labeled primers must be either purchased, e.g., from Integrated DNA Technologies, Coralville, Iowa, or synthesized in the laboratory [19]. Primers 5′-labeled with Cy5, Cy5.5, WellRed D2 (Beckman Coulter), and IRDye800 (Licor)/WellRed D1 (Beckman Coulter) are best suited for the Beckman Coulter 8000 CEQ. Sarstedt PCR tubes (REF 72.735.002) are recommended for this application. For RNA precipitation, use 100 % ethanol (Sigma), 3 M NaOAc (Ambion), and glycogen (10 mg/ml, Roche).

7. Preparation of sequencing ladder: use purified DENV-MINI PCR template. Prepare two sequencing reactions with two different ddNTPs and two differently 5′-labeled primers, e.g., with WellRED D2 or Licor IR800 fluorophores, using USB Cycle Sequencing Kit #78500.

8. Capillary electrophoresis: fractionate the cDNA products dissolved in deionized formamide (SLS, Beckman Coulter, #M608082) on Beckman Coulter CEQ 8000 Genetic Analyzer. Use 96-well sample microtiter plate (Beckman Coulter,

#609801) to load samples, Thermowell® 96-well PCR plate (Costar, #6551) for CEQ™ Separation Buffer (Beckman Coulter), and LPA-1 gel for fractionation (Beckman Coulter).

3 Methods

3.1 PCR for Generating DENV-MINI Template

1. Add the following components to a microcentrifuge tube: 5 µl of 10× High Fidelity PCR Buffer, 1 µl 10 mM dNTP, 2 µl of 50 mM MgSO₄, 0.5 µl of primer 1 and 0.5 µl of primer 2 (Table 1), ~50 ng of purified plasmid, and 0.5 µl of Platinum *Taq*® High Fidelity polymerase. Bring total volume of reaction to 50 µl with H_2O. Vortex well to mix, and then spin briefly in a microfuge.

2. Perform PCR using the following cycle profiles: initial denaturation at 94 °C for 3 min; followed by 30 cycles of denaturation at 94 °C for 1 min, annealing at 62 °C for 1 min, and extension at 72 °C for 1 min; and final extension, incubated at 72 °C for 2 min to produce DENV-MINI DNA template.

3. Verify the integrity of PCR product using standard 1 % agarose gel electrophoresis.

4. Purify the PCR product using PureLink™ Quick PCR Purification Kit according to the manufacturer's protocol.

3.2 In Vitro Transcription

1. Mix the following components for the in vitro transcription reaction: 2 µl of 10× reaction buffer, 2 µl of each dNTP solution, 2 µl of enzyme mix, and 0.5 µg of purified PCR template. Bring total volume of reaction to 20 µl with nuclease-free water. Gently mix the solution by pipetting; then spin briefly to collect the reaction mixture.

2. Incubate the reaction at 37 °C for 4 h.

3. Add 1 µl of TURBO DNase, mix well, and incubate 15 min at 37 °C to remove the template DNA.

4. Verify the integrity of DENV-MINI RNA produced in the in vitro transcription reaction by electrophoresis of a 0.5 µl aliquot on a 1 % agarose gel.

5. Purify the RNA using MEGAclear™ Kit following the manufacturer's protocol.

3.3 RNA Folding

RNA folding can be performed in several ways, e.g., by addition of a ligand or by a change in ionic strength of the buffer. Here, we describe a general method in which the RNA is heated and "snap-cooled" in a low ionic strength buffer to eliminate multiple conformations. RNA folding is commenced by addition of mono and divalent ions to the solution. Alternatively, when RNA is isolated from a biological source, e.g., viruslike particles (VLPs) or cell cultures, the folding step might be omitted to preserve the native RNA structure.

1. Prepare 10 pmol of RNA in 18 μl of water; add 2 μl of 10× renaturation buffer (*see* **Note 3**).

2. Heat the RNA to 85 °C for 1 min and then place on ice for 5 min.

3. Add 100 μl of water and 30 μl of 5× folding buffer to the total volume of 150 μl.

4. Incubate at 37 °C for 15 min.

5. Divide the RNA into two 72 μl aliquots, which will correspond to modified (+) and control (−) reactions.

3.4 Antisense-Interfered SHAPE

Antisense-interfered SHAPE (aiSHAPE) strategy can be applied to study putative *tertiary* RNA–RNA interactions. In this method, displacement of one strand of an RNA duplex by hybridizing an antisense oligonucleotide would disrupt long-distance interactions and would be characterized by enhanced NMIA sensitivity of the displaced nucleotides. Since short (7–10 nt) ai-oligonucleotides are required to prevent major structural distortions in the target RNA, locked nucleic acid (LNA) and 2′-*O*-methyl substitutions are introduced into chimeric oligonucleotides in order to increase the T_m of the resulting duplex. Such modified oligonucleotides have an additional "invasive" property in as much as they can locate and hybridize to their complement in the context of pre-folded RNA, which is an important consideration in many studies.

1. Prepare 10 pmol of RNA in 18 μl of water; add 2 μl of 10× renaturation buffer (*see* **Note 3**).

2. Heat the RNA to 85 °C for 1 min and then place on ice for 5 min.

3. Add 100 μl of water and 30 μl of 5× folding buffer to the total volume of 150 μl.

4. Add the chimeric LNA/DNA oligonucleotides at tenfold excess to the folded RNA keeping the total volume of reaction at 200 μl (*see* **Note 4**).

5. Divide the RNA into two 100 μl aliquots, which will correspond to modified (+) and control (−) reactions.

6. Incubate the sample at 37 °C for 15 min prior to NMIA modification (*see* **Note 5**).

7. To quantify alternations induced by ai-oligonucleotides, raw data are processed as described below.

3.5 RNA Chemical Modification

Chemical modification involves treatment of the RNA under evaluation with an electrophilic SHAPE reagent. Well-characterized, electrophilic SHAPE reagents include isatoic anhydride (IA),

N-methylisatoic anhydride (NMIA), 1-methyl-7-nitro-isatoic anhydride (1M7), and benzoyl cyanide (BzCN). Of these, the most commonly used for high-throughput SHAPE is NMIA. NMIA is easy to handle, its reaction time is relatively short, and it reports RNA structure accurately. The final concentration of the modifying reagent must be optimized to yield single hits over the length of the RNA to be analyzed [20]. Too high NMIA concentration might result in sharp signal decay in the analyzed electropherogram trace, while too low NMIA concentration causes low signal-to-noise ratio in the (+) reaction trace. Use the concentration of reagent that produces a detectable signal while minimizing the difference in signal intensity between long and short cDNA products.

1. Prepare 10× stock of NMIA (20 mM) by adding small amounts of reagent to a 1.5 ml microfuge tube and then adding DMSO (Sigma) (*see* **Note 6**).

2. Add 8 μl of the 10× NMIA to 72 μl aliquot of (+) reaction (final concentration of NMIA is 2 mM) and 8 μl of DMSO to 72 μl aliquot of (−) reaction.

3. Incubate these reactions at 37 °C for 50 min (around five hydrolysis half lives of NMIA).

4. Precipitate the RNA by adding 2.5 volumes of cold 100 % ethanol, 1/10 volumes of 3 M NaOAc, and 0.2 μl of glycogen (10 mg/ml) and incubating at −20 °C for 2 h (*see* **Note 7**).

5. Pellet the RNA by centrifugation at 14,000×g for 30 min.

6. Carefully remove supernatant and air dry the pellet for 1–5 min.

7. Dissolve the RNA in 10 μl of H_2O.

3.6 Reverse Transcription

Sites of RNA modification are detected by reverse transcriptase-mediated primer extension. The SHAPE analysis of DENV-MINI RNA (~720 nt long) requires two sets of 5′-labeled primers: one that binds at the very 3′ end of the DENV-MINI RNA (primer 3) and the other that binds the RNA at nt position 378–400 (primer 4) (Table 1).

1. Prepare (+) and (−) samples for reverse transcription in 0.5 ml microfuge tubes.

2. For the (+) RT reaction, add 6 μl of water and 1 μl of 10 μM Cy5-labeled primer 3 or 4 (depending on the analyzed region of DENV-MINI RNA); for the (−) RT reaction, add 6 μl water and 1 μl Cy5-labeled primer 3 or 4 (10 μM) (*see* **Note 8**).

3. Place tubes in a thermal cycler, and anneal primer to RNA by applying the following program, 85 °C for 1 min, 60 °C 5 min, and 35 °C 5 min, and then raise the temperature to 50 °C.

4. During annealing, prepare 2.5× RT mix by combining 4 μl 5× RT buffer, 1 μl 100 mM DTT, 1.5 μl water, 1 μl 10 mM

dNTPs, and 0.5 µl SuperScript™ III Reverse Transcriptase per RT reaction. Mix well and incubate at 37 °C for 5 min before adding to the annealing reactions.

5. Once the temperature of the annealing reactions reaches 50 °C, add 8 µl of 2.5× RT mix and incubate for 50 min at 50 °C (*see* **Note 9**).

6. Hydrolyze RNA by adding 1 µl 4 M NaOH and incubating at 95 °C for 3 min.

7. Cool the tubes on ice and add 2 µl of 2 M HCl (*see* **Note 10**).

8. Combine (+) and (−) reactions in a single tube and precipitate by adding 10 µl 3 M NaOAc, 0.8 µl glycogen (10 mg/ml), and 300 µl 100 % ethanol.

9. Incubate at −20 °C for 2 h.

10. Pellet the DNA by centrifugation at $14,000 \times g$ for 30 min.

11. Wash pellet twice with 400 µl of cold 70 % ethanol and centrifuge immediately at $14,000 \times g$ for 5 min (*see* **Note 11**).

12. Remove the supernatant and dry the pellet for 3 min by SpeedVac.

13. Dissolve the DNA in 40 µl of deionized formamide.

3.7 Preparation of Sequencing Ladder

Dideoxy sequencing reactions serve as markers to assign the nucleotide reactivity to the RNA sequence. Use PCR-generated DNA template previously applied for in vitro transcription and primers labeled with WellRed D2 or D1/Licor 800. The protocol presented here is scaled up tenfold and while ddA and ddC are used as chain terminators, any terminators may be used to generate the sequencing ladders.

1. Mix 40 µl of the ddA termination mix, 5 pmol of DNA template, 4.6 µL of 10× Sequenase buffer, 10 µl of WellRed D2-labeled primer, and 4.6 µl of Sequenase and H_2O to bring the total volume to 82 µl. Add the Sequenase last.

2. Prepare a second sequencing reaction in the same manner, utilizing ddC and Licor IR800-labeled primer.

3. Proceed to PCR amplification using the following conditions: 25 cycles of 96 °C 20 s, 55 °C 20 s, 72 °C 1 min. Proceed to PCR amplification using conditions recommended in USB Cycle Sequencing Kit (#78500).

4. Combine the ddA and ddC sequencing reactions into one 1.5 ml microfuge tube (164 µl total volume).

5. Precipitate DNA by adding 16 µl 3 M NaOAc (pH 5.2), 16 µl 100 mM EDTA, 1 µl 10 mg/ml glycogen, and 480 µl 95 % ethanol. Mix well, incubate at 4 °C for 2 h, and centrifuge at $14,000 \times g$ for 30 min at 4 °C.

6. Resuspend pelleted DNA in 100 μl of deionized formamide by heating to 65 °C for 10 min, followed by vortexing for at least 30 min.

3.8 Capillary Electrophoresis

High-throughput capillary electrophoresis allows simultaneous fractionation of DNA products from four reactions combined into one sample. Up to eight sets of reactions may be fractionated simultaneously, while as many as 96 samples can be prepared for fractionation over the course of 12 consecutive CE runs on Beckman Coulter CEQ 8000 Genetic Analyzer [21] (*see* **Note 12**).

1. Combine 40 μl of pooled SHAPE reactions with 10 μl of the pooled dideoxy sequencing ladder into one sample and load it to 96-well sample plates.

2. Program and prepare capillary electrophoresis instrument according to the previously published capillary automated footprinting analysis (CAFA) method parameters and initiate run as per manufacturer's instructions [21].

3. Export raw traces into ShapeFinder (*see* Subheading 3.9) [15].

3.9 Data Analysis by ShapeFinder

ShapeFinder is a freely available software that allows the user to visually adjust traces in the output raw electropherogram [15]. Detailed instructions for data handling in ShapeFinder are provided with the software documentation.

1. The raw electropherogram data are corrected for: (1) fluorescent background, (2) spectral overlap between fluorescent traces, (3) mobility shifts to account for the effect of different fluorophores on cDNA mobility, and (4) signal decay resulting from imperfect processivity of reverse transcriptase.

2. The "Setup" function of the "Align and Integrate" tool in ShapeFinder automatically assigns individual peaks to the RNA sequence as defined by user input and the two sequencing ladders.

3. Errors of initial assignments can be corrected manually using the "Modify" function of the same tool.

4. The "Fit" function calculates the areas under the aligned (+) and (−) reaction peaks and charts the reactivity values together with the corresponding nucleotide positions in a tab-delimited text file.

3.10 Data Normalization

To incorporate nucleotide reactivates into the secondary structure prediction algorithm, SHAPE data must be normalized. A consistent approach to normalization is important for comparing data sets and for accurate structure predictions. We normalize SHAPE data to a scale starting at zero (no reactivity). The nucleotide positions with reactivity <0.2, indicative of fully base-paired residues, are defined as the most conformationally constrained, while residues which have a reactivity >0.8 are highly reactive and thus least conformationally constrained (single-stranded).

1. Identify and exclude the highly reactive outliers from subsequent calculations. In general, reactivities greater than 1.5 times the interquartile range are outliers.

2. After eliminating outliers, compute the "effective maximum" reactivity (the mean of the next (highest) 8 % of reactivity values, excluding outliers).

3. Normalize the values by dividing all reactivity values by the "effective maximum."

4. Eliminate values below −0.09, as these are likely the result of RT pausing during cDNA synthesis for reasons other than chemical modification of the template.

5. Save the file with a .shape extension for use with RNAstructure software.

3.11 RNA Modeling

Prediction of experimentally supported RNA secondary structure requires application of RNAstructure software, which uses the pseudo-free energy constraints from SHAPE analysis to enable highly accurate prediction [16]. The software generates graphical representations of the lowest energy 2D RNA structures as well as dot-bracket notation of these structures. The latter can be introduced to RNA structure viewers, e.g., Pseudoviewer [22] or Varna [23], to produce publication-quality images.

1. Use RNA structure, available at: http://rna.urmc.rochester.edu/rnastructure.html.

2. Input SHAPE data as pseudo-free energy constrains.

3. Set the slope and intercept parameters as 2.6 and 0.8, respectively.

4. For a well-supported structure with complete SHAPE data, a small number of similar structures will be obtained.

5. Focus further analysis on the first (lowest free energy) models, which predicted structure agrees the most with experimentally determined SHAPE reactivity values. The good quality structure should contain high reactivity values within single-stranded (flexible) regions, while double-stranded (constrained) regions should have low reactivity. Care must be taken when analyzing the structures, as the software cannot resolve tertiary interaction such as pseudoknot and kissing loops. As a consequence, these structural features must be considered when presenting a final structural model.

4   Notes

1. NMIA concentration is a parameter that should be varied during troubleshooting to obtain one modification for each RNA molecule. For RNA constructs shorter than 100 nt in length, the NMIA solution should be increased (e.g., 100 mM). For RNA constructs longer than 300 nt, a 20–30 mM NMIA solution is preferable. We used 20 mM NMIA stock solution for modification of DENV-MINI RNA (~720 nt).

2. Longer RNAs can be analyzed by high-throughput SHAPE by designing a series of 5′-labeled primers that (1) have annealing sites separated by ~300 nt, (2) are 20–30 nt in length, and (3) have a predicted T_m of >50 °C.

3. This amount of RNA is based on using 5 pmol per (+) and (−) reagent reaction.

4. For the aiSHAPE strategy: locked nucleic acid (LNA)/DNA chimeras can be ordered from Exiqon (http://www.exiqon.com/oligo-tools).

5. Continue the procedure as described in Subheading 3.5, keeping the 2 mM final concentration of NMIA in (+) reaction).

6. Store NMIA and DMSO in a desiccator at room temperature, and prepare stock solutions immediately prior to use in order to minimize exposure to ambient water vapor.

7. In order to minimize co-precipitation of salt with RNA, which can negatively affect peak resolution during fractionation, it is important to minimize refrigeration time and keep the centrifugation parameters as specified in the protocol.

8. From this point on, samples should be protected from light.

9. Incubation of the RT reactions for longer than 50 min may result in aberrant cDNA products.

10. Omission of this step results in poor quality separation of cDNA products.

11. Centrifugation at a higher speed for a longer period results in difficulties in resuspending the pellet(s).

12. The original application of SHAPE utilizes the radioactively 5′-labeled primers for the (+), (−), and two sequencing reactions. Products of these reactions are loaded into adjacent wells in a 5–8 % polyacrylamide slab gel and fractionated by denaturing polyacrylamide gel electrophoresis (PAGE). Quantitative analysis of the gel images produced by conventional SHAPE can be performed using a semiautomated footprinting analysis software (SAFA) described elsewhere [24].

Acknowledgments

This work was funded by the Intramural Research Program of the National Cancer Institute, National Institutes of Health, Department of Health and Human Services.

References

1. Gubler DJ (1998) Dengue and dengue hemorrhagic fever. Clin Microbiol Rev 11(3): 480–496

2. Filomatori CV, Iglesias NG, Villordo SM, Alvarez DE, Gamarnik AV (2011) RNA sequences and structures required for the recruitment and activity of the dengue virus polymerase. J Biol Chem 286(9):6929–6939. doi:10.1074/jbc.M110.162289

3. Manzano M, Reichert ED, Polo S, Falgout B, Kasprzak W, Shapiro BA, Padmanabhan R (2011) Identification of cis-acting elements in the 3′-untranslated region of the dengue virus type 2 RNA that modulate translation and replication. J Biol Chem 286(25):22521–22534. doi:10.1074/jbc.M111.234302

4. Wei Y, Qin C, Jiang T, Li X, Zhao H, Liu Z, Deng Y, Liu R, Chen S, Yu M, Qin E (2009) Translational regulation by the 3′ untranslated region of the dengue type 2 virus genome. Am J Trop Med Hyg 81(5):817–824. doi:10.4269/ajtmh.2009.08-0595

5. Alvarez DE, Lodeiro MF, Luduena SJ, Pietrasanta LI, Gamarnik AV (2005) Long-range RNA–RNA interactions circularize the dengue virus genome. J Virol 79(11):6631–6643. doi:10.1128/JVI.79.11.6631-6643.2005

6. Villordo SM, Gamarnik AV (2009) Genome cyclization as strategy for flavivirus RNA replication. Virus Res 139(2):230–239. doi:10.1016/j.virusres.2008.07.016

7. You S, Padmanabhan R (1999) A novel in vitro replication system for Dengue virus. Initiation of RNA synthesis at the 3′-end of exogenous viral RNA templates requires 5′- and 3′-terminal complementary sequence motifs of the viral RNA. J Biol Chem 274(47):33714–33722

8. Chiu WW, Kinney RM, Dreher TW (2005) Control of translation by the 5′- and 3′-terminal regions of the dengue virus genome. J Virol 79(13):8303–8315.doi:10.1128/JVI.79.13.8303-8315.2005

9. Ackermann M, Padmanabhan R (2001) De novo synthesis of RNA by the dengue virus RNA-dependent RNA polymerase exhibits temperature dependence at the initiation but not elongation phase. J Biol Chem 276(43): 39926–39937. doi:10.1074/jbc.M104248200

10. Nomaguchi M, Ackermann M, Yon C, You S, Padmanabhan R (2003) De novo synthesis of negative-strand RNA by Dengue virus RNA-dependent RNA polymerase in vitro: nucleotide, primer, and template parameters. J Virol 77(16):8831–8842

11. Merino EJ, Wilkinson KA, Coughlan JL, Weeks KM (2005) RNA structure analysis at single nucleotide resolution by selective 2′-hydroxyl acylation and primer extension (SHAPE). J Am Chem Soc 127(12):4223–4231. doi:10.1021/ja043822v

12. Wilkinson KA, Merino EJ, Weeks KM (2006) Selective 2′-hydroxyl acylation analyzed by primer extension (SHAPE): quantitative RNA structure analysis at single nucleotide resolution. Nat Protocols 1(3):1610–1616. doi:10.1038/nprot.2006.249

13. Sztuba-Solinska J, Le Grice SF (2012) Probing retroviral and retrotransposon genome structures: the "SHAPE" of things to come. Mol Biol Int 2012:530754. doi:10.1155/2012/530754

14. Wilkinson KA, Gorelick RJ, Vasa SM, Guex N, Rein A, Mathews DH, Giddings MC, Weeks KM (2008) High-throughput SHAPE analysis reveals structures in HIV-1 genomic RNA strongly conserved across distinct biological states. PLoS Biol 6(4):e96. doi:10.1371/journal.pbio.0060096

15. Vasa SM, Guex N, Wilkinson KA, Weeks KM, Giddings MC (2008) ShapeFinder: a software system for high-throughput quantitative analysis of nucleic acid reactivity information resolved by capillary electrophoresis. RNA 14(10):1979–1990. doi:10.1261/rna.1166808

16. Reuter JS, Mathews DH (2010) RNAstructure: software for RNA secondary structure prediction and analysis. BMC Bioinformatics 11:129. doi:10.1186/1471-2105-11-129

17. Pang PS, Elazar M, Pham EA, Glenn JS (2011) Simplified RNA secondary structure mapping by automation of SHAPE data analysis. Nucleic Acids Res 39(22):e151. doi:10.1093/nar/gkr773

18. Legiewicz M, Badorrek CS, Turner KB, Fabris D, Hamm TE, Rekosh D, Hammarskjold ML, Le Grice SF (2008) Resistance to RevM10 inhibition reflects a conformational switch in the HIV-1 Rev response element. Proc Natl

Acad Sci U S A 105(38):14365–14370. doi:10.1073/pnas.0804461105

19. Legiewicz M, Zolotukhin AS, Pilkington GR, Purzycka KJ, Mitchell M, Uranishi H, Bear J, Pavlakis GN, Le Grice SF, Felber BK (2010) The RNA transport element of the murine musD retrotransposon requires long-range intramolecular interactions for function. J Biol Chem 285(53):42097–42104. doi:10.1074/jbc.M110.182840

20. McGinnis JL, Duncan CD, Weeks KM (2009) High-throughput SHAPE and hydroxyl radical analysis of RNA structure and ribonucleoprotein assembly. Methods Enzymol 468:67–89. doi:10.1016/S0076-6879(09)68004-6

21. Mitra S, Shcherbakova IV, Altman RB, Brenowitz M, Laederach A (2008) High-throughput single-nucleotide structural map-ping by capillary automated footprinting analysis. Nucleic Acids Res 36(11):e63. doi:10.1093/nar/gkn267

22. Byun Y, Han K (2006) PseudoViewer: web application and web service for visualizing RNA pseudoknots and secondary structures. Nucleic Acids Res 34(Web Server issue):W416–W422. doi:10.1093/nar/gkl210

23. Darty K, Denise A, Ponty Y (2009) VARNA: Interactive drawing and editing of the RNA secondary structure. Bioinformatics 25(15):1974–1975. doi:10.1093/bioinformatics/btp250

24. Das R, Laederach A, Pearlman SM, Herschlag D, Altman RB (2005) SAFA: semi-automated footprinting analysis software for high-throughput quantification of nucleic acid footprinting experiments. RNA 11(3):344–354. doi:10.1261/rna.7214405

Chapter 15

Use of Small-Angle X-ray Scattering to Investigate the Structure and Function of Dengue Virus NS3 and NS5

Kyung H. Choi and Marc Morais

Abstract

Small-angle X-ray scattering (SAXS) is a powerful reemerging biophysical technique that can be used to directly analyze many properties related to the size and shape of a macromolecule in solution. For example, the radius of gyration and maximum diameter of a macromolecule can be readily extracted from SAXS data, as can information regarding how well folded a protein is. Similarly, the molecular weight of macromolecular complexes can be directly determined from the complex's scattering profile, providing insight into the oligomeric state and stoichiometry of the assembly. Furthermore, recently developed procedures for ab initio shape determination can provide low-resolution (~20 Å) molecular envelopes of proteins/complexes in their native state. In conjunction with high-resolution structural data, more sophisticated analysis of SAXS data can help address questions regarding conformational change, molecular flexibility, and populations of states within molecular ensembles. Because SAXS samples are easy to prepare and SAXS data is relatively easy to collect, the technique holds great promise for investigating the structure of macromolecules and their assemblies as well as monitoring and modeling their conformational changes. Here we describe typical steps in SAXS sample preparation and data collection and analysis and provide examples of SAXS analysis to investigate the structure and function of dengue virus NS3 and NS5.

Key words Small-angle X-ray scattering, Ab initio shape determination, Radius of gyration, Pair-wise distribution function

1 Introduction

Although small-angle X-ray scattering (SAXS) has only recently begun to garner the attention of nonspecialists, SAXS has a long history. In 1939, Guinier described the measurement of the radius of gyration of particles in solution from their diffuse X-ray scattering [1]. At the time, small-angle techniques provided a direct method for estimating the size of biological macromolecules, yet widespread application of the technique never gained momentum. The subsequent emergence of X-ray crystallography, electron microscopy, and NMR probably overshadowed the fledgling SAXS technique. Despite the considerable successes of these other structural methods, SAXS analysis is undergoing a renaissance.

Radhakrishnan Padmanabhan and Subhash G. Vasudevan (eds.), *Dengue: Methods and Protocols*, Methods in Molecular Biology, vol. 1138, DOI 10.1007/978-1-4939-0348-1_15, © Springer Science+Business Media, LLC 2014

There are several reasons for this reemergence, but perhaps the most important are the growing availability of adequate X-ray sources to collect useful SAXS data and sufficient computational power to fully analyze SAXS data. Another important impetus for the rapid growth of SAXS is the number of recently developed methodologies for extracting structural information.

There are several advantages of SAXS analysis compared to other structural techniques such as X-ray crystallography and NMR. First, protein solutions can be directly analyzed with relatively few restrictions regarding buffer composition, salts, or pH, and thus protein molecules can be investigated in their native state in solution. Second, large protein molecules or virus particles (~500 Å in diameter) that are off-limits to NMR and X-ray crystallography can be studied by SAXS to determine their molecular shape and size. On the other hand, unlike electron microscopy, there is no lower size limit for analyzing structure via SAXS. Hence, SAXS analysis is compatible with virtually the entire range of molecular analyses. Thirdly, relatively small amounts of protein are required for SAXS analysis compared to analytical ultracentrifugation, NMR, or crystallography. Lastly, experimental setup and initial data analysis are fairly straightforward, as long as the protein solution is homogeneous and monodisperse.

In a SAXS experiment, a protein solution is exposed to X-rays, and the diffuse scattering at very small angles from the incident beam is recorded, typically on a 2-D area detector. This region of the scattered radiation contains low-resolution structural information and corresponds to data that would mostly be blocked by the beam stop in an X-ray crystallographic experiment. In contrast to atomic resolution structural techniques like NMR or X-ray crystallography, the high-resolution limit of a typical SAXS experiment is between 10 and 20 Å, and thus SAXS analysis only provides information regarding the global molecular shape of the object. From SAXS data, we can determine many parameters related to protein size such as molecular mass, radius of gyration, hydrated volume, and maximum diameter of the molecule. Additionally, it is possible to construct a low-resolution, 3-D molecular envelope using ab initio shape determination programs. Furthermore, SAXS is useful for evaluating different 3-D models of a protein; SAXS scattering curves can be calculated from a protein model and compared to the experimental scattering curves to validate the model. Similarly, SAXS has been used to evaluate crystal structures of multi-domain proteins that show two or more conformations. In this review, we will focus on the types of structural information that can be extracted from experimental SAXS data as well as the practical considerations that go into conducting a SAXS experiment.

2   Materials

2.1   Instrumentation

1. SAXS data collection requires a monochromatic X-ray source and a compatible detector system. SAXS data are usually collected either at synchrotron radiation sources or using an in-house X-ray generator coupled with a specialized detector/camera for recording small-angle scattering data. Currently, SAXSess (Anton Paar), BioSAXS-1000 (Rigaku), and MICROPix (Bruker AXS) instruments are commercially available for collecting SAXS data on biological samples.

2.2   Protein and Buffer Solutions

1. Prepare protein solutions for SAXS analysis. Protein solutions need to be pure and monodisperse (*see* **Note 1**). A typical protocol involves purifying his-tagged proteins via metal-affinity and size-exclusion chromatography. Several protein concentrations between 1 and 10 mg/mL are initially used for data acquisition to determine the optimal concentration for the final data collection and analysis. Depending on the sample holder of the instrument, between 20 and 80 μL of protein solution is typically needed for each measurement.

2. Prepare buffer solution. Buffer solutions should contain the same exact components in the same concentrations as the protein solution, but without the protein present. Although we are only interested in scattering from our protein, scattering measured from a protein solution is the sum of scattering from the protein and the buffer. Scattering from the protein is calculated by subtraction of buffer scattering from protein solution scattering. It is thus essential that the buffer solution exactly matches the buffer present in the protein solution. It is recommended that the buffer solution is obtained from the protein solution by either buffer exchange or dialysis rather than separately preparing a buffer with the same composition. For example, the protein solution can be concentrated or buffer exchanged using an Amicon Ultra Centrifugal Filter Unit (Millipore) and the filtrate used as a "buffer" solution. Most of buffer conditions are compatible with SAXS data collection, provided there is an adequate difference in electron density between the buffer and the macromolecule of interest.

3. Minimize radiation damage. Radiation damage is a significant source of data deterioration during SAXS data collection. In order to reduce radiation damage during data collection, free radical scavengers such as ~5 % glycerol or 1–5 mM DTT can be added to protein and buffer solutions (*see* **Note 2**).

3 Methods

3.1 Data Collection

1. Plan experiments. SAXS can be measured either at synchrotron sources or using an in-house SAXS instrument. Typical data collection for an individual sample takes 5–10 s at a synchrotron source and 1–4 h using currently available in-house X-ray generators. Temperature-controlled chambers can also be used for long data collection times to preserve sample integrity.

2. Collect scattering data for both protein and buffer solutions. Protein and buffer solutions are placed in a capillary tube and irradiated with the X-ray beam. The same capillary tube should be used for each solution to eliminate scattering differences arising from different capillary tubes. The scattering intensities $I(q)$ are recorded as a function of the scattering vector q ($q = 4\pi \sin\theta / \lambda$, where 2θ is the angle between the incident and scattered radiation, and λ is the radiation wavelength, generally around 0.1 nm). Typical high q values for SAXS measurement are 0.2–0.3 Å^{-1}, which correspond to 30–20 Å resolution (Fig. 1a).

3. Integrate scattering intensities at each q value. Because the particles are randomly distributed (i.e., their positions and orientations are thus uncorrelated), the scattering intensity from a protein solution is continuous and isotropic (Fig. 1a) and is

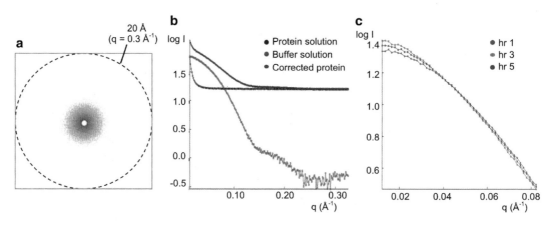

Fig. 1 Small-angle X-ray scattering data collection and processing. (**a**) Typical X-ray scattering pattern of a protein solution. The scattering intensities are radially averaged to yield a 1-D scattering curve of $I(q)$ vs. q. (**b**) 1-D SAXS scattering curves for a protein solution, a buffer solution, and corrected protein. The SAXS data are collected for protein and buffer solutions. The scattering intensity of the buffer solution is then subtracted from that of the protein solution to obtain the corrected scattering intensity for the target protein. This corrected curve is then used for data analysis. (**c**) Example of radiation damage. Protein samples were exposed to X-rays for five consecutive hours and data recorded at 1 h intervals. Only the first, third, and fifth hours are shown for clarity. Increased exposure times led to an increase in log $I(q)$ at the lowest q regions, indicative of protein aggregation due to radiation damage

proportional to the scattering from a single protein molecule averaged over all orientations [2]. Thus, the scattering intensities are radially (circularly) averaged and integrated as one-dimensional (1-D) function, $I(q)$ vs. q, using the SAXS software package provided by either the beamline or SAXS camera manufacturer (Fig. 1b).

4. Calculate protein-only scattering. Scattering intensities are measured from both protein and buffer solutions. Scattering intensities contributed only from protein molecules are obtained by subtracting the scattering intensities of the buffer solution from those of the protein solution at each q value (Fig. 1b). This corrected intensity for protein molecules is then used for further analysis (*see* **Note 3**).

5. Check for radiation damage. Radiation damage can be significant in SAXS data collection, particularly at high flux synchrotron sources, even though scattering data are acquired within 10 s. To determine whether the protein solution suffered from radiation damage, short irradiation (i.e., 0.5 s) of a protein sample can be measured before and after the data collection, and their scattering curves compared. Using a home X-ray source, data collection can take up to several hours. Scattering data can be collected at set time intervals, compared for signs of radiation damage, and then merged with previously collected data (Fig. 1c).

6. Repeat the above data collection procedure at several protein concentrations. Since the scattering intensity is proportional to the amount of scattering objects (electrons for X-ray radiation), higher concentrations of protein solutions will have higher scattering intensities, which increases the signal to noise ratio. However, concentrated solutions may exhibit "interference" between neighboring molecules, and thus SAXS data are collected in several different protein concentrations to determine whether there are concentration-dependent effects on the scattering profile. Protein concentrations that show protein aggregation or interparticle effects are then omitted from subsequent data analysis (*see* **Note 4**).

3.2 Primary Data Analysis

1. Perform initial data analysis. Primary data analysis can be carried out using the program PRIMUS from the ATSAS Suite [3]. All software can be downloaded from http://www.embl-hamburg. de/biosaxs/atsas-online/. The PRIMUS program is used to carry out simple operations on scattering profiles such as buffer subtraction, data averaging, merging, and extrapolation to zero concentrations.

2. Examine the quality of the data first by calculating a Guinier plot, $\ln[I(q)]$ vs. q^2. The scattering curve at low angles near $q = 0$ follows the Guinier approximation

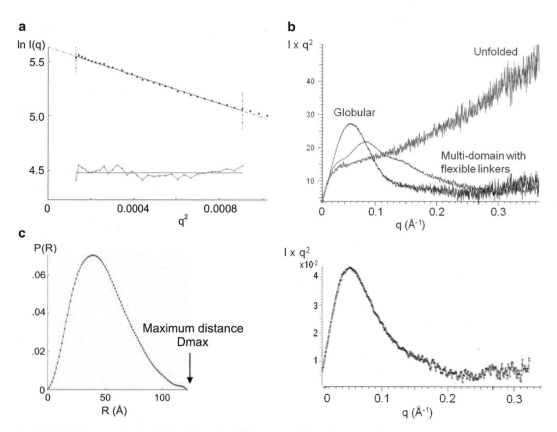

Fig. 2 SAXS data analysis. (**a**) Guinier plot, ln[*I*(*q*)] vs. *q²*. The fit of the Guinier equation is shown in red. The *bottom green line* shows the residual. The radius of gyration *R*$_G$ is calculated from the slope of the Guinier plot. (B) Kratky plot, *I*(*q*) *q²* vs. *q*. *Top*: Examples of a globular protein, an unfolded protein, and a multi-domain protein with flexible linkers are shown (from [4]). *Bottom*: The Kratky plot for this particular protein (dengue NS5) shows a profile typical of a globular protein [5]. (**c**) Pair-wise distance distribution function *P*(*r*). The radius of gyration *R*$_G$ calculated from the *P*(*r*) for the dengue NS5 is 35.6 ± 1 Å, which is in a good agreement with the value of 35.1 ± 1 Å obtained from the Guinier plot [5]. The shape of the *P*(*r*) for the protein has an asymmetric peak with a tailing profile for the higher distances, indicative of a rather elongated shape

$$I(q) \approx I(0)\exp(-q^2 R_G^2 / 3) \text{ or}$$
$$\ln[I(q)] \approx \ln[I(0)] - q^2 R_G^2 / 3 \text{ (for a } qR_G < 1.3)$$

where $I(0)$ is the forward scattering intensity (also called the extrapolated scattering intensity at zero angle) and R_G is the radius of gyration. Radius of gyration is defined as the root mean square average of the distance of the electrons from the center of the particle. Linearity of the Guinier plot near $q=0$ indicates non-aggregated and monodisperse samples (Fig. 2a). For protein samples that obey the Guinier approximation, the radius of gyration is calculated from the slope of the line. If the Guinier plot does not follow a straight line at low angles, the protein is either heterogeneous, aggregated, or there are interparticle repulsions. The data should be discarded for further data analysis unless these behaviors are explicitly accounted for (*see* **Note 4**).

3. Calculate Kratky plot. A Kratky plot, $I(q)\cdot q^2$ vs. q, can provide an indication as to whether the protein of interest is properly folded in solution. As shown in the example in Fig. 2b, folded, partially folded, and unfolded proteins have characteristic Kratky plots [4, 6]. Globular proteins will have a bell-shaped curve. In contrast, extended, semi-stiff polymers such as random coil peptides yield a curve like the unfolded protein shown in Fig. 2b. Partially folded proteins would have a curve that has characteristics of both folded and unfolded proteins. In addition, it is possible to determine whether multiple domains within a protein are arranged as a single unit or whether the domains are structurally independent and connected by a flexible linker. If the two domains are in a fixed arrangement as a single unit, a bell-shaped curve would be obtained. If two domains are connected by a flexible linker, the plot will have a broad multi-peak profile.

3.3 Protein Size and Shape Analysis

1. Determine Porod invariant. The volume and molecular weight of a protein can be estimated from the small-angle X-ray scattering data via the Porod invariant, which can be calculated using the program PRIMUS. The hydrated volume V_P of the particle is computed from the Porod invariant and is used to estimate the molecular mass of a globular protein [3]. The hydrated volume in cubic nanometers (nm^3) is empirically found to be between 1.5 and 2 times the molecular mass in kilodaltons (kDa) [6].

2. Determine forward scattering intensity $I(0)$. The forward scattering intensity $I(0)$ measured at zero angle ($q = 0$) on an absolute scale is equal to the square of the number of electrons in the scattering object and is thus proportional to the molecular mass. Although $I(0)$ cannot be experimentally measured since it is coincident with the direct beam and hence blocked by the beam stop, $I(0)$ can be determined by extrapolation of the scattering curve to $q = 0$. However, this approach still requires that $I(0)$ is on an absolute scale, a condition that is not always practical to meet. In practice, the apparent molecular mass is often determined from the $I(0)$ of a set of standard proteins using the formula

$$MM_p = MM_{st}\{[I(0)_p/c_p]/[I(0)_{st}/c_{st}]\}$$

where MM, $I(0)$, and c are the molecular mass, the forward scattering intensity, and the concentration of the protein of interest (subscript p) or the standard protein (subscript st), respectively [7]. Lysozyme (14.3 kDa), bovine serum albumin (66.2 kDa), and glucose isomerase (172 kDa) are often used as standard proteins.

3. Determine pair-wise distance distribution function, $P(r)$. The pair-wise distance distribution function $P(r)$ describes the probability of finding an electron in the macromolecule separated by distance r from another electron in the particle. In theory, the $P(r)$ can be directly obtained via Fourier transformation of the scattering intensity as a function of q. In practice, the $P(r)$ is calculated from the scattering pattern via indirect Fourier inversion of the scattering intensity $I(q)$, which can be accomplished using the program GNOM [8]. The boundary constraints of $P(r)=0$ at $r=0$ and at the maximum linear dimension, D_{max}, are applied to $P(r)$. This real space representation of the scattering intensity provides information about the particle shape and the maximum dimension D_{max} (Fig. 2c). For example, globular proteins display a bell-shaped curve with maximum at about $D_{max}/2$. Elongated molecules have skewed distributions with a maximum at small distances corresponding to the radius of the cross section and a tailing profile for the longer distances. Proteins consisting of well-separated subunits would have multiple maxima, with the first maximum corresponding to the intra-subunit distances and the others corresponding to the distances between subunits. Radius of gyration R_G can also be determined by $P(r)$ function and should be compared to the R_G determined from the Guinier plot. In some ways, the R_G obtained from the $P(r)$ function may be more reliable than that obtained via the Guinier approximation since all the data is used, whereas only low-resolution data contributes to Guinier analysis.

3.4 Ab Initio Shape Calculations

1. Perform initial shape determination. Although a unique three-dimensional (3-D) structure cannot be retrieved from the one-dimensional (1-D) scattering curve, it is possible to determine approximate molecular shapes, or envelopes, of macromolecules that are consistent with the scattering data. The programs GASBOR and DAMMIN can be used for ab initio molecular shape determination [2, 9]. Both programs use an ensemble of dummy residues (GASBOR) or dummy atoms (DAMMIN) placed in a volume, of which the radius is $D_{max}/2$. Scattering curves from different arrangements of these dummy residues or atoms are calculated and compared to the experimental SAXS data (Fig. 3a). The agreement between a resulting model and the data is determined using the discrepancy χ^2, defined according to Konarev et al. [3]. Since many 3-D shapes can fit a 1-D scattering curve equally well ($\chi^2 < 2.0$), an averaged shape from multiple runs is believed to be a better representative of a protein molecule. Ten to twenty independent calculations are thus performed with or without molecular symmetry imposed (Fig. 3a). The program DAMAVER is used to align the multiple molecular envelopes, select the most

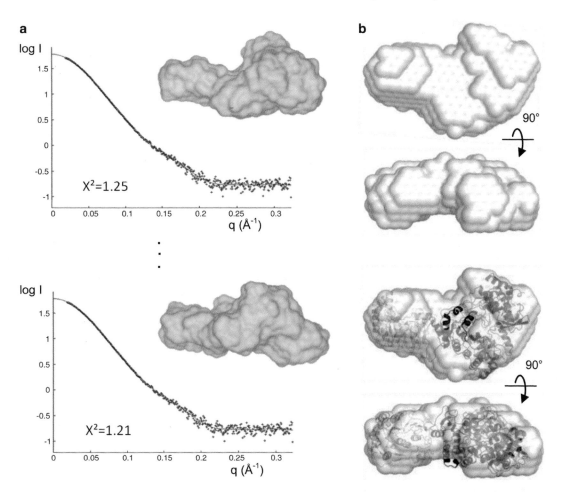

Fig. 3 Applications of SAXS. (**a**) De novo shape determination using GASBOR. Ten individual runs generate ten different models (*black lines*) that fit the experimental scattering curve (*red*) well with X^2 values between 1.2 and 1.4; two of the ten curves and de novo models (*blue surface*) are shown. (**b**) Two views of an averaged molecular envelope obtained from (**a**). In this example, the protein consists of two domains whose independent crystal structures are known. These structures were fitted into the SAXS envelope using SUPCOMB to generate a full-length model of the protein (*bottom*)

typical ones, and build an averaged model [10] (Fig. 3b). If an atomic structure or model structures are available, they can be superimposed to the calculated envelope by programs such as SUPCOMB [11] or Situs [12] to evaluate their agreement (*see* **Note 5**).

3.5 Building Multicomponent Assemblies from Their Partial Structures

1. Model multicomponent assemblies. If individual domain structures are known but not the full-length protein structure, SAXS data can be used to build a full-length protein model from the known subunits [5, 13]. Similarly, multicomponent assemblies can be constructed from the structures of individual components and SAXS data for the entire assembly. Both SASREF and BUNCH programs minimize the discrepancy

between the calculated SAXS curves of the assembled model and the experimental scattering data [14]. The SASREF program uses a simulated annealing protocol to construct an interconnected ensemble of known subunits without steric clashes. The BUNCH program combines a rigid-body modeling for the known subunits and ab initio modeling for the regions of unknown structure and thus can be more useful if the structures of the protein subunits are incomplete (*see* **Note 6**).

3.6 Comparison of Different 3-D Structures with Solution SAXS Data

1. Compare SAXS data with other structural information. SAXS is particularly useful for verification of other structural or modeling data, e.g., whether a crystal structure or a 3-D model likely exists in solution. It is important to realize that it is not necessary to calculate an ab initio molecular envelope in order to compare X-ray structures to SAXS data. It is more appropriate to calculate a scattering curve from a structure for direct comparison with SAXS data. A SAXS profile can be calculated from the 3-D model (PDB coordinates) after adding a hydration shell using the program CRYSOL [15]. The agreement between the calculated SAXS profile of the 3-D model and the experimental SAXS data is determined using the discrepancy χ^2. The lower χ^2 indicates a better fit. Agreement would indicate that the crystallized structure (or model) represents the conformational state of the protein in solution. Disagreement might indicate that crystal packing forces have trapped your protein in a nonbiological conformation.

   This approach has been used to study the flavivirus NS5 and NS3 proteins. Flavivirus NS5 consists of an N-terminal methyltransferase and a C-terminal RNA polymerase domain. Although structures of the two separate domains have been determined, there is no crystal structure for the full-length protein. Hence, various models of the full-length protein have been proposed, which differ in the relative arrangement of the two domains. Comparing scattering curves calculated from the different models to experimental SAXS data indicated which model more likely resembles the structure in solution [5]. Similarly, flavivirus NS3 consists of an N-terminal protease domain and a C-terminal helicase domain. The crystal structures of the full-length NS3 from dengue and Murray Valley encephalitis viruses have been determined in two conformations, which differ in the relative arrangements of the two domains [16, 17]. In order to determine whether NS3 protein exists in either or both conformations in solution, SAXS curves are calculated from the crystal structures and compared to the solution SAXS data [16, 17]. Both structures agree well with the solution SAXS data. Taken together with the crystallographic data, it is concluded that the conformational flexibility of the linker between the two domains exists and both NS3 conformations are likely present in the solution.

3.7 Modeling Multiple Conformational Ensembles

When two or more conformations of the protein are expected to exist in the solution, the relative abundance of each population can be estimated using the program EOM (ensemble optimization method) [18]. The program calculates a pool of potential conformations that are randomly generated ($N > 1,000$). Then subsets of the potential conformations ($N = 50$) are selected to fit the experimental SAXS data. By comparing the profile of the selected conformations to the random conformations, one can also address the protein flexibility.

4 Notes

1. Monodispersity (identical molecules) of protein samples can be addressed by a combination of analytical methods such as dynamic light scattering and/or analytical ultracentrifugation.

2. Radiation damage in data collected at synchrotron sources is frequently observed. We found that the addition of radical scavengers such as 1–5 % glycerol, 1–5 mM DTT, or 1–2 mM TCEP in protein solutions reduces the secondary radiation damage. The use of lower molecular weight salts can help reduce primary radiation since lighter atoms have a smaller cross-sectional scattering area.

3. The scattering intensity difference between protein and buffer solutions is small, especially at high q ranges where intensities are low. Thus, accurate background (buffer) subtraction is essential to obtain accurate measurements of the protein scattering.

4. SAXS data should be collected at a range of protein concentrations (typically 1–10 mg/mL) to determine the optimal concentration for SAXS data analysis. Overlaying the protein samples at different concentrations in the Guinier region should indicate whether there are concentration-dependent effects such as aggregation or interparticle repulsion.

5. SAXS-based ab initio shape calculations assume uniform particle density. Thus, shape calculations could be problematic for a multicomponent system such as protein–nucleic acid and protein–lipid complexes, in which each component has different average electron densities.

6. In some cases, the low-resolution ab initio envelope does not allow accurate positioning of individual subunits. Thus, model building is greatly facilitated if a priori information regarding the structure or assembly is available.

Acknowledgments

This work is supported by NIH grant AI087856 to KHC and GM095516 to MCM. We thank Dr. Mark White and Dr. Cecile Bussetta for providing SAXS examples [19] and for helpful discussions.

References

1. Guinier A (1939) La diffraction des rayons X aux tres petits angles; application a l'etude de phenomenes ultramicroscopiques. Ann Phys (Paris) 12:161–237

2. Svergun DI, Petoukhov MV, Koch MHJ (2001) Determination of domain structure of proteins from X-ray solution scattering. Biophys J 80(6):2946–2953

3. Konarev PV, Volkov VV, Sokolova AV, Koch MHJ, Svergun DI (2003) PRIMUS: a Windows PC-based system for small-angle scattering data analysis. J Appl Crystallogr 36: 1277–1282

4. Putnam CD, Hammel M, Hura GL, Tainer JA (2007) X-ray solution scattering (SAXS) combined with crystallography and computation: defining accurate macromolecular structures, conformations and assemblies in solution. Q Rev Biophys 40(3):191–285

5. Bussetta C, Choi KH (2012) Dengue virus nonstructural protein 5 adopts multiple conformations in solution. Biochemistry 51(30): 5921–5931

6. Mertens HD, Svergun DI (2010) Structural characterization of proteins and complexes using small-angle X-ray solution scattering. J Struct Biol 172(1):128–141

7. Mylonas E, Svergun DI (2007) Accuracy of molecular mass determination of proteins in solution by small-angle X-ray scattering. J Appl Crystallogr 40:S245–S249

8. Svergun DI (1992) Determination of the Regularization Parameter in Indirect-Transform Methods Using Perceptual Criteria. J Appl Crystallogr 25:495–503

9. Svergun DI (1999) Restoring low resolution structure of biological macromolecules from solution scattering using simulated annealing (vol 76, p 2879, 1999). Biophys J 77(5): 2896–2896

10. Volkov VV, Svergun DI (2003) Uniqueness of ab initio shape determination in small-angle scattering. J Appl Crystallogr 36:860–864

11. Kozin MB, Svergun DI (2001) Automated matching of high- and low-resolution structural models. J Appl Crystallogr 34:33–41

12. Wriggers W (2010) Using Situs for the integration of multi-resolution structures. Biophys Rev 2(1):21–27

13. Mastrangelo E, Milani M, Bollati M, Selisko B, Peyrane F, Pandini V, Sorrentino G, Canard B, Konarev PV, Svergun DI, de Lamballerie X, Coutard B, Khromykh AA, Bolognesi M (2007) Crystal structure and activity of Kunjin virus NS3 helicase; protease and helicase domain assembly in the full length NS3 protein. J Mol Biol 372(2):444–455

14. Petoukhov MV, Svergun DI (2005) Global rigid body modeling of macromolecular complexes against small-angle scattering data. Biophys J 89(2):1237–1250

15. Svergun D, Barberato C, Koch MHJ (1995) CRYSOL—a program to evaluate x-ray solution scattering of biological macromolecules from atomic coordinates. J Appl Crystallogr 28:768–773

16. Luo D, Xu T, Hunke C, Gruber G, Vasudevan SG, Lescar J (2008) Crystal structure of the NS3 protease-helicase from dengue virus. J Virol 82(1):173–183

17. Assenberg R, Mastrangelo E, Walter TS, Verma A, Milani M, Owens RJ, Stuart DI, Grimes JM, Mancini EJ (2009) Crystal structure of a novel conformational state of the flavivirus NS3 protein: implications for polyprotein processing and viral replication. J Virol 83(24):12895–12906

18. Bernado P, Mylonas E, Petoukhov MV, Blackledge M, Svergun DI (2007) Structural characterization of flexible proteins using small-angle X-ray scattering. J Am Chem Soc 129(17):5656–5664

19. Zheng J, Gay DC, Demeler B, White MA, Keatinge-Clay AT (2012) Divergence of multimodular polyketide synthases revealed by a didomain structure. Nat Chem Biol 8(7): 615–621

# Chapter 16

Identification of Dengue RNA Binding Proteins Using RNA Chromatography and Quantitative Mass Spectrometry

Alex M. Ward, J. Gunaratne, and Mariano A. Garcia-Blanco

Abstract

A major challenge in dengue virus (DENV) research has been to understand the interaction of the viral RNA with host cell proteins during infection. Until recently, there were no comprehensive studies identifying host RNA binding proteins that interact with DENV RNA (Ward et al. RNA Biol 8 (6):1173–1186, 2011). Here, we describe a method for identifying proteins that associate with DENV RNA using RNA chromatography and quantitative mass spectrometry. The method utilizes a tobramycin RNA aptamer incorporated into an RNA containing the dengue 5′ and 3′ untranslated regions (UTRs) in order to reversibly bind RNA to a tobramycin matrix. The RNA–tobramycin matrix is incubated with SILAC-labeled cell lysates, and bound proteins are eluted using an excess of tobramycin. The eluate is analyzed using quantitative mass spectrometry, which allows direct and quantitative comparison of proteins bound to DENV UTRs and a control RNA–tobramycin matrix. This technique has the advantage of allowing one to distinguish between specific and nonspecific binding proteins based on the ratio of protein preferentially bound to the DENV UTRs versus the control RNA. This methodology can also be used for validation of quantitative mass spectrometry results using conventional Western blotting for specific proteins. Furthermore, though it was specifically developed to identify DENV RNA binding proteins, the RNA chromatography method described here can be applied to a broad range of viral and cellular RNAs for identification of interacting proteins.

Key words RNA affinity chromatography, Dengue virus, RNA binding proteins, Tobramycin, RNA aptamer, Quantitative mass spectrometry, SILAC

1 Introduction

Dengue virus (DENV) is a mosquito-borne positive-strand RNA virus that is endemic in tropical regions and rapidly reemerging in temperate areas [1, 2]. The positive-strand RNA genome is translated into a single polyprotein that is cleaved by a combination of host and viral proteases to give rise to ten viral proteins [3]. In addition to the open reading frame, the viral RNA contains 5′ and 3′ untranslated regions (UTRs) that each form complex secondary structures and also long-range interactions between the two UTRs [4–24]. Several of the secondary structures are absolutely required

Radhakrishnan Padmanabhan and Subhash G. Vasudevan (eds.), *Dengue: Methods and Protocols*, Methods in Molecular Biology, vol. 1138, DOI 10.1007/978-1-4939-0348-1_16, © Springer Science+Business Media, LLC 2014

for DENV replication and/or translation, suggesting that the UTRs are critical assembly sites for both viral and host cell proteins [4–24]. Previous efforts to identify these proteins have relied on molecular mass similarity or mass spectrometry of individual bands from SDS-PAGE gels [15, 21, 25, 26]. We coupled an RNA affinity chromatography technique and quantitative mass spectrometry to obtain an unbiased and comprehensive method for identification of host cell proteins that interact with the 5′ and 3′ UTRs of DENV RNA [4]. The technique relies on the use of a tobramycin RNA aptamer to reversibly attach a bait RNA containing the 5′ and 3′ UTRs from DENV to tobramycin-conjugated sepharose beads. This method was developed to purify spliceosomes [27]. The immobilized RNA is incubated with SILAC-labeled cell lysates and washed, and RNA–protein complexes are eluted with excess tobramycin for quantitative mass spectrometry-based identification. This technique has proven useful not only for identification of associated proteins but also for validation of RNA–protein interactions and binding site mapping.

1.1 RNA Chromatography Using RNA Aptamer Tags

RNA aptamers are sequences selected from a random library of RNAs for their ability to bind to a specific substrate. Some examples of common RNA aptamers used in biochemical purifications include the S1 streptavidin and the J6f1 tobramycin aptamers [27, 28]. The selection process begins with a library of random nucleotide sequences approximately 40 nucleotides in length that also contain known sequences for iterative use in in vitro transcription and reverse transcription (RT) polymerase chain reaction (PCR) amplification. The RNAs are in vitro transcribed and incubated with the substrate of choice, usually a protein or chemical bound to a solid support. Bound RNAs are eluted, amplified, and re-incubated with the substrate for several rounds of selection before undergoing sequencing to identify RNA aptamer sequences. These sequences are used to refine aptamers by minimizing their length or increasing their affinity for the chosen substrate [29]. RNA aptamer sequences are incorporated into an RNA of interest to allow reversible binding of the RNA to a solid substrate for RNA chromatography. RNA aptamers have been successfully used for identification of proteins associated with an RNA of interest as well as for purification of biochemically active complexes such as RNAse P or the spliceosome [4, 27, 28]. Other approaches are available for biochemical purification of RNA binding proteins, including direct covalent conjugation of the RNA to a solid substrate, binding of the RNA via base pairing to an immobilized oligonucleotide, and biotinylation of RNAs followed by purification with anti-biotin antibodies or with streptavidin [30, 31]. Several issues need to be considered when using these alternative approaches, including the difficulty in eluting associated proteins from covalenty linked RNAs and the possible interference of the biotin moiety with protein–RNA interactions. While these issues need not rule out the use of alternative approaches, the ease and adaptability of using an RNA aptamer-based substrate for biochemical purification makes it an attractive method for identification of interacting proteins. Additionally, the ability to gently elute RNA–protein complexes by adding excess aptamer ligand can be very useful in recovering large complexes in their native conformations.

1.2 Quantitative Mass Spectrometry-Based Interactomics

Quantitative mass spectrometry (qMS) is a core technology that complements the plethora of modern molecular biology methods. Among advanced qMS strategies, stable isotope labeling of amino acids in cell culture (SILAC) has emerged as a simple and robust approach [32, 33]. Conceptually, it involves growing two (or sometimes three) populations of cells, one in a medium with amino acids containing common isotopes (e.g., ^{12}C and ^{14}N) ("light") and the other in a medium that contains amino acids with stable heavy isotopes (^{13}C and ^{15}N) ("heavy") before mixing the cells and extracting the proteome for MS analysis. SILAC can quantify both the proteome and its modifications in response to stimuli and perturbations. Among others, one application of this technology is to identify protein partners in RNA–, DNA– and protein–protein interactions [34–38]. Conventionally, protein interaction studies are carried out based on the use of a bait to "fish out" protein partners. These methods have weaknesses, such as difficulty in separating specific interactors from the background and loss of weak interactors. These techniques are now superseded by the SILAC approach that can more readily distinguish "real" from adventitious interactors. Proteins that are biologically and functionally indistinguishable can be readily differentiated by mass spectrometry, and quantitative changes in control and binding proteins can be accurately measured.

2 Materials

2.1 RNA Chromatography

2.1.1 Materials and Basic Buffer Components

1. Diethyl pyrocarbonate (DEPC)-treated ddH$_2$O (*see* **Note 1**).

2. 1 M Tris–HCl, pH 7.0.

3. 1 M Tris–HCl, pH 7.5.

4. 1 M CaCl$_2$.

5. 1 M MgCl$_2$.

6. 1 M KCl.

7. 5 M NaCl.

8. 10 mg/mL yeast RNA.

9. 10 mg/mL bovine serum albumin (BSA).

10. 10 % IGEPAL CA-630 (formerly Nonidet P-40 (NP-40)).

11. Tobramycin, dissolved in coupling buffer to make 100 mM solution.

12. 0.5 M DL-dithiothreitol (DTT).

13. 1 M sodium bicarbonate.

14. 1 mM HCl.

15. Phosphate buffered saline (PBS).

16. Complete mini protease inhibitor tablets.

17. PhosSTOP phosphatase inhibitor tablets.

18. N-Hydroxysuccinimide (NHS)-activated Sepharose 4 Fast Flow.

19. MicroSpin columns.

20. Rotating device for end-over-end rotation of tubes.

21. DENV-2 New Guinea C-strain (DEN2-NGC) RNA.

22. Oligodeoxynucleotide primers.

23. ImProm-II reverse transcription system.

24. GoTaq Flexi DNA polymerase.

25. pGEM T-Easy kit.

26. Chemically competent *E. coli* cells.

27. MEGAscript T7 kit.

28. RNEasy mini kit.

2.1.2 Freshly Prepared Buffers for RNA Chromatography

1. Coupling buffer: 0.2 M sodium bicarbonate, 0.5 M NaCl, pH to 8.3 with NaOH.

2. 4× binding buffer (4× BP): 80 mM Tris–HCl, pH 7.0, 4 mM $CaCl_2$, 4 mM $MgCl_2$.

3. Blocking buffer B: 1× BP, 300 mM KCl, 0.1 mg/mL yeast RNA, 0.5 mg/mL BSA, 0.2 mM DTT, 0.01 % v/v IGEPAL CA-630 (NP-40).

4. RNA binding buffer: 1× BP, 75 mM NaCl, 145 mM KCl, 0.1 mg/mL yeast RNA, 0.2 mM DTT.

5. RNA washing buffer: 1× BP, 75 mM NaCl, 145 mM KCl, 0.1 % v/v IGEPAL CA-630 (NP-40), 0.2 mM DTT.

6. Cell lysis buffer: 50 mM Tris–HCl, pH 7.5, 150 mM NaCl, 1 % v/v IGEPAL CA-630 (NP-40), 1× protease inhibitor, 1× phosphatase inhibitor.

7. Modified 2× protein binding buffer: 2 mM $CaCl_2$, 2 mM $MgCl_2$, 290 mM KCl, 4 mM DTT.

8. Wash buffer: 1× BP, 75 mM NaCl, 145 mM KCl, 3 mM $MgCl_2$, 0.5 % v/v IGEPAL CA-630 (NP-40).

9. Elution buffer: 1× BP, 145 mM KCl, 2 mM $MgCl_2$, 5 mM tobramycin, 0.2 mM DTT.

2.2 Stable Isotopic Labeling of Amino Acids in Cell Culture (SILAC)

2.2.1 Reagents for SILAC

1. Dialyzed fetal bovine serum (FBS).
2. SILAC Dulbecco's Modified Eagle's Medium (DMEM) (Thermo Scientific, 0089985) (*see* **Note 2**).
3. L-Arginine hydrochloride (R0).
4. L-Lysine dihydrochloride (K0).
5. L-Arginine-$^{13}C_6$, $^{15}N_4$ hydrochloride (R10) (Cambridge Isotope, CNLM-539).
6. L-Lysine-$^{13}C_6$, $^{15}N_2$ dihydrochloride (K8) (Cambridge Isotope, CNLM-291).
7. Gentamicin (Life Technologies, 15710-072).
8. Pierce BCA protein assay kit (Thermo Scientific, 23225).

2.2.2 Buffers and Media for SILAC

1. L-Arginine stock (light (R0) or heavy (R10)): 85 mg/mL in PBS.
2. L-Lysine stock (light (K0) or heavy (K8)): 146 mg/mL in PBS.
3. Complete SILAC DMEM (light or heavy media): 10 % (v/v) dialyzed FBS, 0.4 mM L-arginine, 0.8 mM L-lysine, 0.2 % gentamicin.
4. PBS for tissue culture.
5. Trypsin–EDTA solution.

2.3 Quantitative Mass Spectrometry Analysis

2.3.1 Sample Preparation

1. NuPAGE 4× LDS sample buffer (Laemmli sample buffer) (Invitrogen, NP007).
2. NuPAGE sample reducing agent (RA) 10× solution (Invitrogen, NP0009).
3. NuPAGENovex 4–12 % Bis–Tris gel system, MOPS (Invitrogen, NP0321BOX).
4. Colloidal blue staining kit (Invitrogen, LC6025).
5. Sequencing grade modified trypsin (Promega, V5111).
6. DTT and iodoacetamide (IAA) (Sigma, I1149).
7. LiChrosolv® acetonitrile (ACN) gradient grade for LC (Merck, 100029) and Suprapur® formic acid 98–100 % (FA) (Merck, 111670).
8. Ammonium bicarbonate (ABC, Sigma, A6141) buffer: 50 mM ABC in water. To make 10 mL, dissolve 39.6 mg ABC in 10 mL of water and store at RT.
9. DTT: 10 mM DTT in 50 mM ABC. To make 1 mL, dissolve 1.5 mg DTT in 1 mL of 50 mM ABC.
10. IAA: 55 mM IAA in 50 mM ABC. To make 1 mL, dissolve 10.2 mg and IAA in 1 mL of 50 mM ABC.

11. 50 % ACN/25 mM ABC: 50 % 50 mM ABC/50 % ACN.

12. 25 mM ABC: 50 % 50 mM ABC/50 % water.

13. Trypsin stock: 0.2 µg/µL in buffer. To make 100 µL, add 100 µL of sequencing grade modified trypsin resuspension buffer into 20 µg of lyophilized trypsin and store on ice before use.

14. Trypsin digestion solution: 13 ng/µL trypsin in 50 mM ABC. To make 150 µL of trypsin digestion solution, add 10 µL of trypsin stock in 140 µL of cold 50 mM ABC and store at fridge before usage.

15. 5 % FA: To make 20 mL, add 1 mL of FA in 19 mL of water.

16. 0.1 % FA: To make 5 mL, add 100 µL of 5 % FA in 4.9 mL water.

17. Proxeon CB080 perforated microtiter plate.

18. 96-well Nunc plates.

19. LC solvent A: 0.1 % FA in 2 % ACN/water. To make 25 mL, add 25 µL FA and 0.5 mL ACN in water.

20. LC solvent B: 0.1 % FA in 80 % ACN/water. To make 25 mL, add 25 µL FA and 20 mL ACN in water.

2.3.2 Liquid Chromatography–Mass Spectrometry (LC-MS) and Data Analysis

1. A nano LC system such as Proxeon Easy-nLC (Thermo Scientific, LC110) system.

2. Self-packed pre-column of 100 µm (internal diameter) × 2–2.5 cm (length) using MICHROM Bioresources, Inc. MAGIC C18 AQ, 200 A, 5 µm.

3. Fritless nano column from New Objective, PicoFrit™ column, HALO C18, 2.7 µm, 75 µm (internal diameter) × 10 cm (length) × 15 µm (tip).

4. Nano column from Thermo Scientific, Acclaim® PepMap100, C18, 100 A, 3 µm, 75 µm (internal diameter) × 15 cm (length) connected with Thermo Scientific nano-bore ES 542 emitter.

5. LTQ (Linear trap quadrupole) Orbitrap instrument (Thermo Scientific) or similar high resolution instrument.

6. MaxQuant software (an open source at http://www.maxquant.org/) for LTQ Orbitap data analysis or a suitable software for protein quantification compatible with generated raw files (instrument dependent).

7. Human and DENV protein sequence database.

3 Methods

3.1 Generation of RNA Chromatography Matrix

3.1.1 Template Construction and In Vitro Transcription of RNA for Chromatography

We utilize splice overlap extension (SOE) PCR to assemble the T7 promoter sequence, DENV-2 UTRs, and aptamer sequences into a single template for in vitro transcription (oligonucleotide sequences are provided in Table 1).

1. Generate cDNA from DENV-2 NGC viral RNA using random hexamer primers and ImPromII Reverse Transcriptase, according to the manufacturer's instructions (Promega) (*see* **Note 3**).

2. PCR amplify the DENV-2 5′ UTR (nts 1–172) and 3′ UTR (nts 10,242–10,725) from cDNA using GoTaq according to the manufacturer's instructions (Promega) (*see* **Note 4**).

3. Fill in the aptamer fragment from a single oligonucleotide using GoTaq polymerase and an oligonucleotide complementary to the 3′ end of the aptamer fragment (*see* **Note 5**).

Table 1
Oligonucleotides used to generate T7 templates for in vitro transcription of RNA chromatography substrates

Description	DENV2 NGC nucleotides	Sequence (5′ to >3′)
T7-5′UTR forward	1–22	ATA AAG CTT TAA TAC GAC TCA CTA TAG GGA GTA GTT AGT CTA CGT GGA CCG
5′UTR reverse	148–172	GGT CGG TCG TCT TCT CCC GCT AGC ATA AGT CGA CAC GCG GTT TCT CTC
3′UTR forward	10,242–10,261	GGT CGG TCG TCT TCT CCC GCT AGC ATA GGT TAT TCA TCA GAG ATC TGC
3′UTR reverse	10,702–10,725	AGA ACC TGT TGA TTC AAC AGC AC
Aptamer forward		TAT GCT AGC GGG AGA AGA CGA CCG ACC AGA ATC ATG CAA GTG CGT AAG ATA GTC GCG GGC CGG GAA AAA AAA AAG CTT AGT ATG AGC GAG GTT AGC TTA CAC TCG TGC TGA GCC AAA AAA AAA AGA CCG ACC AGA ATC ATG CAA GTG CGT AAG ATA GTC GCG GGC CGG GTA TGT GCG TCT GGA TCC TAT
Aptamer reverse		ATA GGA TCC AGA CGC ACA TAC CGG GCC
T7-5′NS2A forward	3,504–3,524	ATA AAG CTT TAA TAC GAC TCA CTA TAG GGT TTT CAC TAG GAG TCT TGG G
5′NS2A reverse	3,654–3,674	GGT CGG TCG TCT TCT CCC GCT AGC ATA CCG TCA TAG TAG CGC CCA CC
3′NS2A forward	3,739–3,759	CGG GTA TGT GCG TCT GGA TCC TAT GCA GCT GGA CTA CTC TTG AG
3′NS2A reverse	4,201–4,221	GGT CCT GTC ATG GGA ATG TC

Incubate the reaction at 95 °C for 30 s., anneal the primer at 55 °C for 30 s. and extend at 72 °C for 10 min.

4. Fuse the 5′ UTR and aptamer fragment by SOE PCR and gel purify the resulting fragment using the Qiagen Gel Extraction kit according to the manufacturer's instructions.

5. Fuse the 5′ UTR-aptamer fragment to the 3′ UTR fragment using SOE PCR and gel purify as described in **step 4**. As a control for specific binding to the DENV-2 UTRs, generate a control construct by fusing sequences from the DENV-2 NS2A ORF (nts 3,504–3,674 and 3,739–4,221) to the aptamer fragment instead of the 5′ and 3′ UTRs (*see* **Note 6**).

6. Ligate the final fragments into the pGEM T-Easy vector and transform into competent bacteria according to the manufacturer's instructions (Promega).

7. Perform Sanger sequencing on individual clones before PCR amplification of the full-length T7 template to confirm that all the sequences are correct (*see* **Note 7**).

8. PCR amplify T7 templates and analyze on analytical agarose gels prior to purification for use in T7 transcription reactions.

9. Assemble in vitro transcription reactions according the manufacturer's instructions and incubated overnight at 37 °C (Ambion) (*see* **Note 8**).

10. Remove unincorporated nucleotides using the RNEasy mini kit (Qiagen) and calculate yield by measuring absorbance at 260 nm (*see* **Note 9**).

3.1.2 Generation of Tobramycin Matrix, Folding of RNA, and Binding of RNA to Tobramycin Matrix

As described in Hartmuth et al., all spins are at $250 \times g$ [27].

1. Wash 1 mL NHS-activated sepharose beads (50 % slurry) four times with 1 mL 1 mM HCl, using $250 \times g$ spins.

2. Resuspend beads in 950 μL of coupling buffer and add 50 μL 100 mM tobramycin.

3. Incubate overnight at 4 °C with end-over-end rotation.

4. Spin down beads, discard the supernatant fraction, and resuspend the pelleted beads in 1 mL blocking buffer B.

5. Incubate at 4 °C, 1 h.

6. Wash beads 3× with 1 mL PBS and resuspend in 0.5 mL PBS (this makes ~1 mL tobramycin bead slurry). Store at 4 °C until ready for use (*see* **Note 10**).

7. Combine 60–80 pmol of RNA and RNA binding buffer (1:1 mixture of RNA and 2× RNA binding buffer) for a final volume of 250 μL.

8. Heat the RNA to 95 °C for 5 min., and then shift to room temperature for 15 min.

9. Add 30 µL of tobramycin bead slurry and incubate at 4 °C with end-over-end rotation, 1 h.

10. Wash 2× with 0.3 mL RNA washing buffer.

3.2 SILAC and Generation of Cell Lysates

Establish two separate cell populations by culturing cells in either "light" (K0R0) or "heavy" (K8R10) media. After five to six doublings, the cells should be uniformly labeled with light or heavy amino acids (*see* **Note 11**). Depending on the cell type, cells should be observed for changes in growth rate and/or morphology (*see* **Note 12**). Uniformly labeled cells can be frozen and stored in 10 % DMSO for later use. For RNA chromatography, a substantial number of cells will need to be grown to generate lysates. Depending on the cell line used, 5×10^7 to 2×10^8 cells will need to be grown for each sample. Prior to RNA chromatography isotope incorporation percentage and Arg to Pro conversion should be assessed (*see* **Notes 13** and **14**).

3.2.1 Harvesting Adherent Cells and Generating Cell Lysates

1. Wash cells 2× with ice cold PBS.

2. For a T175 flask, add 2 mL trypsin–EDTA to dissociate cells.

3. Inactivate trypsin–EDTA by adding 10–20 mL of complete SILAC DMEM containing 10 % FBS.

4. Transfer cells to a 50 mL conical tube and spin at $500 \times g$, 5 min.

5. Discard the supernatant fraction and resuspend cell pellet in 10–20 mL cold PBS and repeat **steps 4** and **5**.

6. Discard last supernatant fraction and resuspend cell pellet in cell lysis buffer.

7. Incubate on ice, 10 min.

8. Spin lysate at $10,000 \times g$, 10 min.

9. Collect supernatant fraction and measure protein concentration using the Pierce BCA kit.

3.2.2 RNA Affinity Chromatography

1. Preclear cell lysate by combining 10 mg cell lysate with 30 µL of tobramycin matrix and incubating 1 h at 4 °C with end-over-end rotation (*see* **Note 15**).

2. Combine 10 mg precleared cell lysate with 2× modified protein binding buffer (*see* **Notes 16** and **17**).

3. Incubate beads and lysate 1–1.5 h at 4 °C with end-over-end rotation.

4. Spin down beads and save the cell lysate. Resuspend beads in 300 µL of wash buffer and transfer to a MicroSpin column. At this point, the control and experimental beads are combined in the MicroSpin column and processed together.

5. Wash 5× with 300 µL of wash buffer (*see* **Note 18**).

6. Add 30–60 µL of elution buffer and incubate 5 min at RT.

7. Elute proteins by spinning at $250 \times g$ (*see* **Note 19**).

3.3 Analysis of Bound Proteins Using Quantitative LC/MS

To reduce sample complexity, the eluted proteins are fractionated using 1D SDS-PAGE followed by in-gel trypsin digestion essentially as described by Shevchenko et al. [39]. Digested samples are subjected to reversed phase LC, which is directly connected to the mass spectrometer (online).

3.3.1 Sample Preparation

1. Add 1/4 volume of 4× LDS and 1/10 volume of 10× RA to the eluted protein complex sample (final volume should not exceed the volume of the sample well in the gel). Boil the sample for 5 min and spin down at $16,000 \times g$ for 2 min.

2. Load the supernatant fraction into SDS-PAGE leaving one empty lane between each loaded lanes to prevent contamination of each other. Run the gel halfway down (200 V for 20 min).

3. Stain the gel with colloidal Coomassie blue and destain with Milli-Q water (arium® 611UV, Sartorius).

4. Cut the gel into 6–8 slices from each lane with roughly equal amount of proteins (not necessarily be equal in size, e.g., very intense band should be cut separately and bands from blank area can be combined). Cut each protein band on a clean square Petri dish using a gel spot cutters of 2–3 mm. Cut the gel as close to the band as possible to reduce the volume of gel to be processed (*see* **Note 20**).

5. Place the gel pieces into Proxeon CB080 perforated microtiter plate that is placed on 96-well Nunc plate (underneath the perforated plate) (*see* **Note 21**).

6. Wash gel pieces by adding 50 µL of ABC into each well for 5 min.

7. Centrifuge at $175 \times g$ to remove the waste into Nunc plate. Discard the waste into a waste beaker.

8. Dehydrate the gel pieces with 50 µL of ACN for 10 min.

9. Centrifuge at $175 \times g$ to remove the waste into Nunc plate. Discard the waste into a waste beaker.

10. Repeat **step 6** for another 5 min and discard the waste by centrifuging.

11. Cover the gel pieces with DTT and incubate for 30 min at 56 °C. Remove and discard the waste.

12. Dehydrate the gel pieces with 50 µL of ACN for 10 min. Remove and discard waste.

13. Repeat **step 10** for 5 min and discard waste.

14. Cover the gel pieces with IAA and incubate for 20 min in the dark at RT. Remove and discard the waste.

15. Repeat **steps 6–10**.

16. Add 50 µL of ABC or more to cover the gel pieces. Sample can be stored at 4 °C overnight.

17. Add same volume of ACN into each well and incubate for 15 min at RT to destain Novex® Colloidal blue-stained gel pieces. Remove and discard the waste.

18. Repeat **steps 16** and **17** with 50 % until dark blue color disappears.

19. Dehydrate the gel pieces with 50 µL of ACN for 10 min. Remove and discard waste.

20. Repeat **step 19** for 5 min and discard waste.

21. Cover the gel pieces with cold trypsin (typically requires 30 µL) (*see* **Note 22**) digestion solution.

22. Allow the gel pieces to swell for 30 min in the refrigerator (~10 °C).

23. Inspect the gel pieces and if there is not enough solution to cover the swollen gel pieces, add more trypsin digestion solution.

24. Incubate in at 4 °C for another 60 min.

25. Add 25 mM ABC to cover the gels (typically 10–15 µL).

26. Incubate for overnight at 37 °C with a new 96-well Nunc plate underneath.

27. Centrifuge the plate at $175 \times g$ to collect the digested peptides (flow-through) from perforated plate into Nunc plate.

28. Add 20 µL of 5 % FA to gel pieces containing wells (top plate) and incubate for 10 min. Collect the flow-through by centrifuging at $175 \times g$.

29. Add 20 µL of ACN to gel pieces containing wells (top plate) and incubate for 10 min. Collect the flow-through by centrifuging at $175 \times g$.

30. Repeat **step 28**.

31. Repeat **step 29** two times.

32. SpeedVac to dryness. Peptides can be stored at –20 °C.

33. Reconstitute in 12 µL of 0.1 % FA in water.

3.3.2 Liquid Chromatography–Mass Spectrometry (LC-MS)

1. Inject 5 µL of peptide mixture in 0.1 % FA in water to Proxeon EASY-LC system (or similar system) directly connected to a high resolution mass spectrometer (Orbitrap Classic or XL). To further separate peptides use a New Objective, PicoFrit™ column, HALO C18, 2.7 µm, 75 µm (internal diameter) × 10 cm (length) with 15 µm (tip) or Thermo Scientific, Acclaim® PepMap100 column, C18, 100 A, 3 µm, 75 µm (internal diameter) × 15 cm (length) connected with Thermo Scientific

nano-bore ES 542 emitter using union, and a gradient of 5–40 % LC solvent B for 120 min at a flow rate of 300 nL/min.

2. Eluting peptides are directly injected to mass spectrometry that is on data-dependent mode with one full scan in the Orbitrap ($m/z = 310$–1,400; resolution = 60,000; target value = 1×10^6, top ten intense peaks, normalized collision energy of 35).

3. Analyze the raw data by MaxQuant software using default settings [40].

4. Possible interacting partners can generally be identified by considering proteins detected with SILAC ratio >2 with P (ratio significance B) <0.05 and at least 2 ratio counts. The data should be verified by a reverse experiment (*see* **Note 23**) and biological replicates.

4   Notes

1. DEPC (Sigma, D5758) is used to inactivate RNases that may be present in ddH$_2$O. Add DEPC to 0.1 % and stir for 4–6 h, and then autoclave for 20 min at 120 °C. Where possible, reagents should be made using DEPC-treated ddH$_2$O and autoclaved.

2. SILAC DMEM is the standard DMEM formulation lacking L-lysine or L-arginine.

3. DENV-2 NGC RNA can be obtained from either infected cells or virus stocks. The RNA is reverse transcribed using random hexamer primers and the resulting cDNA can be used for amplification using primers specific for the UTRs or ORF.

4. The oligonucleotides for amplification of the 5′ UTR contain the T7 promoter sequence prior to the start of the 5′ UTR or overlapping sequences with the aptamer fragment at the end of the 5′ UTR. The oligonucleotides for amplification of the 3′ UTR contain overlapping sequence with the aptamer fragment prior to the start of the 3′ UTR and a NotI restriction site at the end of the 3′ UTR.

5. The aptamer sequence used in our studies contains two streptavidin aptamer sequences flanking a single tobramycin aptamer. It was designed for tandem affinity purification utilizing both the streptavidin and tobramycin aptamers. Hartmuth et al. used the tobramycin aptamer alone for purification of spliceosome complexes [27]. Taq polymerase was used for the fill-in reaction to make a dsDNA aptamer fragment because the higher temperatures were required to melt the secondary structures formed by the aptamer sequences. Previous attempts to fill in using *E. coli* DNA Pol I Klenow Fragment at 37 °C led to deletion of the aptamer sequences.

6. Care should be taken when designing controls for RNA chromatography. For the DENV-2 control, we considered length, GC-content and the presence of complementary sequences for base pairing. This allows for any base-specific preference and also for nonspecific binding to dsRNA regions. Sequence from NS2A was a convenient source for sequence with similar length and GC-content. We added complementary sequences to the NS2A control RNA flanking the aptamer sequence to mimic the complementary sequences (involved in cyclization of the viral genome) that occur in the DENV 5′ and 3′ UTRs.

7. We used PCR templates for T7 transcription. Cloning of the final SOE PCR product was useful for preserving the construct, preventing the need to repeat the SOE PCR and also allowing confirmation of the sequence prior to beginning experiments. In addition, having the T7 promoter fused directly to the beginning of the 5′ UTR and amplifying from the end of the 3′ UTR incorporates fewer extraneous nucleotides that are not present in the DENV sequence.

8. Shorter incubations (4–6 h) are also possible for efficient in vitro transcription. Also, nuclease-resistant nucleotides can be included in the reaction, though this results in lower RNA yields. Incorporation of 2′-fluoro-UTP and 2′-fluoro-CTP (Epicentre, R2F110U and R2F110C) does not interfere with binding to the tobramycin matrix. However, additional experiments have demonstrated that we can achieve similar results from RNA chromatography and qMS without the need for nuclease-resistant nucleotides (Ward et al., unpublished results).

9. Use the RNA cleanup protocol included in the kit instructions and perform two elutions to maximize yield.

10. Hartmuth et al. describe a protocol for testing efficiency of RNA binding to the tobramycin matrix using a radioactively labeled RNA [27]. An alternative approach to this is the use of RNA binding buffer lacking yeast RNA, allowing accurate measurement of the input and flow-through concentrations of RNA before and after binding to the tobramycin matrix. This will give a rough idea of whether the tobramycin matrix is able to bind the aptamer RNA. Ideally, 50–70 % of the input RNA should be bound to the tobramycin matrix.

11. Prior to SILAC experiments the cell lines that will be used for RNA chromatography should be checked for how SILAC affects their viability and morphological characteristics. This is because the media used contains dialyzed FBS that is using prepared using 10,000 MW cutoff filters where some small molecules essential for some cell lines may be lost.

12. We have used HeLa and HuH-7 cells in our studies. These cell lines have not exhibited any significant changes in morphology or growth rate when cultured in SILAC media.

13. To make sure complete isotope incorporation to the proteome SILAC incorporation test must be carried out prior to large-scale SILAC experiments. Cell lysates generated from growing cells (after about 5–6 doublings) in "light" and "heavy" media should be subjected to MS analysis as described in Subheading 3.3. Lysates are generated as described in Subheading 3.2.1. Incorporation plots can be generated by using MaxQuant software. SILAC incorporation should be at least >96 %. It may not reach to 100 % (ideal) as the purity of commercially available stable isotope is not 100 %. To further confirm this incorporation equal protein amounts from "light" and "heavy" can be mixed followed by LC-MS analysis. The SILAC ratio (non-normalized) of majority of proteins should be 1:1 if incorporation is >96 %.

14. In some cell lines arginine-to-proline conversion is considerable. During SILAC incorporation test, this conversion can be checked. Select "heavy"-labeled proline-containing peptides with highest intensities. Check the mass spectra (you can open MS raw file using XCalibur software to visualize chromatogram and MS spectra). The mass shift of 6 Da can be observed when $^{13}C^{15}N$-Arg is used if the conversion occurred. This interferes with accurate quantification of any proline-containing peptides. Hence, it should be controlled or minimized. Arginine-to-proline conversion can be controlled by reducing the arginine concentration or using the media supplemented with unlabeled proline [41, 42].

15. This step is necessary to remove proteins that nonspecifically bind to the tobramycin affinity matrix [43].

16. This is only a guideline for how much cell lysate to use. If additional manipulations are made in preparing the cell lysate (nuclear or cytoplasmic extract, organelle preparations), then the concentration of cell lysate used may require further optimization. Typically, using a higher concentration of cell lysate results in identification of more proteins by qMS.

17. Typically, the "light" lysate is incubated with the control RNA chromatography matrix and the "heavy" lysate is incubated with the experimental matrix. It is easier to eliminate keratin contamination from the subsequent analysis when the experimental matrix is incubated with the "heavy" lysate, since any contamination would be "light." Potential interacting partners should be further verified and prioritized using a reverse experiment. Typically "heavy"-labeled cells for bait containing sample and "light"-labeled cells for control sample are used in forward

experiment, whereas "heavy"-labeled cells for control and "light"-labeled cells for bait sample in reverse experiment. Specific interactors should appear with high SILAC ratio in forward experiment and lower (inverse) ratio in reverse experiment.

18. The final wash should contain little detectable protein by silver staining, whereas the first elution should contain significantly more protein. Washes can be concentrated and analyzed by silver staining to determine if there is nonspecifically bound protein remaining on the matrix. The salt or detergent concentrations in the washing buffer can be optimized if there is a significant problem with nonspecific binding to your RNA of interest. Nonspecific binding proteins can be identified in the dataset based on the light–heavy ratio. In a typical forward experiment, a protein that binds specifically to the RNA of interest will have an elevated SILAC ratio (generally twofold or more). In order to increase the confidence level, Ratio Significance B value also can be considered. Nonspecific binding proteins and proteins that preferentially bind to the control RNA have light–heavy ratio closer to 1:1 (pulled down equally by the bait and the control bead). Additionally, even when not detected as preferentially binding to the experimental RNA, abundant RNA binding proteins can interfere with the analysis by masking unique peptides from proteins bound specifically to the RNA of interest.

19. Additional elutions can be performed to ensure that all bound protein has been removed from the matrix.

20. All following steps must be performed a laminar flow hood. Wear gloves at all times to avoid keratin contamination. Always use clean spatulas, pre-packed pipette tips (manually packed tips are not recommended) and LoBind tubes (designed for storing peptides at low concentrations).

21. Here we describe "microplate version" of MS sample preparation. Thus, a mini gel setup is suitable. For large-scale experiment, a microtube version can also be used [39].

22. Trypsin is commonly used enzyme to generate peptides for the bottom-up proteomics, which is a widely used method to identify proteins and posttranslational modifications by proteolytic digestion of proteins prior to analysis by mass spectrometry. Sequencing grade modified trypsin (Promega) is recommended as it is in high purity and prevents autolysis. Trypsin cleaves at arginine(R)-X and lysine (K)-X if X is not proline. Thus, labeling with both arginine and lysine is essential to obtain high confident quantification and high coverage when trypsin is used for digestion.

23. The reverse experiment is helpful in increasing confidence of the SILAC data by confirming possible contaminants (always

low ratio) and eliminating any labeling-bias identification. In this experiment, the control construct is incubated with the "heavy" (K8R10) lysate, while the UTR construct is incubated with the "light" (K0R0) lysate. All subsequent steps are performed as described and the data from a forward and reverse experiment are compared. Specific interactors should be identified with high ratio in the forward experiment, whereas low ratio (reciprocal to forward ratio) in the reverse experiment.

Acknowledgments

The authors would like to acknowledge Walter Blackstock for guidance with mass spectrometry and experimental design and Sheena Wee and Siok Ghee Ler in Quantitative Proteomics Group IMCB for technical assistance in mass spectrometry.

References

1. Weaver SC, Reisen WK (2010) Present and future arboviral threats. Antiviral Res 85(2):328–345

2. Mackey TK, Liang BA (2012) Threats from emerging and re-emerging neglected tropical diseases (NTDs). Infect Ecol Epidemiol 2

3. Lindenbach B, Thiel HJ, Rice CM (2007) Flaviviridae: the viruses and their replication. Vol. I. Field's virology, 5th edn. Lippincott, Williams and Wilkins, Philadelphia, PA

4. Ward AM, Bidet K, Yinglin A, Ler SG, Hogue K, Blackstock W, Gunaratne J, Garcia-Blanco MA (2011) Quantitative mass spectrometry of DENV-2 RNA-interacting proteins reveals that the DEAD-box RNA helicase DDX6 binds the DB1 and DB2 3′ UTR structures. RNA Biol 8(6):1173–1186

5. Alvarez DE, De Lella Ezcurra AL, Fucito S, Gamarnik AV (2005) Role of RNA structures present at the 3′UTR of dengue virus on translation, RNA synthesis, and viral replication. Virology 339(2):200–212

6. Alvarez DE, Filomatori CV, Gamarnik AV (2008) Functional analysis of dengue virus cyclization sequences located at the 5′ and 3′UTRs. Virology 375(1):223–235

7. Alvarez DE, Lodeiro MF, Luduena SJ, Pietrasanta LI, Gamarnik AV (2005) Long-range RNA–RNA interactions circularize the dengue virus genome. J Virol 79(11):6631–6643

8. Anwar A, Leong KM, Ng ML, Chu JJ, Garcia-Blanco MA (2009) The polypyrimidine tract-binding protein is required for efficient dengue virus propagation and associates with the viral replication machinery. J Biol Chem 284(25):17021–17029

9. Brinton MA, Dispoto JH (1988) Sequence and secondary structure analysis of the 5′-terminal region of flavivirus genome RNA. Virology 162(2):290–299

10. Brinton MA, Fernandez AV, Dispoto JH (1986) The 3′-nucleotides of flavivirus genomic RNA form a conserved secondary structure. Virology 153(1):113–121

11. Chiu WW, Kinney RM, Dreher TW (2005) Control of translation by the 5′- and 3′-terminal regions of the dengue virus genome. J Virol 79(13):8303–8315

12. Clyde K, Harris E (2006) RNA secondary structure in the coding region of dengue virus type 2 directs translation start codon selection and is required for viral replication. J Virol 80(5):2170–2182

13. Emara MM, Liu H, Davis WG, Brinton MA (2008) Mutation of mapped TIA-1/TIAR binding sites in the 3′ terminal stem-loop of West Nile virus minus-strand RNA in an infectious clone negatively affects genomic RNA amplification. J Virol 82(21):10657–10670

14. Filomatori CV, Lodeiro MF, Alvarez DE, Samsa MM, Pietrasanta L, Gamarnik AV (2006) A 5′ RNA element promotes dengue virus RNA synthesis on a circular genome. Genes Dev 20(16):2238–2249

15. Garcia-Montalvo BM, Medina F, del Angel RM (2004) La protein binds to NS5 and NS3 and to the 5′ and 3′ ends of Dengue 4 virus RNA. Virus Res 102(2):141–150

16. Grange T, Bouloy M, Girard M (1985) Stable secondary structures at the 3'-end of the genome of yellow fever virus (17 D vaccine strain). FEBS Lett 188(1):159–163

17. Hahn CS, Hahn YS, Rice CM, Lee E, Dalgarno L, Strauss EG, Strauss JH (1987) Conserved elements in the 3' untranslated region of flavivirus RNAs and potential cyclization sequences. J Mol Biol 198(1):33–41

18. Holden KL, Harris E (2004) Enhancement of dengue virus translation: role of the 3' untranslated region and the terminal 3' stem-loop domain. Virology 329(1):119–133

19. Holden KL, Stein DA, Pierson TC, Ahmed AA, Clyde K, Iversen PL, Harris E (2006) Inhibition of dengue virus translation and RNA synthesis by a morpholino oligomer targeted to the top of the terminal 3' stem-loop structure. Virology 344(2):439–452

20. Olsthoorn RC, Bol JF (2001) Sequence comparison and secondary structure analysis of the 3' noncoding region of flavivirus genomes reveals multiple pseudoknots. RNA 7(10):1370–1377

21. Paranjape SM, Harris E (2007) Y box-binding protein-1 binds to the dengue virus 3'-untranslated region and mediates antiviral effects. J Biol Chem 282(42):30497–30508

22. Polacek C, Foley JE, Harris E (2009) Conformational changes in the solution structure of the dengue virus 5' end in the presence and absence of the 3' untranslated region. J Virol 83(2):1161–1166

23. Takegami T, Washizu M, Yasui K (1986) Nucleotide sequence at the 3' end of Japanese encephalitis virus genomic RNA. Virology 152(2):483–486

24. Yocupicio-Monroy M, Padmanabhan R, Medina F, del Angel RM (2007) Mosquito La protein binds to the 3' untranslated region of the positive and negative polarity dengue virus RNAs and relocates to the cytoplasm of infected cells. Virology 357(1):29–40

25. Lei Y, Huang Y, Zhang H, Yu L, Zhang M, Dayton A (2011) Functional interaction between cellular p100 and the dengue virus 3' UTR. J Gen Virol 92(Pt 4):796–806

26. Li W, Li Y, Kedersha N, Anderson P, Emara M, Swiderek KM, Moreno GT, Brinton MA (2002) Cell proteins TIA-1 and TIAR interact with the 3' stem-loop of the West Nile virus complementary minus-strand RNA and facilitate virus replication. J Virol 76(23):11989–12000

27. Hartmuth K, Vornlocher HP, Luhrmann R (2004) Tobramycin affinity tag purification of spliceosomes. Methods Mol Biol 257:47–64

28. Srisawat C, Engelke DR (2001) Streptavidin aptamers: affinity tags for the study of RNAs and ribonucleoproteins. RNA 7(4):632–641

29. Klug SJ, Famulok M (1994) All you wanted to know about SELEX. Mol Biol Rep 20(2):97–107

30. Caputi M, Mayeda A, Krainer AR, Zahler AM (1999) hnRNP A/B proteins are required for inhibition of HIV-1 pre-mRNA splicing. EMBO J 18(14):4060–4067

31. Stellato C, Gubin MM, Magee JD, Fang X, Fan J, Tartar DM, Chen J, Dahm GM, Calaluce R, Mori F, Jackson GA, Casolaro V, Franklin CL, Atasoy U (2011) Coordinate regulation of GATA-3 and Th2 cytokine gene expression by the RNA-binding protein HuR. J Immunol 187(1):441–449

32. Mann M (2006) Functional and quantitative proteomics using SILAC. Nat Rev Mol Cell Biol 7(12):952–958

33. Ong SE, Blagoev B, Kratchmarova I, Kristensen DB, Steen H, Pandey A, Mann M (2002) Stable isotope labeling by amino acids in cell culture, SILAC, as a simple and accurate approach to expression proteomics. Mol Cell Proteomics 1(5):376–386

34. Butter F, Scheibe M, Morl M, Mann M (2009) Unbiased RNA–protein interaction screen by quantitative proteomics. Proc Natl Acad Sci U S A 106(26):10626–10631

35. Mittler G, Butter F, Mann M (2009) A SILAC-based DNA protein interaction screen that identifies candidate binding proteins to functional DNA elements. Genome Res 19(2):284–293

36. Ong SE, Mann M (2006) A practical recipe for stable isotope labeling by amino acids in cell culture (SILAC). Nat Protoc 1(6):2650–2660

37. Schulze WX, Mann M (2004) A novel proteomic screen for peptide–protein interactions. J Biol Chem 279(11):10756–10764

38. Selbach M, Mann M (2006) Protein interaction screening by quantitative immunoprecipitation combined with knockdown (QUICK). Nat Methods 3(12):981–983

39. Shevchenko A, Tomas H, Havlis J, Olsen JV, Mann M (2006) In-gel digestion for mass spectrometric characterization of proteins and proteomes. Nat Protoc 1(6):2856–2860

40. Cox J, Mann M (2008) MaxQuant enables high peptide identification rates, individualized p.p.b.-range mass accuracies and proteome-wide protein quantification. Nat Biotechnol 26(12):1367–1372

41. Lossner C, Warnken U, Pscherer A, Schnolzer M (2011) Preventing arginine-to-proline conversion in a cell-line-independent

manner during cell cultivation under stable isotope labeling by amino acids in cell culture (SILAC) conditions. Anal Biochem 412(1): 123–125

42. Rigbolt KT, Blagoev B (2010) Proteome-wide quantitation by SILAC. Methods Mol Biol 658:187–204

43. Trinkle-Mulcahy L, Boulon S, Lam YW, Urcia R, Boisvert FM, Vandermoere F, Morrice NA, Swift S, Rothbauer U, Leonhardt H, Lamond A (2008) Identifying specific protein interaction partners using quantitative mass spectrometry and bead proteomes. J Cell Biol 183(2):223–239

Chapter 17

Analysis of Affinity of Dengue Virus Protein Interaction Using Biacore

Yin Hoe Yau and Susana Geifman Shochat

Abstract

Surface plasmon resonance (SPR) biosensors have become the mainstream method for biomolecular interaction analysis. It offers many advantages over conventional methods by its label-free, real-time monitoring, low sample consumption, high throughput, and remarkable sensitivity. We have examined dengue virus protein interactions in the context of antibody affinity measurement, protein–protein interaction, and in the screening of small molecule inhibitors as well as the characterization of the interactions between the small molecule binders and the relevant dengue protein. Here we describe the basic methods involved in performing SPR assays as well as in data processing and evaluation using some examples of dengue proteins.

Key words Surface plasmon resonance, SPR, Affinity, Kinetics, Biacore, Biomolecular interaction analysis, Dengue virus, Flavivirus

1 Introduction

Surface plasmon resonance (SPR) biosensors are label-free, fast, sensitive, and high-throughput instruments designed for monitoring biomolecular interactions [1–3]. Some of the applications of these versatile biosensors are binding characterization, epitope/binding domain mapping, mechanism studies [4], probing membrane systems [5, 6], assay development, and with advances in the technology, ligand/antibody screening, fragment-based drug design [7–9], small ligand characterization [10, 11], active concentration determination [12, 13], thermodynamic measurements, and many others. In the context of this chapter, we will discuss the basic applications of simple bimolecular interaction pertinent to dengue research. The important steps described in this chapter include (a) surface preparation, (b) assay optimization, (c) running experiments, (d) data processing, and (e) evaluation for the determination of the kinetic constants and affinity for the interaction.

Radhakrishnan Padmanabhan and Subhash G. Vasudevan (eds.), *Dengue: Methods and Protocols*, Methods in Molecular Biology, vol. 1138, DOI 10.1007/978-1-4939-0348-1_17, © Springer Science+Business Media, LLC 2014

Leading SPR biosensors in the market (like Biacore 3000/T100/T200 and ProteOn XPR) consist of three major parts—injection port, interfluidic cartridge (IFC), and detection system. The functionality of these parts is controlled by a simple integrative software system. A decline in performance in any of these parts will likely cause a drop in data quality. The injection port and needle(s) must be clean and unblocked, free from salt precipitation, and free from carry-over samples/chemicals from the last assay. The IFC is the most delicate part of all. Any of its microchannels can easily be blocked by dust particles or sample precipitation; therefore, maintenance of equipment is crucial. In our facility the equipment is flushed continuously with ultrapure water when not in use.

1.1 SPR Biosensors in Dengue Research

The first publication in dengue research using SPR biosensors in 2009 was from our platform in which we characterized several monoclonal antibodies (mAbs) and single-chain Fvs (scFvs) recognizing domain III of flavivirus envelope glycoprotein (E) from different dengue virus (DENV) serotypes and West Nile virus (WNV) [14]. Since then, many groups working in this field have used SPR biosensors to study pH effect on precursor membrane (pr) protein with E [15, 16], to characterize interaction between E protein domain(s) and antibodies and small molecules [17–19], to evaluate the interaction between nonstructural protein 1 (NS1) and anti-NS1 IgM/IgG in a rapid test kit [20], to select mouse mAb that strongly neutralize all four dengue virus serotypes [14], to determine the affinity of human antibody fragments to dengue NS3 [21], to gain structural insights into binding mechanism between wild-type and mutant dengue proteins [22], and also to compare several methods in discriminating specific and nonspecific inhibitors of dengue virus protease [23]. As exemplified in these publications, SPR biosensor technology has wide application in dengue research and can be used to study the interaction of viral proteins, develop therapeutic antibodies, and also screen for small molecule inhibitors. The low sample consumption and the rapid turnover time for measuring the interaction affinity can contribute immensely to studies directed towards therapeutic discovery.

Here we describe the standard working protocols used in our facility for measuring affinity. Table 1 lists the dengue articles where SPR biosensor technology has been used (with keywords of *SPR*, *dengue*, and *flavivirus*).

2   Materials

All protocols described herein are based on Biacore 3000 system (Biacore, GE Healthcare). However these protocols can be easily adapted to other SPR biosensors.

Table 1
List of Scientific papers on Dengue which used SPR biosensor technology

	SPR eq.	Chip	Immob	Interaction	Regen	Ref.
Protein-Ab/Fab						
1	B3000	SA	Biotinylation	*DENV4 NS2B$_{18}$NS3–Fab	15 mM HCl	[21]
2	B3000	CM5	Amine coupling	*D1/D2/D3/WN Ed3–mAb/scFv	50 mM HCl	[14]
3	T100	CM5	Anti-mouse IgG	*Anti-NS1–DENV-1 NS1	10 mM glycine-HCl, pH 1.7	[20]
4	BX	CM5	?	*DENV2 Ed3–mAbs	10 mM glycine-HCl/0.2 M NaCl, pH 3.0	[18]
5	B3000	?	?	*E protein–mAbs	?	[19]
Protein–protein						
6	B3000	CM5	Amine coupling	*DENV4 NS5/mutants–DENV4 full-length NS3/helicase	15 mM HCl	[22]
7	B2000	CM5	Amine coupling	*DENV pr–DENV E	50 mM NaOH/1 M NaCl	[15]
8	B3000	SA	Biotinylation	*DENV pr–DENV E	50 mM NaOH	[16]
Protein–small molecule						
9	B3000	CM5	Cysteine coupling	*DENV2 NS3–small molecule	Nil	[23]
10	B3000	CM5	?	*EdI/EdII–small molecule	Nil	[17]

SPR eq.=SPR equipment used; chip=sensor chip used; immob=immobilization method; interaction=interaction studied; regen=regeneration solution; ref=reference article; BX=Biacore X; B2000=Biacore 2000; B3000=Biacore 3000; T100=Biacore T100; CM5=carboxymethylated dextran surface; SA=streptavidin dextran surface; ?=not reported or unclear
*The immobilized interactant

2.1 Sensor Chip

1. CM5 (Biacore) [BR-1000-14 or BR-1005-30].

2.2 Common Buffers

1. HBS-EP (HEPES-buffered saline with EDTA and P-20): 10 mM HEPES, 150 mM NaCl, 0.34 mM EDTA, 0.005 % P-20, pH 7.4 [BR-1001-88].

2. PBS-P (phosphate-buffered saline with P-20): 10 mM phosphate, 150 mM NaCl, 0.005 % P-20, pH 7.4 [BR-1006-72].

2.3 Biacore Consumables

1. Biacore plastic vials and caps: 0.8 ml 7 mm rounded polypropylene microvials [BR-1002-12] and 7 mm thin polyethylene snap caps [BR-1002-13] (Fig. 1a).

2. Biacore glass vials and caps: 1.8 ml 9 mm borosilicate vial [BR-1002-07] and (rubber cap type 3) penetrable ventilated kraton G cap [BR-1005-02] (Fig. 1b) and 4.0 ml 16 mm borosilicate vial [BR-1002-09] and (rubber cap type 5) penetrable ventilated kraton G cap [BR-1006-55] (Fig. 1c).

2.4 Biacore Cleaning and Maintenance Solutions

1. BIAdesorb solution 1: 0.5 % sodium dodecyl sulfate (SDS).

2. BIAdesorb solution 2: 50 mM glycine, pH 9.5.

3. BIAnormalizing solution: 70 % (w/w) glycerol.

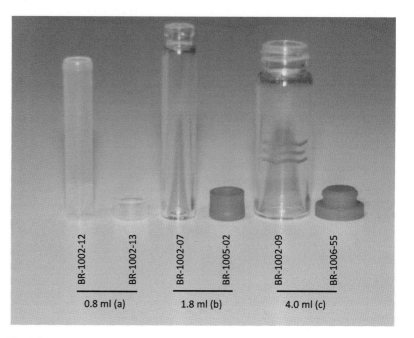

Fig. 1 Consumables (vials and caps) for Biacore. (**a**) Polypropylene microvial and snap cap that holds a maximum of 0.8 ml of solution. (**b**) Borosilicate vial and penetrable ventilated kraton G cap that holds a maximum of 1.8 ml of solution. (**c**) Borosilicate vial and penetrable ventilated kraton G cap that holds a maximum of 4.0 ml of solution

2.5 Biacore Immobilization Solutions

1. NHS: 0.5 M *N*-hydroxysuccinimide (*see* **Note 1**).
2. EDC: 0.1 M *N*-ethyl-*N*-(3-dimethylaminopropyl)-carbodiimide hydrochloride.
3. EA: 1 M ethanolamine-hydrochloride, pH 9.5.
4. NaOAc: 0.1 M sodium acetate, pHs 4.0, 4.5, 5.0, 5.5, 6.0, and 6.5.

2.6 Common Regeneration Solutions

1. Acid regeneration solution: 5–150 mM HCl.
2. Base regeneration solution: 5–150 mM NaOH.
3. SDS regeneration: 0.01–0.5 %.

3 Methods

3.1 Preparation and Cleaning of Equipment

It is essential to ensure that the equipment is clean and in a good working condition (*see* **Note 2**).

1. Dock a *maintenance chip* into the system [*Command → Dock*; *then click OK*].
2. Insert a bottle of filtered ultrapure water. Empty the waste bottle.

3. Place 3 ml of *BIAdesorb solution 1* and 3 ml of *BIAdesorb solution 2* into glass vials and cap (Fig. 1c), separately.

4. Activate *desorb* [*Tools→ Working Tools→ Desorb*; *then click Start*] and follow the on-screen instructions. This procedure takes about 6 min 20 s.

5. When the *desorb* procedure is complete, undock the maintenance chip [*Command→ Undock*; *then click OK*].

6. Dock a new s*ensor chip CM5* (any functionalized sensor chip will do) and replace ultrapure water with HBS-EP buffer.

7. Place 0.5 ml *BIAnormalizing solution* into a glass vial and cap (Fig. 1b), then activate *normalize* [*Tools→ Working Tools→ Normalize*; *then click Start*], and follow the on-screen instructions. This procedure takes about 7 min.

8. When the normalize procedure is completed, start a manual run [*Run→ Run sensorgram*; *then select the channel, click OK*; *set flow rate, click OK*] with minimal flow rate (e.g., 10 ml/min) to monitor the signal quality (Fig. 2).

3.2 Immobilization of Ligand

Biomolecules can be immobilized onto a variety of sensor chip surfaces with many different chemistries [24]. Immobilized biomolecules are referred to as the ligand in the context of SPR biosensors. Standard amine-coupling procedure [25], which constitutes more than half of reported SPR applications, will be described here.

1. Start a manual run [*Run→ Run sensorgram*]. Enter a filename and click OK. Alternatively, start an automated *Wizard* [*Run→ Run Application Wizard→ Surface preparation→ Start→ Immobilization*] and then follow the on-screen instructions.

2. *Surface activation*: Prepare a 110 ml mixture of 1:1 (v/v) of 0.5 M NHS: 0.1 M EDC and inject for 7 min at 10 ml/min (*see* **Note 3**).

3. *Sample coupling*: Dilute protein sample in 10 mM sodium acetate, pH 5.0 (or any other low salt buffer at an appropriate pH) (*see* **Note 4**). Inject the sample and monitor the increase in response. Stop injection when the desired response level is achieved.

4. *Blocking*: Block remaining activated surface with a 7 min injection of EA.

5. Take note of the final response (protein immobilized) (Fig. 3).

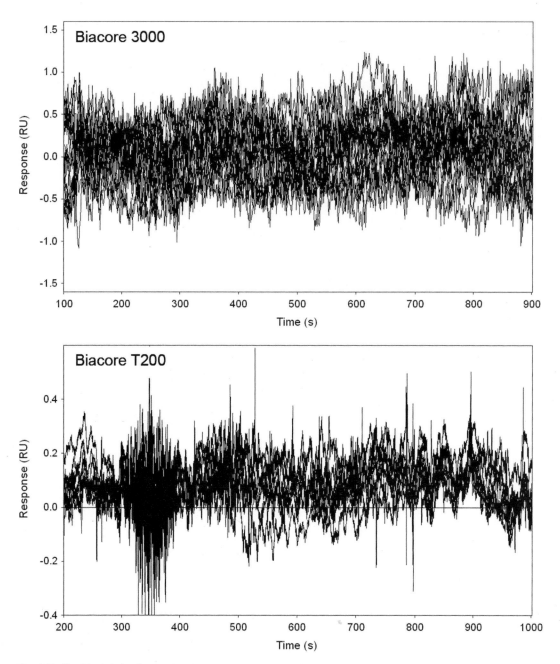

Fig. 2 Buffer blank injection cycles demonstrate the cleanliness and functionality of Biacore 3000 (*top*) and Biacore T200 (*bottom*) in our platform. Well-maintained equipment should have blank responses close to equipment sensitivity

6. Prepare a reference channel with an unrelated protein, with similar physical properties as the test protein, immobilized similarly to the same level. In the event where there is no suitable alternative, an unmodified channel can be used.

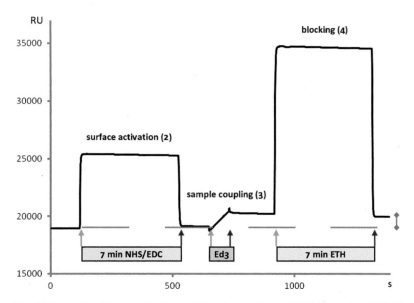

Fig. 3 A typical amine-coupling immobilization profile on CM5 chip using HBS-EP buffer showing the three distinctive phases of surface activation (2), sample coupling (3), and blocking (4). In this example, dengue 2 envelope protein (D2Ed3) was pre-diluted into 10 mM sodium acetate, pH 5.0, and immobilized onto the surface of a CM5 chip to 1000 RU [16]. The total amount of protein immobilized (*blue arrow*) is given by the difference between final stable response and initial buffer baseline (*dotted line*). Injection start is denoted by a *green arrow* and injection end by a *red arrow*

3.3 Test Binding of Analyte and Scouting of Regeneration Condition

Test binding of analyte, prior to setting up a full-scale analysis, serves three purposes: (1) to ensure the conditions (immobilization level, analyte concentration range, buffer composition and pH, etc.) for interaction are suitable, (2) to ensure any interaction between ligand and analyte is in good measurable response, and (3) whether regeneration is necessary and if so, what regeneration solution to use.

1. Start an analysis run and then follow the on-screen instructions [*Run → Run Application Wizard → Binding analysis → Start*].

2. If analyte is not prepared in the same buffer as intended SPR running buffer, exchange buffer and concentrate the analyte to the highest concentration possible (e.g., 300 mM) (*see* **Note 5**).

3. Prepare tenfold dilutions of analyte (from stock concentration) in running buffer to 1:1, 1:10, 1:100, 1:1,000, and 1:10,000.

4. Inject from the lowest to the highest concentration at a flow rate intended for kinetic assay (e.g., 30 ml/min). The concentration that shows saturation response or signs of saturation will be optimal for kinetic study (Fig. 2) (*see* **Note 6**).

An automated run can be set up as follows:

Cycle 1	Buffer blank
Cycle 2	Buffer blank
Cycle 3	Buffer blank
Cycle 4	Analyte 1:10,000
Cycle 5	Analyte 1:1,000
Cycle 6	Analyte 1:100
Cycle 7	Analyte 1:10
Cycle 8	Analyte 1:1
Cycle 9	Buffer blank
Cycle 10	Buffer blank
Cycle 11	Buffer blank

5. Estimate the rate of dissociation (k_d) from the result by calculating the linear decline in response after injection end. The rate of dissociation can be converted to half-life ($t_{1/2}$) through the equation $\sqrt{d}$. $t_{1/2}$ the half life will determine if regeneration is necessary for assay set up (*see* **Note 7**).

6. In the event where regeneration is necessary, find a good regeneration solution starting with the following interactant list—antibodies: acid regeneration; proteins: acid and/or SDS regeneration; peptides: SDS regeneration; and nucleic acids: base regeneration. Always start from low concentration and work upwards and vary contact time (5–30 s) and frequency (1–2 regeneration injections). *For dengue virus proteins, most antibody-protein and protein–protein interactions can be regenerated with a short pulse of mild acid* [14, 21, 22].

3.4 Setting Up a Binding Analysis/ Kinetic Run

In order to run an experiment in which the kinetic and affinity constants can be determined, the following conditions need to be satisfied: (1) the selected concentration range provides a good spread of responses from low binding to close to saturation, (2) the injection time is sufficiently long to achieve equilibrium for most concentrations, (3) the dissociation time is sufficiently long to allow for visible decline in response, (4) the regeneration condition is efficient in regenerating the surface without damaging the immobilized ligand(s) (at least five identical injections should be performed to ensure that the binding signal should be exactly the same), and (5) the buffer baseline is stable for the next cycle run. If all these conditions are reached, it is possible to run a multi-cycle kinetic experiment which is described below. A properly run kinetic assay with at least five analyte concentrations yields the kinetic parameters, i.e., k_a, k_d, and the binding affinity K_D (Fig. 4).

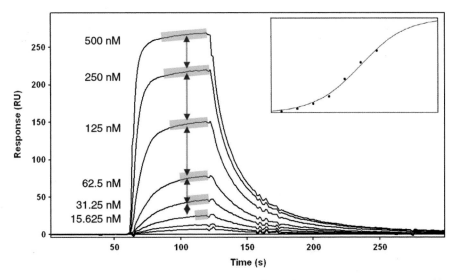

Fig. 4 Corrected sensorgrams of twofold serially diluted Fab 3F11 injected across immobilized DENV4 NS2B$_{18}$NS3 in HBS-EP buffer at 25 °C [17]. Sensorgram responses in the association phase should go from close-distant-close manner for a full-concentration range in increasing analyte concentration (*red arrow*). Saturation response can be calculated from a dose-response plot (inset). Green bar denotes the equilibrium response for each of the corresponding analyte concentrations where rate of complex formation balances with rate of complex dissociation

1. Start a *Binding analysis* run [*Run → Run Application Wizard → Binding analysis/Kinetic analysis → Start*] and then follow the on-screen instructions.

2. Set up a threefold serial-dilution sample run in the concentration range and conditions that have been predetermined in *Test binding of analyte and scouting of regeneration condition* above. For example, 2 min buffer stabilization; 1, 3, 11, 33, 100, 300 mM analyte; 2 min injection; 15 min dissociation; 30 s 10 mM HCl regeneration, 2 min buffer stabilization after regeneration.

3.5 Data Processing and Model Fitting

Model fitting is kept to the most basic 1:1 Langmuir bimolecular interaction model (A + B → AB), which can describe most of the biological interactions and reports the rates of reaction (k_a and k_d) and the affinity constant (K_D).

1. Visually check through all binding analysis/kinetic run sensorgrams and exclude any incorrect responses in the analysis [*File → Open → Select .blr result file → Import curves → highlight curves → Overlay plot*].

2. Step 1: Load all the sensorgrams from the test flow channel (where interacting ligand is immobilized) and a reference flow channel (where an unrelated protein is immobilized) into *BIAevaluation* software (Biacore).

3. Step 2: Subtract all reference responses (sensorgrams from reference channel) from the corresponding test responses (sensorgrams from test channel) to obtain specific binding responses (*see* **Note 8**).

4. Step 3: Perform a Y-alignment (zero buffer baseline at *Y*-axis) on the *channel-referenced* responses at a point before injection start where buffer baseline response is stable [*Calculate→ Y-transform→ Zero at Average of Selection*] (Fig. 5a).

5. Step 4: Perform an X-alignment (zero buffer baseline at *X*-axis) at the point of injection start [*Calculate→ X-transform*] (Fig. 5b).

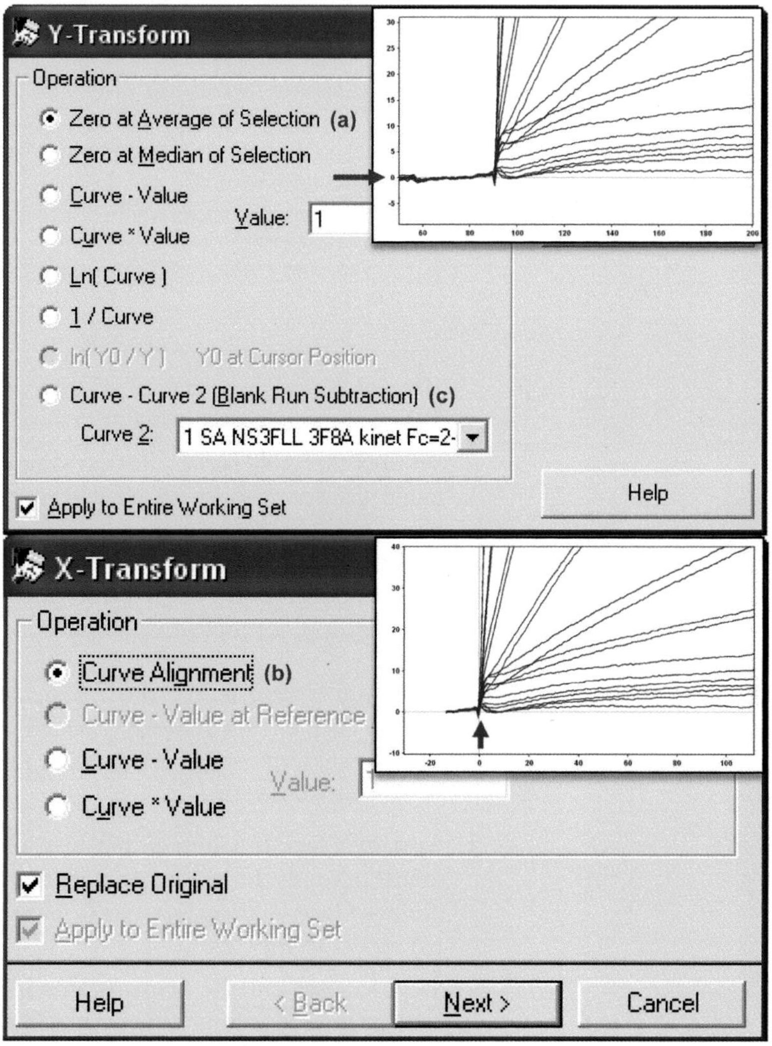

Fig. 5 Screenshots of BIAevaluation functions: Y-transform and X-transform. Functions mentioned in text are marked (**a**), (**b**), and (**c**) in *red*. Insets show results of the respective zero transforms at point indicated by *red arrows*

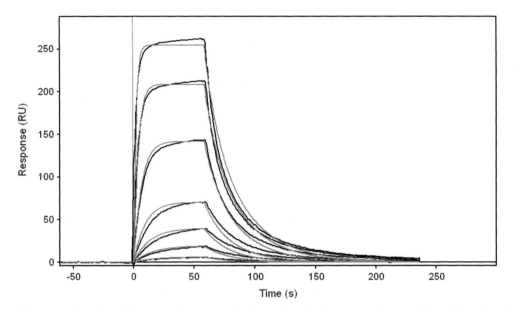

Fig. 6 Sensorgrams from Fig. 4 was fit to a 1:1 simple bimolecular interaction model (*red lines*) to yield $k_a = 8.48 \times 10^5$ M^{-1} s^{-1}, $k_d = 8.02 \times 10^{-2}$ s^{-1}, and $K_D = 95.6$ nM [17]

6. Step 5: Subtract a blank buffer cycle from ALL the aligned sensorgrams. In publications, this is commonly referred to as *buffer referencing* (*see* **Note 9**) [*Calculate* → *Υ-transform* → *Curve–curve 2 (Blank Run Subtraction)*] (Fig. 5c).

7. Step 6: Fit corrected sensorgrams to a 1:1 simple bimolecular interaction model to obtain the rates of association (k_a), dissociation (k_d), and affinity constant (K_D) (Fig. 6) (*see* **Note 10**).

8. Replicates of these kinetic parameters from independent experiments are used for calculation of standard errors (*see* **Note 11**).

3.6 Evaluation of Affinity Result

Evaluation is a final comprehensive task in any quantitative SPR study to assess, basically, if the value of the measured parameters of the interaction is biologically relevant (for instance, it is suspicious to have an affinity constant of 1 mM for a set of interactants which concentrations can never reach a few mM in cells!) and is not an artificial product influenced by an incorrect experimental set up or physical limitation in equipment. It is absolutely necessary to assess if the values obtained are relevant to the biological system.

3.7 Figure Preparation for Publication

Publication of SPR results has greatly matured over the last 10 years mainly through advocating from David Myszka's group [26, 27]. It is recommended that in every SPR figure published, it is important to have:

(a) Low overall binding responses (within 50 RU for sensitive equipment and within 100 RU for less sensitive ones).

(b) One or two highest concentration(s) that reached saturation (*see* **Note 12**).

(c) Overlay of analyte's replicates shown.

(d) Overlay with a 1:1 simple bimolecular interaction model shown.

(e) Measurable rate of dissociation.

(f) A wide concentration range (from saturation to near zero response).

(g) Buffer blanks.

4   Notes

1. Prepare all solutions using ultrapure water (treated deionized water to achieve a resistance of 18 MΩ at 25 °C) and then filter through a 0.22 mm membrane (Millipore) and degas (sonication, vacuum suction, etc.) for at least 5 min. Centrifuge all injected samples (henceforth termed *analyte*) for 5 min at 10,000 × g to pellet down any insoluble material and transfer the supernatant to a clean tube prior to any experimentation.

2. Sulfo-NHS (Pierce) is a good alternative for less charged biomolecules intended for higher immobilization.

3. Biacore recommends to run *desorb* once a week and *sanitize* once a month.

4. These concentrations of reagents will yield a response of about 6000–7000 RU when run under these conditions.

5. pH below 4 should never be used for amine-coupling procedure. Dengue nonstructural proteins tend to have a fair amount of nonspecific interaction with the negatively charged surface of sensor chip CM5. We have found the nonspecific interaction to be in increasing order from [NS3 helicase/NS3 protease/NS5 RNA-dependent RNA polymerase/NS5 methyltransferase] < full-length NS3 < full-length NS5 in ionic buffer. Hence, when monitoring the interaction between NS5 and NS3 proteins, it is preferable to immobilize the NS5 as ligand while passing through NS3 as the analyte [22]. The more diluted the analyte sample is, the closer its buffer composition is to the running buffer. Tenfold is usually sufficient to significantly remove any refractive index mismatches unless high-refractive index compounds like DMSO, glycerol, and high salt are present.

6. Sometimes near saturation or saturation cannot be achieved due to factors like sample availability, sample solubility, low affinity interaction, and inherent nature of the interaction. Under these circumstances, the concentration of analyte that generates the highest possible response will do.

7. The time to wait for dissociation in buffer before initiating a regeneration is very much a discretion of each researcher.

One should take into consideration factors like ligand/analyte stability, duration of the entire experimental assay, availability of equipment time, availability of efficient regeneration cycle, and whether the overall binding response has reduced by at least 10 %. We typically regenerate any dissociation that takes longer than 30 min. For interactions that no effective regeneration can be found, titration kinetics can be used [28]. NaOH should not be used as a regeneration solution because even a concentration of 50 mM was found to inactivate the surface ligands, NS3 and NS5.

8. In publications, this is commonly referred to as *channel referencing*.

9. *Channel referencing* and *buffer referencing* are collectively referred to as *double referencing* in publication.

10. Interaction with fast association and fast dissociation kinetics will only yield affinity constant (K_D), not k_a and k_d.

11. Usually a minimum of $n = 3$ is required for standard error determination.

12. Saturation is different from equilibrium. Each analyte concentration can have its own equilibrium but, collectively, only one saturation point (R_{max}).

Acknowledgements

This work was partly supported by research grants awarded to Dr. Subhash Vasudevan (DUKE-NUS, Singapore) and Dr. Julien Lescar (School of Biological Sciences, Nanyang Technological University, Singapore).

References

1. Rich RL, Myszka DG (2008) Survey of the year 2007 commercial optical biosensor literature. J Mol Recognit 21(6):355–400

2. Rich RL, Myszka DG (2009) Grading the commercial optical biosensor literature-Class of 2008: "The Mighty Binders". J Mol Recognit 23(1):1–64

3. Rich RL, Myszka DG (2011) Survey of the 2009 commercial optical biosensor literature. J Mol Recognit 24(6):892–914

4. Myszka DG, Arulanantham PR, Sana T, Wu Z, Morton TA, Ciardelli TL (1996) Kinetic analysis of ligand binding to interleukin-2 receptor complexes created on an optical biosensor surface. Protein Sci 5(12):2468–2478

5. Rich RL (2002) Kinetic analysis of estrogen receptor/ligand interactions. Proc Natl Acad Sci 99(13):8562–8567

6. Besenicar M, Macek P, Lakey JH, Anderluh G (2006) Surface plasmon resonance in protein-membrane interactions. Chem Phys Lipids 141(1–2):169–178

7. Navratilova I, Hopkins AL (2010) Fragment screening by surface plasmon resonance. ACS Med Chem Lett 1(1):44–48

8. Retra K, Irth H, van Muijlwijk-Koezen JE (2010) Surface plasmon resonance biosensor analysis as a useful tool in FBDD. Drug Discov Today Tech 7(3):e181–e187

9. Elinder M, Geitmann M, Gossas T, Kallblad P, Winquist J, Nordstrom H, Hamalainen M,

Danielson UH (2011) Experimental validation of a fragment library for lead discovery using SPR biosensor technology. J Biomol Screen 16(1):15–25

10. Papalia GA, Leavitt S, Bynum MA, Katsamba PS, Wilton R, Qiu H, Steukers M, Wang S, Bindu L, Phogat S, Giannetti AM, Ryan TE, Pudlak VA, Matusiewicz K, Michelson KM, Nowakowski A, Pham-Baginski A, Brooks J, Tieman BC, Bruce BD, Vaughn M, Baksh M, Cho YH, Wit MD, Smets A, Vandersmissen J, Michiels L, Myszka DG (2006) Comparative analysis of 10 small molecules binding to carbonic anhydrase II by different investigators using Biacore technology. Anal Biochem 359(1):94–105

11. Geitmann M, Unge T, Danielson UH (2006) Interaction kinetic characterization of HIV-1 reverse transcriptase non-nucleoside inhibitor resistance. J Med Chem 49(8):2375–2387

12. Richalet-Sécordel PM, Rauffer-Bruyère N, Christensen LL, Ofenloch-Haehnle B, Seidel C, Van Regenmortel MH (1997) Concentration measurement of unpurified proteins using biosensor technology under conditions of partial mass transport limitation. Anal Biochem 249(2):165–173

13. Chavane N, Jacquemart R, Hoemann CD, Jolicoeur M, De Crescenzo G (2008) At-line quantification of bioactive antibody in bioreactor by surface plasmon resonance using epitope detection. Anal Biochem 378(2):158–165

14. Rajamanonmani R, Nkenfou C, Clancy P, Yau YH, Shochat SG, Sukupolvi-Petty S, Schul W, Diamond MS, Vasudevan SG, Lescar J (2009) On a mouse monoclonal antibody that neutralizes all four dengue virus serotypes. J Gen Virol 90(Pt 4):799–809

15. Zheng A, Umashankar M, Kielian M (2010) In vitro and in vivo studies identify important features of dengue virus pr-E protein interactions. PLoS Pathog 6(10):e1001157

16. Zhang Q, Hunke C, Yau YH, Seow V, Lee S, Tanner LB, Guan XL, Wenk MR, Fibriansah G, Chew PL, Kukkaro P, Biukovic G, Shi PY, Shochat SG, Grüber G, Lok SM (2012) The stem region of premembrane protein plays an important role in the virus surface protein rearrangement during dengue maturation. J Biol Chem 287(48):40525–40534

17. Schmidt AG, Lee K, Yang PL, Harrison SC (2012) Small-molecule inhibitors of dengue-virus entry. PLoS Pathog 8(4):e1002627

18. Li PC, Liao MY, Cheng PC, Liang JJ, Liu IJ, Chiu CY, Lin YL, Chang GJJ, Wu HC (2012) Development of a humanized antibody with high therapeutic potential against dengue virus type 2. PLoS Negl Trop Dis 6(5):e1636

19. Balsitis SJ, Williams KL, Lachica R, Flores D, Kyle JL, Mehlhop E, Johnson S, Diamond MS, Beatty PR, Harris E (2010) Lethal antibody enhancement of dengue disease in mice is prevented by Fc modification. PLoS Pathog 6(2):e1000790

20. Fry SR, Meyer M, Semple MG, Simmons CP, Sekaran SD, Huang JX, McElnea C, Huang CY, Valks A, Young PR, Cooper MA (2011) The diagnostic sensitivity of dengue rapid test assays is significantly enhanced by using a combined antigen and antibody testing approach. PLoS Negl Trop Dis 5(6):e1199

21. Moreland NJ, Tay MYF, Lim E, Paradkar PN, Doan DNP, Yau YH, Geifman Shochat S, Vasudevan SG (2010) High affinity human antibody fragments to dengue virus non-structural protein 3. PLoS Negl Trop Dis 4(11):e881

22. Zou G, Chen YL, Dong H, Lim CC, Yap LJ, Yau YH, Shochat SG, Lescar J, Shi PY (2011) Functional analysis of two cavities in flavivirus NS5 polymerase. J Biol Chem 286(16):14362–14372

23. Bodenreider C, Beer D, Keller TH, Sonntag S, Wen D, Yap L, Yau YH, Shochat SG, Huang D, Zhou T, Caflisch A, Su XC, Ozawa K, Otting G, Vasudevan SG, Lescar J, Lim SP (2009) A fluorescence quenching assay to discriminate between specific and nonspecific inhibitors of dengue virus protease. Anal Biochem 395(2):195–204

24. O'Shannessy DJ, Brigham-Burke M, Peck K (1992) Immobilization chemistries suitable for use in the BIAcore surface plasmon resonance detector. Anal Biochem 205(1):132–136

25. Johnsson B, Löfås S, Lindquist G (1991) Immobilization of proteins to a carboxymethyldextran-modified gold surface for biospecific interaction analysis in surface plasmon resonance sensors. Anal Biochem 198(2):268–277

26. Rich RL, Myszka DG (2003) A survey of the year 2002 commercial optical biosensor literature. J Mol Recognit 16(6):351–382

27. Center for Biomolecular Interaction Analysis. http://www.cores.utah.edu/interaction/index.php.

28. Karlsson R, Katsamba PS, Nordin H, Pol E, Myszka DG (2006) Analyzing a kinetic titration series using affinity biosensors. Anal Biochem 349(1):136–147

# Chapter 18

Functional Genomics Approach for the Identification of Human Host Factors Supporting Dengue Viral Propagation

Nicholas J. Barrows, Sharon F. Jamison, Shelton S. Bradrick, Caroline Le Sommer, So Young Kim, James Pearson, and Mariano A. Garcia-Blanco

Abstract

Dengue virus (DENV) is endemic throughout tropical regions of the world and there are no approved treatments or anti-transmission agents currently available. Consequently, there exists an enormous unmet need to treat the human diseases caused by DENV and block viral transmission by the mosquito vector. RNAi screening represents an efficient method to expand the pool of known host factors that could become viable targets for treatments or provide rationale to consider available drugs as anti-DENV treatments. We developed a high-throughput siRNA-based screening protocol that can identify human DENV host factors. The protocol herein describes the materials and the procedures necessary to screen a human cell line in order to identify genes which are either necessary for or restrict DENV propagation at any stage in the viral life cycle.

Key words RNA interference (RNAi), Dengue virus, Yellow fever virus, Whole-genome RNAi screening, Whole-genome siRNA screening, Dengue virus host factors, Flavivirus

1 Introduction

Dengue virus (DENV) is endemic throughout tropical regions around the world as is its primary mosquito vector, *Aedes aegypti*. The number of countries reporting dengue-related diseases to the World Health Organization increased at a rate of 13 countries per decade from 1955 to 2004, demonstrating an emerging threat to populations worldwide [1]. Currently, there is no effective vaccine that can protect against all four DENV serotypes, and only palliative care is available for infected patients suffering from severe disease. The only proven way to combat urban outbreaks requires a combination of public health measures, including promotion of mosquito control and personal protection [2]. Consequently, there exists an enormous unmet need to treat the human diseases caused by DENV and block viral transmission by the mosquito vector.

Radhakrishnan Padmanabhan and Subhash G. Vasudevan (eds.), *Dengue: Methods and Protocols*, Methods in Molecular Biology, vol. 1138, DOI 10.1007/978-1-4939-0348-1_18, © Springer Science+Business Media, LLC 2014

Rapid identification of host factors and biological processes that either promote or restrict virus propagation may advance development of antiviral therapeutics. The recent utilization of whole-genome RNA interference (RNAi) screens identified hundreds of human and insect host factors that are required for efficient DENV propagation [3, 4]. RNAi is a cellular process in which short interfering RNAs (siRNA) bind perfectly complementary mRNA targets and induce their endonucleolytic cleavage [5, 6]. The consequence of RNAi is the depletion of the targeted mRNA and their protein products. Genome-scale RNAi uses either chemically synthesized siRNA duplexes or expressed short hairpin RNAs (shRNA) to target mRNAs. These libraries of siRNAs and shRNAs are used to systematically evaluate the roles that genes play in various biological processes [7]. RNAi screening represents an efficient method to expand the pool of known host factors that could become viable targets for treatments or provide rationale to consider available drugs as anti-DENV treatments.

We developed a high-throughput siRNA-based screen to identify human DENV host factors. Our strategy used two independent siRNA libraries which are simultaneously tested. Host factor candidates are identified only when the two independent tests provide mutually supportive results (*see* **Note 1**) [8, 9]. The protocol herein describes the materials and the procedures used to screen 2,240 genes or one batch of siRNA containing plates. An entire genome-scale screen is divided into eleven nearly equal batches. Reverse transfection of siRNAs into cells, infection of the transfected cells with DENV, and fixation and staining of the infected cells are detailed (*see* **Notes 2** and **4**).

2 Materials

2.1 Tissue Culture, siRNA Transfection, and Virus Infection

1. Tissue culture medium: DMEM, 10 % FBS, PS 100 U/mL.
    (a) Dulbecco's modified Eagle medium (DMEM).
    (b) Penicillin-Streptomycin (PS), 10,000 U/mL.
    (c) Heat-inactivated fetal bovine serum (FBS).
2. Complete screening medium: DMEM, 5 % FBS, 0.01 M HEPES, PS 100 U/mL.
    (a) Dulbecco's modified Eagle medium (DMEM).
    (b) Penicillin-Streptomycin (PS), 10,000 U/mL.
    (c) Heat-inactivated fetal bovine serum (FBS).
    (d) HEPES, 1 M Buffer Solution.
3. Phosphate buffered saline without calcium or magnesium (PBS).
4. Trypsin/EDTA, 0.25 %.
5. Opti-MEM® Reduced Serum Medium, no Phenol Red (Opti-MEM) (Gibco# 11058-021).

6. HuH-7 cells (*see* **Note 11**).

7. Qiagen siRNA duplex targeting green fluorescent protein (Qiagen# SI04380467).

8. Qiagen siRNA duplex targeting ATP6V0C (ATP6V0C siRNA) (Qiagen# SI00307384).

9. Qiagen AllStars Negative Control siRNA (Qiagen # 1027280).

10. Qiagen Human Whole-Genome siRNA set, Library v1.0 (*see* **Note 5**).

11. 384 Well Flat Clear Bottom Black Polystyrene TC-Treated Microplates (microtiter assay plate) (Corning# 3712).

12. 75 cm² tissue culture flasks, vented caps (T-75 flask).

13. Lipofectamine RNAiMax (Invitrogen# 13778-150).

14. Corning® 125 mL Octagonal PET Storage Bottles, Sterile (Corning # 431731).

15. Corning® 500 mL Octagonal PET Storage Bottles, Sterile (Corning # 431733).

16. Dengue 2 virus, New Guinea C strain (*see* **Note 14**).

17. 384ST 70 µl Tips, Sterile, (ST70 tips) (Agilent Technologies # 19133-012).

18. 1, 5, 10, 25, and 50 mL serological pipettes.

19. Omni Trays (Thermo # 242811).

20. Matrix WellMate Tubing Assemblies (Matrix cartridge) (Thermo # 201-30002).

21. Thermo Scientific Matrix WellMate Flexible, High-Speed, 8-Channel Microplate Dispenser.

22. Rainin manual multichannel pipette (Rainin# L12-20XLS; P20).

23. Sorvall Legend RT + centrifuge with Sorvall S1102 swinging bucket rotor.

2.2 Fixation and Viral Envelope Protein Staining

1. 4 % Paraformaldehyde (w/v) (4 % PFA):

    (a) Make at least 3,300 mL 4 % PFA diluted in PBS immediately prior to the genomic screen. Filter through a 0.45 µm filter to remove particulates which disrupt image analysis and store at 4 °C.

2. 4,500 mL PBS with 0.1 % Tween-20 (PBS-Tween):

    (a) Add 450 mL 10× Dulbecco's Phosphate Buffered Saline to 4,050 mL 0.45 µm filtered H₂O. Filtering is necessary to remove particulates which disrupt image analysis.

    (b) Add 4.5 mL Tween-20.

    (c) Mix well.

3. 10× Dulbecco's Phosphate Buffered Saline.

4. Tween-20, Sigma# P1370-100ML.

5. bisBenzamide H 33342 trihydrochloride (Hoechst):

    (a) Dilute 10 mg/mL in dH$_2$O, store at 4 °C protected from light.

6. Triton X-100.

7. Pan-flavivirus anti-DENV envelop protein mouse monoclonal antibody (4G2) [10].

8. Goat anti-mouse AlexaFluor 488 (Alexa488).

9. Normal Donkey Serum (Millipore# S30-100 mL).

10. Bottle top Vacuum filter, 1 L.

11. Rainin LTS pipette tips 20 µL.

12. Rainin LTS pipette tips 200 µL.

13. Matrix pipette tips 1,250 µL (Matrix# 8042).

14. Adhesive plate sealing film (Fisher# AB-0580).

15. Reagent reservoirs.

16. Ethanol 70 %.

17. Ethanol 100 %.

18. Distilled water.

19. Kimtech Science Kimwipes.

20. Velocity11/Agilent Bravo Liquid Handling Platform.

21. Biotek ELx405 96 well plate washer.

22. Cellomics Target Activation software package and associated ArrayScan^{VTI} instrumentation and robotic arm.

23. VP179 aspiration manifold, V&P Scientific.

24. Matrix Impact2 Multichannel Electronic Pipette.

25. Baker BioPROtect II biological safety cabinet.

---

3   Methods

3.1 Tissue Culture, Day 0

1. Prepare five T-75 flasks each with 2.0×10^6 HuH-7 cells in 14 mL tissue culture medium (*see* **Note 3**).

3.2 Reverse Transfection, Day 2 Time Started:

1. Remove 14 microtiter assay plates from the –80 °C freezer (*see* **Notes 4** and **5**). Set on the bench to defrost at room temperature. Record microtiter assay plate ID labels.

2. Warm the incomplete screening medium (DMEM, 5 % FBS, PS), 0.25 % Trypsin, PBS, Opti-MEM, 1 M HEPES in water bath set to 37 °C. The HEPES is added to the incomplete screening medium during subsequent steps to create the complete screening medium.

3. Sanitize the laminar flow cabinet with 70 % ethanol.

4. Aliquot 200 mL 70 % ethanol into a T-75 flask. Retrieve the Matrix cartridge (*see* **Note 6**). Run ethanol through the Matrix cartridge in order to sanitize it. Let sit until needed with ethanol in the line.

5. When microtiter assay plates are fully defrosted, centrifuge the plates at room temperature at $500 \times g$ for 1 min in a centrifuge with swinging bucket rotor. Transfer plates to the laminar flow cabinet.

6. Sanitize the outside of the micotiter assay plates with 70 % ethanol.

7. Ensure that there is at least 300 µL of 0.2 µM ATP6V0C siRNA and 0.2 µM AllStars Negative Control siRNA solutions (*see* **Note 7**).

8. Remove the plate seals from all microtiter assay plates carefully.

9. For each microtiter assay plate, use the P20 multichannel pipette to:

   (a) Remove all dH$_2$O from wells E2, G2, I2, K2 & F23, H23, J23, L23.

   (b) Add 5 µL 0.2 µM ATP6V0C siRNA to wells E2, G2, J23, L23.

   (c) Add 5 µL 0.2 µM AllStars Negative Control siRNA to wells I2, K2, F23, H23.

   (d) Any remaining siRNA can be frozen and used the following day.

10. Centrifuge the microtiter assay plates at room temperature at $500 \times g$ for 1 min in order to make sure the control siRNAs are concentrated at the bottom of the well.

11. Make the transfection mixture by mixing 360 µL of RNAimax with 71.6 mL of Opti-MEM in the 125 mL storage bottles. Mix well but do not create bubbles (*see* **Note 8**).

12. Run the 70 % ethanol partially through the Matrix cartridge thereby creating an air gap between the ethanol and the end of the Matrix cartridge line. Submerge the Matrix cartridge end in PBS and run 15 mL of PBS through the line to rinse out the ethanol. Run the PBS partially out of the line creating a new air gap.

13. Submerge the end of the cartridge into the transfection mixture and minimally prime the line.

14. Set the Matrix WellMate to dispense 10 µL per well for all columns.

15. Load one assay plate. Be sure "A1" is in the upper left. Remove the lid. Press start (*see* **Note 9**).

16. Replace lid on the first plate and set aside. Fill the remaining plates.

17. Centrifuge the microtiter assay plates at $800 \times g$ for 10–20 s to ensure that the transfection mixture reaches the siRNA solution (*see* **Note 10**).

18. Program termination:

    (a) Run the remaining transfection mixture out.

    (b) Run 15 mL PBS through the cartridge.

    (c) Run 15 mL 70 % ethanol through the line.

    (d) Let the cartridge line sit submerged in 70 % ethanol while you prepare the HuH-7 cells for dispensing.

19. Add 3.75 mL 1 M HEPES to 375 mL of DMEM, 5 % FBS, PS to create the complete screening medium.

20. Retrieve 3 out of the 5 assigned T-75 flasks containing that day's HuH-7 cells (*see* **Note 11**).

21. Remove the growth media from the flasks and rinse the HuH-7 cells with 4 mL PBS. Remove the PBS. Then rinse the HuH-7 cells with 1.5 mL of Trypsin/EDTA, 0.25 %. Remove the Trypsin/EDTA, 0.25 % and discard.

22. Add 1.5 mL fresh Trypsin/EDTA, 0.25 % to each flask, and return to the 37 °C incubator for 5 min.

23. Remove the Trypsin-treated flasks and tap the flasks in order to dislodge the cells.

24. Add 10 mL of the complete screening medium to each flask. Triturate vigorously several times to dissociate the HuH-7 cells from each other. Pool the cell suspension from all three flasks and remove 0.5 mL of cell suspension for counting on a hemocytometer.

25. Calculate how much cell suspension is necessary to complete dispensing (*see* **Note 12**).

26. Prepare the solution in the sterile 500 mL storage bottles.

27. Run the ethanol out of the cartridge and rinse with 15 mL PBS.

28. Prime the cartridge with 25 mL of cell suspension.

29. Check that the dispense volume is set to 50 μL per well.

30. Place the first microtiter assay plate into the loading tray with A1 in the upper left.

31. Remove lid and press start. After dispensing the cell suspension, replace lid on the microtiter assay plate and set aside. Continue filling all 14 plates in this manner (*see* **Note 9**).

32. Set microtiter assay plates on the stainless steel floor of the laminar flood hood and incubate for 5 min at room temperature prior to placing all assay plates on a single shelf of the 37 °C tissue culture incubators.

33. Time completed _____.

34. Program termination:

    (a) Run the cell suspension out of the line.

    (b) Run 15 mL PBS through the cartridge.

    (c) Run 15 mL 70 % ethanol through the cartridge.

    (d) Let the cartridge line sit submerged in 70 % ethanol for 10 min. Then run the ethanol out from the line.

3.3 DENV Infection, Day 4

1. Addition of DENV 52 h after siRNA treatment using the Bravo Liquid Handling Platform from Velocity11 for the delivery of 20 μL of viral-containing complete screening medium into the microtiter assay plates (*see* **Notes 6** and **13**).

2. The Velocity11 and Bravo unit are housed within a Baker BioPROtect II biological safety cabinet approved for biosafety level 2 projects.

3. Warm at least 200 mL of complete screening media at 37 °C.

4. Set an Omni Tray onto position 1 on Bravo deck then fill with 50 mL 70 % ethanol.

5. Set an Omni Tray onto position 2 on Bravo deck and fill with 50 mL 100 % ethanol.

6. Set an Omni Tray with fresh Kimwipes onto position 3 on Bravo deck.

7. Set an Omni Tray onto position 4 on Bravo deck then fill with 50 mL sterile dH_2O.

8. Set a box of sterile ST70 tips onto position 5 on Bravo deck (*see* **Note 18**).

9. Set an empty Omni Tray onto position 7 on Bravo deck.

10. Don all appropriate biohazard clothing. The appropriate personal protection equipment will be determined by each institution and may include biosafety level 2 protections including a gown, double-layer non-latex gloves and eye protection.

11. Make complete screening medium by addition of 1.25 mL 1 M HEPES to 123.8 mL of the DMEM, 5 % FBS, PS into a sterile 125 mL bottle.

12. Defrost 4 mL of the DENV at 37 °C (*see* **Note 14**).

13. Mix the DENV viral stock gently with a 5 mL pipette.

14. Add 3.4 mL of DENV stock to the 125 mL of complete screening medium.

15. Mix the diluted viral stock gently 3× with a 25 mL pipette.

16. Add at least 50 mL of diluted DENV stock into the empty Omni Tray at position 7 on the Bravo deck. Replenish the viral-containing media in the Momentary as needed throughout the procedure.

17. Load viral delivery program (*see* **Note 15**).

18. Set assay for 14 replications.

19. Obtain microtiter assay plates from the 37 °C incubator one plate at a time.

20. Take 1st uninfected microtiter assay plate from the 37 °C incubator and transfer it to the Baker BioPROtect II biosafety cabinet and place it on position 8 on the Bravo deck (*see* **Note 16**).

21. Remove all lids.

22. Run viral delivery program (*see* **Note 16**).

23. At the "pause" replace lid on the infectious assay plate on position 8. It is important to note that the plate is now infectious and any spills should be treated using the appropriate biosafety cleanup protocol.

24. Remove the plate from position 8 in the Baker BioPROtect II biosafety cabinet and return to the 37 °C incubator. Care must be taken to follow all established biosafety protocols. Return to **step 20**. Repeat for each assay plate.

25. Replenish the viral-containing media as needed in the Omni Tray at position 7 by addition of 50 mL.

26. ST70 tips at position 5 on Bravo deck are reused throughout the screen. The tips are decontaminated inside the Baker BioPROtect II. The viral delivery program ends by washing the ST70 tips in dH$_2$O, then in 70 % ethanol, and then in 100 % ethanol and finally blotting dry on sterile Kimwipes. Tips are left at position 5 on Bravo deck for use in subsequent days.

27. When done, discard all viral-containing or exposed liquids into bleach with a final concentration of bleach not to be less than 10 %. Each virus-contaminated Omni Tray is exposed to 10 % bleach and then discarded into biohazard bags.

3.4 Fixation of HuH-7 Cells 42 h After Infection with DENV, Day 6

1. Make 300 mL of PBS with 0.5 % Triton-X100.

2. Make sure you have on hand ~4,500 mL PBS-Tween.

3. Obtain 300 mL ice cold 4 % PFA.

4. Empty the collection bottles associated with the Biotek ELx405 plate washer (*see* **Note 6**).

5. Fill the dispensing bottles associated with the Biotek ELx405 plate washer with PBS-Tween.

6. Prime the Biotek ELx405 plate washer by running 100 mL of PBS-Tween wash buffer through the system to remove previously used buffer and air bubbles before washing the first plate each day.

7. Add 1.65 mL normal donkey serum to each of three aliquots of 165 mL PBS-Tween. Label the bottles as indicated and store on ice or at 4 °C:

   (a) Blocking buffer.

   (b) Primary antibody.

   (c) Secondary antibody and Hoechst.

8. Don all appropriate biohazard clothing. The appropriate personal protection equipment will be determined by each institution and may include biosafety level 2 protections including a gown, double-layer non-latex gloves, and eye protection.

9. Install the VP179 aspiration manifold in the laminar flow hood (*see* **Note 17**).

10. Remove seven microtiter assay plates from the 37 °C incubator.

11. Remove all except 15 μL of virus-containing media from each well using the VP179 aspiration manifold in the laminar flow hood (*see* **Note 17**).

12. Add 50 μL of ice cold 4 % PFA to each well using the Matrix electronic pipette.

13. Incubate each microtiter assay plate for 12 min.

14. After fixation, using the Biotek ELx405, wash each microtiter assay plate. The Biotek ELx405 is programmed to add and remove 50 μL of PBS-Tween to each well 3 times.

15. Permeabilize the cells by adding 50 μl of PBS with 0.5 % Triton X-100 to each well using the Matrix Impact2 Multichannel Electronic Pipette.

16. Incubate each plate for 15 min.

17. After permeabilization, repeat wash **step 14**.

18. Add 30 μL of the blocking buffer containing PBS-Tween and 1 % normal donkey serum using the Matrix Impact2 Multichannel Electronic Pipette to each well. Let this first set of plates incubate in this blocking buffer while you treat the second set of seven plates.

19. Return to **step 9**, and repeat **steps 9–18** with the second set of seven microtiter assay plates.

   (a) The catch-up point is **step 18**.

   (b) After addition of the blocking solution to the first plate of the second set, start timer to incubate for 50 min.

20. During the incubation period, add 165 μL of 4G2 antibody to the 165 mL of PBS-Tween, 1 % normal donkey serum labeled "primary antibody" bottle prepared earlier.

21. After 50 min, wash each microtiter assay plate using the Biotek ELx405 as in **step 14**.

22. Add 30 µL of "primary antibody" solution to each well. Start timer to incubate 1 h right after addition to the first microtiter assay plate.

23. During the incubation period, add 82.5 µL of Alexa488 and 165 µL Hoechst to PBS-Tween, 1 % normal donkey serum labeled "Secondary Antibody and Hoechst" bottle prepared earlier.

    (a) Work with least ambient light possible to protect the fluorescent dyes on Alexa488.

24. After 1 h, wash each microtiter assay plate using the Biotek ELx405 as in **step 14**.

25. Add 30 µL of "Secondary Antibody and Hoechst" mixture to each well. Start timer to incubate 1 h right after the first plate is done. Incubate under a piece of aluminum foil to protect the plates from light.

26. After 1 h, wash each microtiter assay plate using the Biotek ELx405 as in **step 14**.

27. Add 50 µL PBS.

28. Hermetically seal, clean imaging surface with 95 % ethanol, and set up the Array Scan.

4   Notes

1. Ideally, each siRNA treatment leads to profound depletion of the host protein(s) encoded by the targeted transcripts and does not affect the levels of any other protein in the cell. Practically, however, siRNAs exhibit variable potency against their intended target mRNAs and can downregulate expression of unintended ones—the so-called off-target effects [11]. As such, the standard of the field requires that any RNAi-based result is deemed reliable only when multiple independent siRNAs provide evidence that supports the hypothesis. Moreover, it is desirable that RNAi-mediated phenotypes be rescued by re-expression of a siRNA-resistant mRNA that encodes the protein of interest, and additionally it is judicious to confirm RNAi phenotypes by other genetic or biochemical methods. At the completion of a genome-scale screen using an RNAi library design similar to the design we employ, the screener(s) generates a hit list in which each hit has at least two siRNAs that produce the interrogated phenotype.

2. We always test our proposed procedures prior to beginning any large-scale screen. First, the entire written procedure is applied using a single microtiter assay plate from the siRNA library. The outcome of this practice is to determine if there exists a significant error in the written or planned procedure

which would lead to failure to produce high-quality data. We recommend that unexpected observations made during this test are explored prior to moving forward with a pilot screen. Second, the procedure is tested against a set of microtiter assay plates during a pilot screen. The goal of the pilot screen is to determine if there is any reason not to move forward with the complete screen. The pilot screen represents the last opportunity to identify problems or reconsider the experimental design. The ideal pilot screen assay plates permit the screening team to complete every step of the procedure for a single batch of microtiter assay plates including limited data analysis with the intention of producing a short "hit" list. Further, the same set of microtiter assay plates will be tested again during the screen, and the data from the pilot screen and genomic screen can be compared directly. It is expected that the results from individual wells will behave similarly between the pilot and genomic screen plates, which demonstrates that the protocol is applied effectively.

3. Three T-75 flasks of HuH-7 cells are used for the siRNA reverse transfection each day. Two T-75 flasks of HuH-7 cells are used for tissue culture in order to generate the next five T-75 flasks of HuH-7 cells needed for another batch of microtiter assay plates. In order to avoid passaging the same cells every day, we set up parallel cell passages. Passage A was used to start screens on days even numbered days, while Passage B was used to start screens on odd-numbered days.

4. The number of microtiter assay plates which are prepared each day is dependent on how many plates can be efficiently processed by the screening team and dependent on the capacity to image the plates before the next batch is ready the following day. We chose to process 14 microtiter assay plates each day. Therefore, there are 11 batches for an entire genomic screen with two separate libraries formatted on a total of 144 plates. We arranged the screening schedule such that a new batch was started each day so the entire screen takes less than 3 weeks from initiating cell culture to producing the first genomic database of screening data.

5. Our protocol applied the Qiagen Human Whole-Genome siRNA set, Library v1.0 toward interrogating 22,909 predicted mRNAs in which each mRNA is targeted by four unique chemically synthesized siRNA duplexes labeled A, B, C, or D. Ideally, each of the four siRNAs targets a unique, non-overlapping sequence which does not have homology with a different gene; however, this is not always the case. The siRNAs are arrayed in columns 3–22 and rows A-P of the microtiter assay plate. Each well contains 1pmol total siRNA in 5 μL dH$_2$O. The siRNAs are arrayed prior to the project and stored

at −80 °C. The final concentration for the siRNAs, after addition of Opti-MEM, RNAimax, and cell suspension, is 15.4nM. Our library format uses a 2×2 pooled arrangement such that siRNA duplexes A and B targeting gene X are arrayed in one well of a 384 well microtiter assay plate, while siRNA duplexes C and D targeting gene X are arrayed on a separate 384 well microtiter assay plate. Both siRNA pools are located at analogous positions within each paired 384 well microtiter assay plate creating two independent libraries which differ only by siRNA sequence identity. The AB and CD libraries are screened and high-confidence targets are identified when positive in both libraries thereby limiting false-positive results incurred by testing a single pool of siRNAs multiple times [8]. Indeed siRNA-induced phenotypes are very precise and repetition of single pools of siRNAs is unlikely to improve on accuracy. While our approach minimizes false-positives, one caveat associated with this strategy is that it likely elevates the rate of false-negatives due to ineffective siRNA sequences.

6. Consistent implementation of the screening protocol maximizes the distinction between a negative control population and valuable hits. Our group utilizes several useful technologies which facilitate screening:

(a) The Thermo Scientific Matrix WellMate Flexible, High-Speed, 8-Channel Microplate Dispenser evenly dispensed the transfection reagent and HuH-7 cells into the microtiter assay plates.

(b) The consistent distribution of DENV-containing media within wells on each plate, between plates, and across several days in a screen is essential in order to identify wells with low infection. The Velocity11/Agilent Bravo Liquid Handling Platform delivers and gently mixes viral-containing media in the 384 microwell plate.

(c) Incomplete microwell washing could prevent identification of interesting hits. The wash steps during immuno-fluorescence staining procedure must be performed consistently so that the fluorescent signal from each well accurately reflects the DENV infection within the well. The Biotek ELx405 96 well plate washer is used to perform all wash steps.

(d) Finally, the Cellomics Target Activation software package is applied to images gathered by the ArrayScan^{VTI}. While the liquid handling apparatus improve consistent application of the protocol, the imaging and analysis of the well directly determines which statistics will contribute to the final analysis; therefore, great care must be taken when developing the computation portion of the project.

7. The siRNA duplex targeting green fluorescent protein (GFP siRNA) and AllStars Negative Control siRNA served as the negative controls, while siRNA duplex targeting the vacuolar ATPase served as the positive control during assay development and RNAi screening. The vacuolar ATPase is a host enzyme necessary for acidification of the endosome and essential for DENV infection. Ideally any negative control(s) used during assay development provides an adequate estimation of the variability of the diverse genomic population, although during assay development this is unlikely to be known. Qiagen included the GFP siRNAs as a negative control on every microtiter assay plate and we added the AllStars Negative Control siRNA during the screen. Post-screen analysis determines that the combined distribution for the negative controls mirrors the genomic distribution, but each negative control behaves differently from the other. The ideal positive control would be an endogenous gene which, when targeted by siRNA, results in a highly reproducible, strong change in the phenotype, relative to the negative control. For assay development, the Qiagen siRNA duplex targeting green fluorescent protein served as a negative control, while Qiagen siRNA duplex targeting the endogenous gene ATP6V0C, a subunit of the vacuolar ATPase, was used as a positive control.

8. 53.76 mL of Opti-MEM and transfection reagent are needed to fill 14 microtiter assay plates. An additional 7.5 mL is needed to prime the Matrix cartridge line, 7.5 mL is needed to ensure that line stays filled until the end, and 7.0 mL is necessary to ensure that the end of the Matrix cartridge line remains submerged throughout the dispensing step. Importantly, the exact conditions we describe may not result in the best transfections for every cell type.

9. The Matrix WellMate will move the microtiter assay plate across the deck. Ensure that the vacuum hose does not get tangled with moving parts as it could disrupt delivery of the transfection mixture. Also, it is important to observe the microtiter assay plate orientation because the operator can identify if row or column errors may be introduced due to an unexpected instrument malfunction.

10. According to the manufactures recommendation, the siRNA and transfection mixture should be allowed to incubate for at least 20 min. Generally, the trypsin, cell counting, and dilution of the HuH-7 cells can be completed easily within this time limit.

11. Two qualities possessed by HuH-7 cells makes the cell line an ideal choice for siRNA screens [12]. First, the HuH-7 cell line is susceptible to DENV infection and produces infectious

progeny virus, essential to our assay, which evaluates all stages in viral lifecycle. Second, HuH-7 cells grow as an evenly distributed monolayer with well-separated and regularly ovoid nuclei facilitating automated imaging and analysis.

12. For one batch prepare 330 mL of a 2.4×10^4 cells/mL solution in a 500 mL tissue culture grade disposable bottle. Each well within the microtiter assay plate will receive 0.050 mL cell suspension with 1,200 HuH-7 cells per well.

13. The duration of and calculated multiplicity of infection (MOI) for the initial infection of DENV was optimized to provide consistent and maximal differentiation between the negative and positive control populations. The calculated MOI uses the estimated number of cells in a well and the Vero-derived viral titer (*see* **Note 14**). The reverse transfection conditions, cell plating density, and RNAi duration have been established [3, 8, 9]. DENV viral production from HuH-7 cells can be observed 20 h postinfection (data not shown). In order for the genomic screen to identify gene products necessary for any step(s) in the viral life cycle, we permitted DENV to replicate for 42 h. The Z' factor of 0.87 was observed when an MOI of 0.2 was applied to the cells suggesting an "excellent" assay [13]. All further assay development and genomic screening used this set of conditions.

14. The HuH-7 cells were infected with MOI 0.2 using dengue 2, New Guinea C strain viral stock produced by the *Aedes albopictus* derived C6/36 cell line. The viral titer is determined by infection of Vero (African green monkey kidney) cell monolayers followed by quantification by a foci formation assay.

15. The Velocity11 programming software BenchWorks v3.0.0 permits the user to assign predesigned tasks to each position on the deck. The user adjusts the height, volume, speed, and order of the tasks within the software controlling the Bravo unit. The development of the software program is completed well before the screen and must be carefully evaluated for errors by individuals trained to operate the system. At this time, Agilent has discontinued BenchWorks v3.0.0 and replaced it with VWorks.

16. We programmed the Velocity11 software BenchWorks v3.0.0 to dispense 20 µL of virus-containing media to each well of the assay plate and mix twice slowly in a volume of 40 µL. Finally the tips are touched lightly to the inside of each well to remove any liquid before re-lidding and moving the next assay plate onto position 8 for infection. The assay plates are infected in a predetermined order so that systematic errors may be easier to identify during assay analysis.

17. The VP179 aspiration manifold is a simple lever system that consistently removes a predetermined volume of virus-containing

media from all 384 wells of a single assay plate. This system is used to remove all except 15 μL of media to ensure that cells do not dry out before adding 4 % PFA.

18. Remove the ST70 tips from column 1. Wells in column 1 from the microtiter assay plate will not receive virus. The uninfected wells are useful for setting background corrections when using automated image analysis, which was not discussed in this protocol.

References

1. WHO's Communicable Disease Global Atlas (2005) World Health Organization. www.who.int/denguenet. Accessed 9 Sept 2012.

2. Guzman MG, Halstead SB, Artsob H, Buchy P, Farrar J, Gubler DJ, Hunsperger E, Kroeger A, Margolis HS, Martinez E, Nathan MB, Pelegrino JL, Simmons C, Yoksan S, Peeling RW (2010) Dengue: a continuing global threat. Nat Rev Microbiol 8(Suppl 12):S7–S16. doi:10.1038/nrmicro2460

3. Sessions OM, Barrows NJ, Souza-Neto JA, Robinson TJ, Hershey CL, Rodgers MA, Ramirez JL, Dimopoulos G, Yang PL, Pearson JL, Garcia-Blanco MA (2009) Discovery of insect and human dengue virus host factors. Nature 458(7241):1047–1050. doi:10.1038/nature07967

4. Krishnan MN, Ng A, Sukumaran B, Gilfoy FD, Uchil PD, Sultana H, Brass AL, Adametz R, Tsui M, Qian F, Montgomery RR, Lev S, Mason PW, Koski RA, Elledge SJ, Xavier RJ, Agaisse H, Fikrig E (2008) RNA interference screen for human genes associated with West Nile virus infection. Nature 455(7210):242–245. doi:10.1038/nature07207

5. Elbashir SM, Harborth J, Lendeckel W, Yalcin A, Weber K, Tuschl T (2001) Duplexes of 21-nucleotide RNAs mediate RNA interference in cultured mammalian cells. Nature 411(6836):494–498. doi:10.1038/35078107

6. Bian G, Shin SW, Cheon HM, Kokoza V, Raikhel AS (2005) Transgenic alteration of Toll immune pathway in the female mosquito Aedes aegypti. Proc Natl Acad Sci U S A 102(38):13568–13573. doi:10.1073/pnas.0502815102

7. Echeverri CJ, Perrimon N (2006) High-throughput RNAi screening in cultured cells: a user's guide. Nat Rev Genet 7(5):373–384. doi:10.1038/nrg1836

8. Barrows NJ, Le Sommer C, Garcia-Blanco MA, Pearson JL (2010) Factors affecting reproducibility between genome-scale siRNA-based screens. J Biomol Screen 15(7):735–747. doi:10.1177/1087057110374994

9. Le Sommer C, Barrows NJ, Bradrick SS, Pearson JL, Garcia-Blanco MA (2012) G protein-coupled receptor kinase 2 promotes flaviviridae entry and replication. PLoS Negl Trop Dis 6(9):e1820. doi:10.1371/journal.pntd.0001820

10. Henchal EA, Gentry MK, McCown JM, Brandt WE (1982) Dengue virus-specific and flavivirus group determinants identified with monoclonal antibodies by indirect immuno-fluorescence. Am J Trop Med Hyg 31(4):830–836

11. Jackson AL, Bartz SR, Schelter J, Kobayashi SV, Burchard J, Mao M, Li B, Cavet G, Linsley PS (2003) Expression profiling reveals off-target gene regulation by RNAi. Nat Biotechnol 21(6):635–637. doi:10.1038/nbt831

12. Nakabayashi H, Taketa K, Miyano K, Yamane T, Sato J (1982) Growth of human hepatoma cells lines with differentiated functions in chemically defined medium. Cancer Res 42(9):3858–3863

13. Zhang JH, Chung TD, Oldenburg KR (1999) A simple statistical parameter for use in evaluation and validation of high throughput screening assays. J Biomol Screen 4(2):67–73

Chapter 19

Investigating Dengue Virus Nonstructural Protein 5 (NS5) Nuclear Import

Johanna E. Fraser, Stephen M. Rawlinson, Chunxiao Wang, David A. Jans, and Kylie M. Wagstaff

Abstract

Dengue virus (DENV) nonstructural protein 5 (NS5) plays a central role in viral replication in the cytoplasm of infected cells. Despite this, NS5 is predominantly located in the nucleus of infected cells where it is thought to play a role in suppression of the host antiviral response. We have investigated the nuclear localization of NS5 using immunofluorescent staining for NS5 in infected cells, showing that NS5 nuclear localization is significantly inhibited by Ivermectin, a general inhibitor of nuclear transport mediated by the cellular nuclear transport proteins importin α/β (IMPα/β). Experiments in living mammalian cells transfected to express green fluorescent protein (GFP)-tagged NS5 protein confirm that NS5 is predominantly nuclear and that this localization is inhibited by Ivermectin, demonstrating that NS5 contains an Ivermectin-sensitive IMPα/β-recognized nuclear localization signal [Pryor et al. Traffic 8:795–807, 2007]. Consistent with this observation, mutation of critical residues within the nuclear localization signal (the A2 mutant; [Pryor et al. Traffic 8:795–807, 2007]) results in an 80 % reduction in nuclear localization of NS5. Finally we demonstrate direct, high-affinity binding of NS5 to IMPα/β using an AlphaScreen protein–protein binding assay.

Key words Dengue, Nuclear import, Importin, GFP, NS5, Nuclear localization sequence

1 Introduction

1.1 Dengue Virus

Dengue virus (DENV), of which there are four immunologically distinct serotypes (DENV 1–4), is the causative agent of dengue fever and its more severe dengue hemorrhagic fever form [2]. As a member of the *Flavivirus* genus, DENV possesses a single-stranded, positive-sense RNA genome that is translated as a single polyprotein prior to cleavage/processing into three structural and seven nonstructural proteins [3]. Despite the fact that DENV replication takes place entirely in the cytoplasm, the nonstructural protein 5 (NS5) polymerase, which plays a central role in replication [4], is predominantly located inside the nucleus during infection, where it is thought to play a role in suppression of the host antiviral response [1, 5, 6].

Radhakrishnan Padmanabhan and Subhash G. Vasudevan (eds.), *Dengue: Methods and Protocols*, Methods in Molecular Biology, vol. 1138, DOI 10.1007/978-1-4939-0348-1_19, © Springer Science+Business Media, LLC 2014

1.2 Nuclear Protein Import

Nuclear transport of molecules >45 kDa requires the recognition of a specific nuclear localization signal (NLS) in the cargo protein by members of the importin (IMP) superfamily of nuclear transport receptors, of which multiple α- and β-types exist [7, 8]. Classical nuclear transport occurs when an NLS is recognized by either the IMPα/β heterodimer or IMPβ alone (or a homologue thereof), followed by translocation through the nuclear envelope-embedded nuclear pore complexes. Once inside the nucleus, binding of RanGTP to IMPβ dissociates the complex and the IMPs are recycled back to the cytoplasm [7, 9]. Nuclear protein export is an analogous process, whereby nuclear export signals (NESs) in the cargo protein are recognized by exportin proteins (IMPβ homologues, of which CRM-1 is the best understood) in the nucleus, in a trimeric state with RanGTP, prior to export to the cytoplasm and dissociation of the complex [10].

1.3 DENV Nonstructural Protein 5 (NS5) Nuclear Transport

Nucleocytoplasmic transport of DENV NS5 protein is regulated by multiple targeting signals located in an interdomain region between the N-terminal methyltransferase domain and the C-terminal RNA-dependent RNA polymerase domain. The NS5 interdomain region contains two NLS signals: the aNLS (recognized by IMPα/β) and the bNLS (recognized by IMPβ1 alone), of which the aNLS appears to be dominant [1]. In addition, the interdomain region also contains a CRM-1-recognized NES and the binding site for NS5 interaction with NS3, an essential part of the viral replication complex [5, 6]. Mutation of any of these signals has dramatic effects on the viability of the virus [1, 6]. In particular, mutation of 5 lysine (K) residues within the aNLS to alanine (A) results in such severe effects on viral replication that the virus is unable to be recovered. Partial mutation of this region (DENV A2 mutant; KKK387-389AAA) produces a severely impaired virus with dramatically reduced virus growth kinetics [1, 6].

Clearly, understanding the nucleocytoplasmic shuttling of NS5 protein is critical to understanding DENV in general. Here we describe methods for investigating NS5 nuclear localization. We determine infectious virus titer using plaque assay and quantitative RT-PCR (qRT-PCR) and analyze NS5 localization in infected cells by immunostaining and quantitative confocal laser scanning microscopy (CLSM)/image analysis. Quantitative CLSM is also performed in live GFP-NS5 transfected cells, to demonstrate that the NS5 A2 mutant displays inhibited nuclear localization. In addition, we use the compound Ivermectin, a general inhibitor of IMPα/β-mediated nuclear import [11, 12] to demonstrate inhibition of NS5 nuclear localization in infected and GFP-NS5 transfected cells. Finally we confirm the NS5-IMPα/β association using an AlphaScreen-based protein–protein binding assay [13].

2   Materials

2.1   Growth of Aedes albopictus, C6/36, Cells

1. Cells of the *Aedes albopictus* C6/36 line (ATCC catalog number: CRL-1660).

2. Basal Medium Eagle (BME) medium supplemented with 10 % fetal calf serum (FCS): BME medium supplemented with 10 % heat-inactivated FCS, 2 mM L-glutamine, 50 U/L penicillin, 0.05 mg/L streptomycin, and 1× nonessential amino acids.

3. Humidified incubator with 5 % CO_2 atmosphere at 28 °C.

4. 25 and 175 cm² sterile cell culture flasks (Nunc).

5. 1× PBS: (standard 1× PBS contains 8 g NaCl, 0.2 g KCl, 1.44 g of Na_2HPO_4, 0.24 g KH_2PO_4 in 1 L of water; adjust pH to 7.4).

6. 0.05 % trypsin.

7. 6-well cell culture plates (Nunc).

2.2   Generation of DENV Stocks

1. 175 cm² sterile cell culture flasks.

2. BME (+10 % FCS).

3. BME (+2 % FCS).

4. Humidified incubator with 5 % CO_2 atmosphere at 28 °C.

5. 1× PBS.

6. Dengue virus stock: DENV-2, New Guinea C strain.

7. Sterile 1.5 ml microcentrifuge tubes (screw cap and safe for storage at −80 °C; Sarstedt or similar), sterilized by autoclave.

8. Dry ice.

9. −80 °C freezer.

2.3   Determination of Viral Titer by Quantitative RT-PCR (qRT-PCR)

1. Plasmid encoding DENV-2 (pDVWSK601; available from Dr. Kylie Wagstaff upon request).

2. In vitro transcription kit, mMESSAGE mMACHINE T7 (Ambion), or equivalent.

3. RNA purification kit, RNeasy Mini Kit (Qiagen), or equivalent.

4. Spectrophotometer.

5. RNA standards (prepared from in vitro transcribed DENV RNA).

6. Virus to be titered (collected from infected cell supernatant).

7. QIAamp Viral RNA Mini Kit (Qiagen).

8. RNase-free pipettes.

9. Nuclease-free H_2O.

10. TaqMan Fast Virus 1-Step Master Mix (Applied Biosystems).

11. Forward and reverse primers (50 μM) directed towards DENV-2 NS3 (forward, 5'TCGGAGCCGGAGTTTACAAA; reverse, 5'TCTTAACGTCCGCCCATGAT, modified from [14]).

12. TaqMan probe directed towards a region between forward and reverse primers, fluorescently labeled at the 5' end with 6-carboxyfluorescein (FAM) as the reporter dye and a minor groove binder (MGB) quencher at the 3' end (10 μM): 5'-FAM-ATTCCATACAATGTGGCA-MGB-3' (custom made; Applied Biosystems).

13. 96-well PCR plate (4titude, or equivalent).

14. Transparent optical adhesive plate seal (4titude, or equivalent).

15. Real-time PCR machine (Eppendorf Mastercycler ep realplex 4 gradient S, or equivalent).

2.4 Determination of Viral Titer by Plaque Assay

1. Cells of the *Aedes albopictus* C6/36 line grown in 6-well cell culture plates.

2. BME (+10 % FCS).

3. BME (+2 % FCS).

4. 2× BME: BME powder made up in half the recommended volume of sterile H_2O (2× solution) and supplemented with 4 % heat-inactivated FCS, 4 mM L-glutamine, 100 U/L penicillin, 0.1 mg/L streptomycin, and 2× nonessential amino acids.

5. Humidified incubator with 5 % CO_2 atmosphere at 28 °C.

6. Sterile 1.5 ml microcentrifuge tubes (sterilized by autoclave).

7. Dengue virus samples to titer (collected from infected cell supernatant).

8. 1× PBS.

9. Laboratory water bath, set to 55 °C.

10. 2 % SeaPlaque™ agarose (Lonza) made up in sterile MilliQ H_2O (sterilized by autoclave).

11. Microwave to melt SeaPlaque™ agarose solution.

12. 0.33 % neutral red solution (Sigma-Aldrich).

2.5 Preparation of Vero Cells for Infection Assays

1. Cells of the African green monkey kidney Vero cell line (ATCC catalog number: CCL-81).

2. DMEM complete: Dulbecco's Modified Eagle's Medium (DMEM) supplemented with 10 % heat-inactivated FCS, 2 mM L-glutamine, 50 U/L penicillin, and 0.05 mg/L streptomycin.

3. DMEM complete (+2 % FCS).

4. Humidified incubator with 5 % CO_2 atmosphere at 37 °C.

5. 25 cm^2 sterile cell culture flasks (Nunc).

6. 1× PBS.

7. 0.05 % Trypsin.

8. 6-Well and 12-well cell culture plates.

9. Autoclave-sterilized 15×15 mm square glass coverslips and 24 mm round glass coverslips (ProSciTech, or equivalent).

2.6 Infection of Vero Cells for Immunostaining Assays

1. Vero cells grown on 15 mm×15 mm square glass coverslips in 12-well tissue culture plates.

2. Ivermectin (Sigma-Aldrich).

3. DMEM complete.

4. Humidified incubator with 5 % CO_2 atmosphere at 37 °C.

5. Dengue virus samples with known titer.

6. 1× PBS.

2.7 Methanol/ Acetone Fixation of Infected Cells

1. Infected cells in a 12-well tissue culture plate.

2. 1× PBS.

3. Methanol/acetone solution: 50 % methanol, 50 % acetone.

4. 4 °C refrigerator.

2.8 Immunofluo- rescent Staining of Fixed Cells for NS5 Localization

1. Fixed cells in a 12-well tissue culture plate.

2. PBS/0.2 % Triton-X-100: Triton-X-100 added to 1× PBS after sterilization to a final concentration of 0.2 %.

3. 1× PBS.

4. PBS/1 % BSA: bovine serum albumin (BSA) added to 1× PBS after sterilization to a final concentration of 1 %.

5. Anti-NS5 primary antibody; we used an anti-NS5 polyclonal antibody raised in rabbit against residues 397–772 of DENV-2 NS5 (gift from Dr. Keng Teo). An alternative would be poly-clonal anti-NS5 antibody raised in rabbits against residues 879–892 of DENV-2 NS5 (Genetex; catalog number GTX103350).

6. Goat-anti-rabbit Alexa Fluor 488 green fluorescent dye conju-gated secondary antibody (Molecular Probes, Leiden, The Netherlands).

7. Glass microscope slides.

8. 2 % *n*-propyl gallate/glycerol solution.

2.9 Confocal Laser Scanning Microscopy (CLSM)

1. Olympus FluoView FV1000 CLSM microscope (or equivalent).

2.10 Image Analysis

1. ImageJ software (available as freeware from http://rsbweb. nih.gov/ij/).

2.11 Maintenance and Preparation of Cos-7 Cells for Transfection

1. Cells of the African green monkey kidney Cos-7 cell line.
2. DMEM complete.
3. Humidified incubator with 5 % CO_2 atmosphere at 37 °C.
4. 25 cm² sterile cell culture flasks.
5. 1× PBS.
6. 0.05 % trypsin.
7. 6-well cell culture plates.
8. Autoclave-sterilized 24 mm round glass coverslip.

2.12 Transfection of Cos-7 Cells Using Lipofectamine2000

1. Cos-7 cells grown on 24 mm round coverslips in 6-well tissue culture plates (*see* Subheading 2.11).
2. Sterile 1.5 ml microcentrifuge tubes (sterilized by autoclave).
3. DNA encoding GFP-NS5, GFP-NS5 A2 mutant, and GFP alone (available from Dr. Kylie Wagstaff upon request).
4. Lipofectamine2000 (Invitrogen).
7. DMEM Serum-Free: DMEM supplemented with 2 mM L-glutamine.
8. Humidified incubator with 5 % CO_2 atmosphere at 37 °C.

2.13 Live Cell Imaging of Transfected Cos-7 Cells

1. Attofluor cell chamber (Invitrogen).
2. *DMEM phenol red-free:* DMEM without phenol red, supplemented with 10 % heat-inactivated FCS, 2 mM L-glutamine, 50 U/L penicillin, 0.05 mg/L streptomycin, and 20 mM HEPES.
3. Ivermectin made up in 100 % DMSO.
4. Humidified incubator with 5 % CO_2 atmosphere at 37 °C.
5. KimWipes (Kimberly-Clark).

2.14 Preparation of His$_6$-NS5, GST-IMPα, and GST-IMPβ

1. *Escherichia coli (E. coli)* strain BL21(DE3) competent cells (Invitrogen).
2. DNA encoding His$_6$-NS5, GST-IMPα, and GST-IMPβ (available from Dr. Kylie Wagstaff upon request).
3. *Luria-Bertani (LB) media*: 10 g/L bacto-tryptone, 5 g/L bacto-yeast extract, 5 g/L NaCl; sterilized by autoclave.
4. *LB agar (LB media containing 2 % agar)*: sterilized by autoclave and cooled to 60 °C; ampicillin added to 100 μg/ml; 30 ml added per Petri dish and allowed to set at room temperature for 1 h.
5. Bacterial incubator set at 37 °C.
6. Bacterial shaker incubator set at 37 °C.
7. Centrifuge (Sorvall Evolution RC superspeed centrifuge, or equivalent).

8. Ampicillin; 100 mg/ml stock solution.

9. Isopropyl-beta-D-thiogalactoside (IPTG); 1 M stock solution.

10. Lysis buffer: 50 mM Na_2HPO_4 and 500 mM NaCl were made up and adjusted using a pH meter to pH 7.4. Glycerol and imidazole then added, pH 7.4, 500 mM NaCl, 10 % v/v glycerol, 40 mM imidazole, 1 mM DTT (added fresh), 1 protease inhibitor cocktail tablet/250 ml (Roche; added fresh), 4 mg/ml lysozyme (added fresh), 2 µg DNase I (added fresh).

11. −20 °C freezer.

12. −80 °C freezer.

13. Sonicator.

14. 0.2 µm syringe filters, 50 ml syringes.

15. HiTrap FF columns (GE Life Sciences).

16. FPLC machine (GE Life Sciences ÄKTA or equivalent).

17. Binding buffer: 50 mM Na_2HPO_4, pH 7.4, 500 mM NaCl, 10 % v/v glycerol, 40 mM imidazole, 1 mM DTT (added fresh).

18. Wash buffer: 50 mM Na_2HPO_4, pH 7.4, 500 mM NaCl, 10 % v/v glycerol, 80 mM imidazole, 1 mM DTT (added fresh).

19. Elution buffer: 50 mM Na_2HPO_4, pH 7.4, 500 mM NaCl, 10 % v/v glycerol, 500 mM imidazole, 1 mM DTT (added fresh).

20. 8 % polyacrylamide gels, PAGE apparatus.

21. 6× SDS loading dye: 0.05 % w/v bromophenol blue, 7.5 % v/v glycerol, 2.5 % w/v SDS, 0.35 M Tris, pH 6.8, 600 mM DTT.

22. SDS running buffer: 25 mM Tris, pH 8.3, 0.19 M glycine, 0.1 % SDS w/v.

23. Pre-stained molecular weight markers.

24. Coomassie blue staining solution: 0.25 % w/v Coomassie Brilliant Blue, 45 % v/v methanol, 7 % v/v acetic acid.

25. Destaining solution: 45 % v/v methanol, 7 % v/v acetic acid.

26. 50 kDa MWCO dialysis tubing (Amersham).

27. Dialysis buffer: 50 mM Na_2HPO_4, pH 7.4, 500 mM NaCl, 10 % v/v glycerol, 1 mM DTT (added fresh).

28. 75 kDa MWCO centrifugal filter devices (Amicon, Millipore; or equivalent).

29. Sephacryl H-200 26/60 size-exclusion chromatography column (GE Life Sciences).

30. Size-exclusion chromatography (SEC) buffer: 50 mM Na_2HPO_4, pH 7.4, 500 mM NaCl, 10 % v/v glycerol, 1 mM EDTA, 1 mM DTT (added fresh).

31. BSA.

32. Bradford protein-binding reagent (Bio-Rad).

33. Spectrophotometer.

34. GST Buffer A: 50 mM Tris, pH 8.0, 100 mM NaCl.

35. GST Buffer B: 50 mM Tris, pH 8.0, 100 mM NaCl, 10 mM glutathione.

36. Glutathione Sepharose matrix slurry (GE Life Sciences).

37. Protein columns (Bio-Rad).

38. PD-10 desalting columns (GE Life Sciences).

39. 1× PBS.

2.15 Biotinylation of GST-IMPα

1. Sulfo-NHS-Biotin (Pierce).

2. GST-IMPα.

3. NAP-5 columns (GE Life Sciences).

4. 1× PBS.

5. 75 kDa MWCO centrifugal filter devices (Amicon, Millipore; or equivalent).

6. −80 °C freezer.

2.16 AlphaScreen Protein–Protein Binding Assay to Determine the Binding Affinity of His$_6$-NS5 to IMPα/β

1. His$_6$-NS5 protein, biotinylated-GST-IMPα, and GST-IMPβ.

2. 1 M dithiothreitol (DTT) stock.

3. 4× intracellular buffer stock (4× IB): 440 mM KCl, 20 mM NaHCO$_3$, 20 mM MgCl$_2$, 4 mM EGTA, 0.4 mM CaCl$_2$, 80 mM HEPES, pH 7.4.

4. Sterile microcentrifuge tubes.

5. 1× PBS.

6. 384-well white microtiter plates (OptiPlates, PerkinElmer).

7. AlphaScreen Histidine (nickel chelate) Detection Kit (PerkinElmer).

8. 2.5 % w/v BSA.

9. PerkinElmer EnSpire plate reader fitted with Alpha technology (or equivalent).

10. SigmaPlot graphing program (or equivalent).

3 Methods

3.1 Maintenance and Preparation of C6/36 Insect Cells

C6/36 cells are cultured in BME (+10 % FCS) media to 90 % confluency by passaging at a ratio of 1:8 or 1:12 every 3 or 4 days, respectively, in 25 cm^2 sterile tissue culture flasks.

1. Once cells have reached 90 % confluency in a 25 cm^2 sterile tissue culture flask, remove the existing medium and wash cells gently three times with 2 ml of sterile PBS.

2. Remove the sterile PBS and replace with 500 µl 0.05 % trypsin. Rock the flask gently by hand to ensure an even coverage over the cell monolayer and incubate at 28 °C in a 5 % CO_2 humidified incubator no longer than 5 min to release the cells.

3. Once cells have been released (*see* **Note 1**), add 4.5 ml BME (+10 % FCS) to neutralize the trypsin and gently resuspend the cells.

4. Place 5 ml BME (+10 % FCS) into a 25 cm² sterile tissue culture flask.

5. For a 1:8 split, add 0.625 ml of resuspended cells or for a 1:12 split add 0.417 ml (*see* **Note 2**).

6. Disperse cells evenly by rocking the plate back and forth and side to side three times. Incubate cells in a 28 °C humidified 5 % CO_2 incubator until required.

3.2 Preparation of Cells for Generation of Viral Stocks

1. Place 30 ml BME (+10 % FCS) into each of two 175 cm² sterile tissue culture flasks.

2. Add 2.5 ml of resuspended cells to each flask.

3. Disperse cells evenly by rocking the plate back and forth and side to side three times. Incubate cells in a 28 °C humidified 5 % CO_2 incubator for 3 days or until 80 % confluent.

3.3 Preparation of Cells for Plaque Assay

1. Place 2 ml BME (+ 10 % FCS) into each well of each 6-well plate (*see* **Note 3**).

2. Add 833 µl of resuspended cells to each well (*see* **Note 4**).

3. Disperse cells evenly by rocking the plate back and forth and side to side three times. Incubate cells in a 28 °C humidified 5 % CO_2 incubator overnight for infection the next day.

3.4 Generation of DENV Stocks

1. Prepare two 175 cm² tissue culture flasks of C6/36 cells at 80 % cell confluency (*see* Subheading 3.1, **step 2**).

2. Remove BME (+10 % FCS) and wash the cells twice with 10 ml sterile PBS.

3. Add a previously titered viral stock (e.g., New Guinea C strain of DENV-2; *see* **Note 5**) to a multiplicity of infection (MOI; *see* **Note 6**) of 0.1 to each flask (*see* **Note 7**) and incubate at 28 °C for 1.5 h, rocking the flasks gently by hand every 10–15 min.

4. Remove the virus and wash cells with 10 ml sterile PBS.

5. Add 30 ml BME (+ 2 % FCS) to each flask and incubate at 28 °C, 5 % CO_2.

6. Observe cells daily for signs of cytopathic effects (CPE; *see* **Note 8**).

7. When widespread CPE is observed (~5 days), collect the media and aliquot into 0.5 ml aliquots in sterile microcentrifuge tubes.

8. Freeze immediately on dry ice and store at −80 °C.

9. Determine viral titer by qRT-PCR or plaque assay (*see* Subheadings 3.3 and 3.4, respectively).

3.5 Determination of Viral Titer by Quantitative RT-PCR (qRT-PCR)

1. Generate in vitro transcribed RNA for qRT-PCR standards (*see* **Note 9**):

   (a) Prepare 1 µg of highly pure linearized DNA encoding DENV-2 (e.g., pDVWSK601 cut with *XbaI*; plasmid available from Dr. Kylie Wagstaff upon request).

   (b) Set up an in vitro transcription reaction using mMESSAGE mMACHINE T7 Kit (use RNase-free pipettes): combine 1 µg linearized DNA with kit-supplied NTPs, capping analog, reaction buffer, T7 RNA polymerase, DTT, and RNase inhibitor in a final volume of 20 µl, as per the manufacturer's instructions. Incubate at 37 °C for 4 h.

   (c) To remove template DNA, add 1 U DNase I per reaction (supplied with the mMESSAGE mMACHINE T7 Kit). Incubate at 37 °C for 30 min.

   (d) Purify the RNA to remove unincorporated NTPs using the RNeasy Mini Kit (Qiagen), as per the manufacturer's instructions.

   (e) Quantify purified RNA using a spectrophotometer.

   (f) Prepare 4×100-fold dilutions of RNA in nuclease-free H_2O, beginning at 10^8 copies/µl for pDVWSK601, 1 ng/µl $= 1.7 \times 10^8$ copies per µl). To do this, label three separate Eppendorf tubes 10^6, 10^4, and 10^2. Add 495 µl of nuclease-free H_2O to each tube. Label a fourth tube 10^8 and dilute the in vitro transcribed RNA in nuclease-free H_2O as appropriate to achieve 10^8 copies per µl (ensure the volume is greater than 50 µl). Take 5 µl of the 10^8 standard and add it to the 10^6 tube. Mix well and then take 5 µl from the 10^6 tube and add it to the 10^4 tube. Continue this serial dilution to obtain the 10^2 standard.

   (g) Aliquot RNA standards into 40 µl samples and store at −80 °C until required.

2. Isolate DENV RNA from the supernatant of infected C6/36 cells using a QIAamp Viral RNA Mini Kit as per the manufacturer's instructions.

3. Prepare a qRT-PCR master mix (*see* **Note 10**). To make the master mix, combine the following components on ice (final concentrations are given):

   • 900 nM forward primer.

   • 900 nM reverse primer.

Table 1
PCR cycle conditions for qRT-PCR

Step	Temperature (°C)	Time	No. of cycles
Reverse transcription	50	5 min	1
RT inactivation	95	20 s	1
Denaturation	95	5 s	40
Amplification	60	30 s	

- 200 nM TaqMan probe.
- 1× TaqMan Fast Virus 1-Step Master Mix.
- H$_2$O to give 15 μl per reaction.
  - (a) Aliquot 15 μl of the master mix into each of the required triplicate wells of a 96-well plate.
  - (b) Carefully add 10 μl of RNA (standard or unknown, or H$_2$O for "no template control" (NTC)) to each triplicate set of wells and mix gently by pipetting.
  - (c) Seal the plate using transparent optical adhesive plate seal and transfer immediately to the PCR machine.
  - (d) Open the Mastercycler ep realplex software and complete the plate layout, including the location and number of RNA copies in each standard, the location of the NTC, and the location of virus samples to be quantified (this enables auto-analysis at the end of the run cycle).
  - (e) Assay using the cycle conditions, using the default ramping and data collection settings (*see* Table 1).
  - (f) Use the auto-analysis function on the Mastercycler ep realplex software to create a standard curve and obtain extrapolated values for viral samples.

DENV collected from infected C6/36 cells was titered using qRT-PCR. The amplification curves produced by the Mastercycler ep realplex software show the RNA standards crossing the threshold fluorescence between cycles 11 and 31 (Fig. 1a) with the highest standard (1×10^8 copies) crossing the threshold (red line) at cycle 11 and the lowest standard (1×10^2) crossing the threshold at cycle 31. Thus, a sample with more RNA copies will produce a fluorescent signal at a lower cycle number. By correlating the number of RNA copies with the cycle number at which the fluorescent threshold is crossed, a standard curve can be produced (as generated by the Mastercycler ep realplex software; Fig. 1b). Extrapolating from this curve, and correcting for the input sample volume, indicates that the number of copies of viral RNA present in the viral stock is 2.25×10^8 RNA copies/ml (Fig. 1c). This value can then be used to normalize virus concentrations between samples.

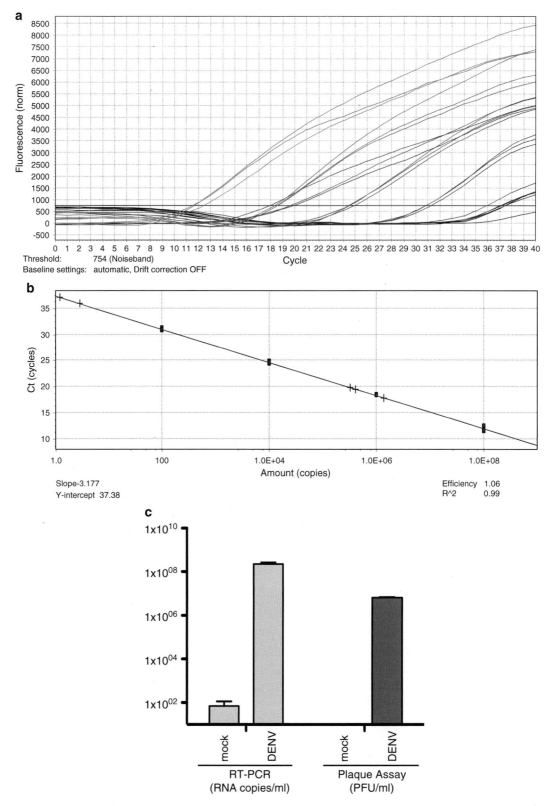

Fig. 1 DENV viral titer determined by qRT-PCR and plaque assay. Wild-type DENV or mock supernatant was analyzed by qRT-PCR and plaque assay to determine the viral titer. (**a**) Typical RT-PCR amplification plot performed as described in Subheading 3.3. (**b**) Standard curve generated from the amplification plots in **a**. (**c**) Virus titer determined by qRT-PCR and plaque assay (RNA copies vs. plaque-forming units; *see* Subheading 3.4). Results are for the mean ± SEM (*n* = 3)

3.6 Determination of Viral Titer by Plaque Assay

1. Prepare C6/36 cells in 6-well plates 1 day prior to assay (*see* Subheading 3.1, **step 3**). Two wells are required for each dilution to be assayed. Two additional wells are required for mock infection.

2. Prepare viral dilutions:

   (a) Label 6 microfuge tubes with the dilution to be generated (e.g., 10^{-2} to 10^{-7}).

   (b) Add 900 µl of BME (+2 % FCS) to each of 6 microcentrifuge tubes.

   (c) Add 100 µl virus stock to the first tube.

   (d) Mix by gently pipetting up and down.

   (e) Generate a 1:10 serial dilution series by adding 100 µl of this solution to the next tube and repeat until all dilutions have been created.

3. Remove the media from the cells and wash with 2 ml sterile PBS per well.

4. Remove PBS and add 400 µl of each virus dilution to each of the two wells (duplicates). Add 400 µl BME (+2 % FCS) to the two mock infection wells. Incubate at 28 °C for 1.5 h, rocking gently by hand every 20 min.

5. Meanwhile, melt the 2 % SeaPlaque™ agarose (2.5 ml/well) in the microwave and store at 55 °C in a water bath to prevent setting. Warm an equal quantity of 2× BME (+4 % FCS) to 28 °C.

6. When virus incubation is complete, remove virus by aspiration and wash cells once with 2 ml sterile PBS.

7. Remove PBS, and then quickly, but gently, mix the 2× BME (+4 % FCS) with the melted SeaPlaque™ agarose.

8. Add 5 ml of the SeaPlaque™ agarose/2× BME (+4 % FCS) solution gently to each well and allow to set for 20 min (*see* **Note 11**).

9. Turn plates upside down and incubate at 28 °C for 5 days.

10. After 5 days, melt 2 % SeaPlaque™ agarose (1.25 ml/well) in a microwave and maintain at 55 °C in a water bath until ready. Warm an equal quantity of 2× BME (+4 % FCS) to 28 °C. Mix the 2× BME (+4 % FCS) with the melted SeaPlaque™ agarose and add a 1:20 dilution of 0.33 % neutral red solution.

11. Overlay 2.5 ml of the agarose/BME/neutral red mixture into each well. Allow to set for 20 min, and then incubate upside down at 28 °C overnight.

12. The following morning, examine the bottom of the plates for plaque formation. Count the number of plaques in a dilution where well-defined individual plaques can be observed (>10 but <100). Viral titer is expressed as plaque-forming

units (PFU)/ml, taking into account the dilution factor and volume plated.

Quantification by plaque assay determines the number of infectious particles as 6.5×10^6 pfu/ml (Fig. 1c). This confirms that qRT-PCR can be used to estimate the amount of virus present in a sample. Differences in the number of infectious particles per ml determined by the two methods can be attributed to the increased sensitivity of qRT-PCR relative to plaque assay and the fact that qRT-PCR measures all RNA, not only RNA incorporated into infectious particles [15].

3.7 Maintenance and Preparation of Vero Cells for Infection Assays

Vero cells are cultured in DMEM complete and are grown to 90 % confluency by passaging at a ratio between 1:10 and 1:20 every 3 or 4 days, respectively, in 25 cm² sterile tissue culture flasks.

1. Once cells have reached 90 % confluency in a 25 cm² sterile tissue culture flask, remove the existing DMEM complete and wash cells three times gently with 2 ml of sterile PBS.

2. Remove the sterile PBS and replace with 500 µl 0.05 % trypsin. Rock the flask gently by hand to ensure an even coverage over the cell monolayer and incubate at 37 °C in a 5 % CO_2 humidified incubator no longer than 5 min to release the cells.

3. Once cells have been released (*see* **Note 12**), add 4.5 ml DMEM complete to neutralize the trypsin and gently resuspend the cells.

3.7.1 Maintenance of Vero Cell Line

1. Place 5 ml DMEM complete into a 25 cm² sterile tissue culture flasks.

2. For a 1:10 split, add 0.5 ml of resuspended cells or add 0.25 ml for a 1:20 split (*see* **Note 13**).

3. Disperse cells evenly by rocking the plate back and forth and side to side three times. Incubate cells in a 37 °C humidified 5 % CO_2 incubator until required.

3.7.2 Infection and Immunostaining Assays

1. Place a sterile 15 mm square glass coverslip into each well of a 12-well cell culture plate.

2. Overlay 1 ml DMEM complete into each well (*see* **Note 14**).

3. Add 150 µl of the cell suspension to each well (*see* **Note 15**), and mix gently by rocking the plate back and forth and side to side three times. Incubate cells in the 37 °C humidified 5 % CO_2 incubator overnight for infection the following day and fixation/immunostaining as appropriate following that.

3.7.3 Infection of Vero Cells for Immunostaining Assays ± Ivermectin Treatment

1. Seed Vero cells into 12-well cell culture plates containing glass coverslips according to Subheading 3.7.2, and grow to 80 % confluency.

2. Pretreat appropriate cells with Ivermectin (1–25 μM) for 3 h prior to infection.

3. Calculate the amount of virus to be used:

    (a) A multiplicity of infection (MOI) of 4 is required (*see* **Note 16**).

    (b) There are ~2.5×10^5 Vero cells in one well of a 12-well plate at 80 % confluency.

4. Dilute virus in DMEM complete (+2 % FCS) so that 200 μl/well can be used to obtain the desired MOI.

5. Remove the media/drug from the cells and wash once with 1 ml sterile PBS per well.

6. Add 200 μl diluted virus to each well required for infection and 200 μl of DMEM complete (+2 % FCS) for a mock infected control well. Incubate at 37 °C for 1 h, rocking gently by hand every 10 min.

7. Remove the virus by aspiration and wash the cells once with 1 ml PBS.

8. Add 1 ml of DMEM complete (+2 % FCS) to each well and incubate at 37 °C overnight.

9. At 24 h post infection, fix cells using methanol/acetone prior to immunostaining.

3.8 Methanol/ Acetone Fixation of Infected Cells

1. Remove media from infected cells in 12-well tissue culture plate.

2. Wash twice with 1 ml room temperature PBS.

3. Remove PBS and add 1 ml of methanol/acetone (1:1 solution) (*see* **Note 17**). Incubate 2 min.

4. Remove the fixation media and wash four times with 1 ml PBS.

5. Store cells at 4 °C in PBS until immunofluorescence staining.

3.9 Immunofluo- rescent Staining of Fixed Cells for NS5 Localization

1. Remove PBS from fixed cells in 12-well tissue culture plates.

2. Add 1 ml PBS/0.2 % Triton-X-100 to permeabilize cells. Incubate for 20 min at room temperature.

3. Remove PBS/0.2 % Triton-X-100 from cells and wash once with 1 ml PBS.

4. Block cells by adding 1 ml PBS/1 % BSA to each well. Incubate 20 min at room temperature or overnight at 4 °C.

5. Dilute primary antibody 1:1,000 in PBS/1 % BSA. Add 300 μl to each well and incubate for 1 h.

6. Remove antibody and wash five times with 1 ml PBS.

7. Dilute secondary antibody (goat-anti-rabbit Alexa Fluor 488 green fluorescent dye conjugate) 1:1,000 in PBS/1 % BSA. Incubate for 60 min in the dark.

8. Wash cells five times with 1 ml PBS.

9. Mount cells onto slides using 2 % *n*-propyl-gallate glycerol solution.

10. Examine cells using confocal laser scanning microscopy (CLSM).

3.10 Confocal Laser Scanning Microscopy (CLSM)

1. Place the slide onto the microscope.

2. Locate the cells using a 60× objective to find green fluorescent cells using the UV filters (other objectives such as a 100× may also be used).

3. Place the cells in the center of the microscopic field, turn off the UV, and expose cells to the laser beam.

4. Use the imaging software to focus and capture images (*see* **Note 18**).

5. Images are saved for later analysis using the ImageJ software.

3.11 Image Analysis

1. Open each CLSM image using the NIH ImageJ software (*see* **Note 19**).

2. Using the software, analyze each transfected cell in the image for fluorescence in the nucleus (Fn) and the cytoplasm (Fc), by selecting an appropriate region of interest and using the measure tool in the ImageJ program. Also measure background fluorescence (Fb) by measuring autofluorescence in a non-transfected cell.

3. The nucleocytoplasmic ratio (Fn/c) for each cell is calculated using the equation $Fn/c = (Fn - Fb)/(Fc - Fb)$.

4. Pool the Fn/c values for all cells in a given sample and determine statistical significance of the differences using a statistical software package (InStat software or equivalent).

CLSM analysis of DENV-infected Vero cells pretreated with increasing concentrations of Ivermectin and immunostained for NS5 (Fig. 2a) demonstrates that NS5 is significantly more cytoplasmic in the presence of Ivermectin. This observation is confirmed by determination of the Fn/c values, whereby treatment with as little as 3.125 μM Ivermectin results in a 40 % reduction in NS5 nuclear accumulation in infected cells (Fig. 2b). This demonstrates that NS5 possesses an IMPα/β-mediated, Ivermectin-inhibited NLS [12].

3.12 Maintenance and Preparation of Cos-7 Cells for Transfection Assays

Cos-7 cells are cultured in DMEM complete and are grown to 90 % confluency by passaging at a ratio between 1:10 and 1:20 every 3 or 4 days, respectively, in 25 cm² sterile tissue culture flasks.

1. Once cells have reached 90 % confluency in a 25 cm² sterile tissue culture flask, remove the existing DMEM complete and wash cells three times gently with 2 ml of sterile PBS.

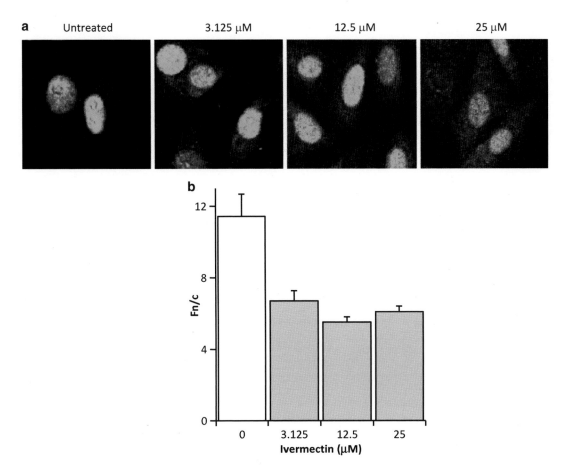

Fig. 2 Nuclear accumulation of NS5 in infected cells is decreased in response to the general inhibitor of IMPα/β-mediated nuclear transport, Ivermectin. (**a**) Vero cells were pretreated for 3 h with the indicated concentrations of Ivermectin, prior to infection with DENV (MOI of 5). Cells were fixed with methanol/acetone 24 h post infection (p.i.), immunostained for NS5, and imaged using CLSM. (**b**) Image analysis performed on digitized images such as those in **a** to determine the nuclear to cytoplasmic fluorescence ratio (Fn/c) as described in Subheading 3.10. Results are for the mean ± SEM ($n > 50$)

2. Remove the sterile PBS and replace with 500 µl 0.05 % trypsin. Rock the flask gently by hand to ensure an even coverage over the cell monolayer and incubate at 37 °C in a 5 % CO_2 humidified incubator no longer than 5 min to release the cells.

3. Once cells have been released (*see* **Note 20**),, add 4.5 ml DMEM complete to neutralize the trypsin and gently resuspend the cells.

3.12.1 Maintenance of the Cos-7 Cell Line

1. Place 5 ml DMEM complete into a 25 cm² sterile tissue culture flasks.

2. For a 1:10 split, add 0.5 ml of resuspended cells or add 0.25 ml for a 1:20 split (*see* **Note 21**).

3. Disperse cells evenly by rocking the plate back and forth and side to side three times. Incubate cells in a 37 °C humidified 5 % CO_2 incubator until required.

3.12.2 Transfection Assays

1. Place a sterile 24 mm round glass coverslip into each well of a 6-well cell culture plate.

2. Overlay 1 ml DMEM complete into each well (*see* **Note 22**).

3. Add 350 µl of the cell suspension to each well (*see* **Note 23**), and mix gently by rocking the plate back and forth and side to side three times. Incubate cells in the 37 °C humidified 5 % CO_2 incubator overnight for transfection the following day and imaging 18 h post transfection.

3.13 Transfection of Cos-7 Cells Using Lipofectamine2000

1. Prepare Cos-7 cells for transfection (*see* Subheading 3.12.1).

2. Eighteen hours after seeding cells, prepare DNA for transfection by adding 4 µg plasmid DNA encoding GFP-NS5, GFP-NS5-A2 mutant, or GFP alone, to separate sterile Eppendorf tubes. Make up to 250 µl with DMEM serum-free.

3. Dilute Lipofectamine2000 1:25 in DMEM serum-free and add 250 µl to each DNA sample. Mix gently and incubate at room temperature for 20 min.

4. Add the 500 µl DNA/Lipofectamine2000 mix dropwise to the cells and incubate at 37 °C, 5 % CO_2 for imaging 18 h later.

3.14 Live Cell Imaging of Transfected Cos-7 Cells ± Ivermectin Treatment

1. Prior to imaging, pre-warm DMEM phenol red-free to 37 °C (*see* **Note 24**) and warm the Attofluor cell chamber by placing in a 37 °C incubator for 5 min.

2. Treat appropriate samples with 2 µM (final concentration) Ivermectin for 1 h prior to imaging (*see* **Note 25**). Also add an equivalent volume of DMSO to the untreated samples for 1 h prior to imaging to control for diluent effects.

3. To image cells, remove coverslips from the wells (*see* **Note 26**), drain remaining media by gently touching the side of the coverslip to a KimWipe lying flat on the bench, and then place into the bottom half of the cell chamber. Ensure that the coverslip is sitting flush with the bottom of the chamber.

4. Place the top half of the cell chamber onto the bottom and screw closed gently (*see* **Note 27**). Add 1 ml of warm DMEM phenol red-free to the top of the chamber.

5. Place the chamber into the heating block on the microscope (*see* **Note 28**).

6. Locate the cells using a 60× objective to find green fluorescent cells using the UV filters (other objectives such as a 100× may also be used).

7. Place the cells in the center of the microscopic field, turn off the UV, and expose cells to the laser beam.

8. Use the imaging software to focus and capture images (*see* **Note 29**).

9. Images are saved for later analysis using the ImageJ software (*see* Subheading 2.10).

CLSM imaging of cells transfected to express GFP-NS5 or GFP alone indicates that NS5 localizes strongly to the nucleus of Cos-7 cells (Fig. 3a, b). GFP alone shows roughly equivalent levels in both the cytoplasm and nucleus (Fn/c ~ 1.4), indicative of free diffusion of the protein between the two compartments.

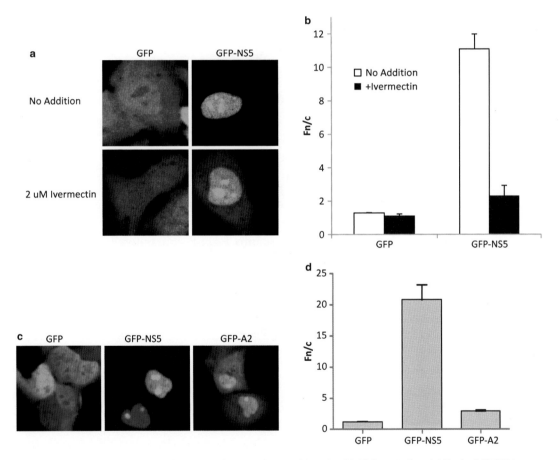

Fig. 3 Nuclear accumulation of NS5 is inhibited by Ivermectin or the A2 NLS mutation. (**a**) Typical CLSM images of Cos-7 cells transfected to express the indicated GFP-fusion proteins in the absence/presence of 2 μM Ivermectin. (**b**) Image analysis performed on digitized images such as those in **a** to determine the Fn/c ratio, as described in Subheading 3.10. Results are for the mean ± SEM (*n* > 48). (**c**) Typical CLSM images of GFP-NS5, GFP-NS5 A2 mutant, or GFP alone. (**d**) Image analysis was performed on digitized images such as those in **c** to determine the Fn/c ratio, as described in Subheading 3.10. Results are for the mean ± SEM (*n* > 45)

This is as expected as GFP does not contain an active NLS. Treatment with Ivermectin significantly ($p < 0.001$) reduces the nuclear accumulation of GFP-NS5, while GFP localization remains unaffected. These results confirm that NS5 contains an Ivermectin-sensitive IMPα/β recognized NLS.

Analysis of cells transfected with the GFP-NS5 A2 mutant indicates that this mutant is more cytoplasmic compared to WT GFP-NS5 (Fig. 3c). Quantification of the Fn/c ratios confirms that the GFP-NS5 A2 NLS mutations cause an ~80 % reduction in nuclear localization compared to WT GFP-NS5 ($p < 0.0001$; Fig. 3d). This suggests that the A2 mutations render the NS5 NLS significantly less functional.

3.15 Preparation of His$_6$-NS5, GST-IMPα, and GST-IMPβ

Transform clones encoding His$_6$-NS5, GST-IMPα, and GST-IMPβ into *E. coli* BL21(DE3) competent cells (*see* **Note 30**):

1. Add 100 ng of the expression plasmid to 50 µl of BL21(DE3) competent cells, mix gently, and incubate on ice for 30 min.

2. Heat shock cells at 42 °C for 45 s, place immediately on ice, and chill 2 min.

3. Add 1 ml LB and incubate at 37 °C for 1 h.

4. Plate 100 µl cells onto LB agar containing 100 µg/ml ampicillin and incubate overnight at 37 °C.

3.15.1 Preparation of His$_6$-NS5

1. Inoculate 12 ml LB containing 100 µg/ml ampicillin with a single colony of His$_6$-NS5 transformed BL21(DE3) and grow overnight at 37 °C in a bacterial shaker incubator set at 220 rpm.

2. Transfer the culture into 1 L LB containing 100 µg/ml ampicillin (*see* **Note 31**) and incubate in a bacterial shaker incubator at 37 °C, 220 rpm until an OD$_{600}$ of 0.6 is reached.

3. Induce protein expression with the addition of 1 mM IPTG (final concentration) and incubate in a bacterial shaker incubator at 16 °C, 250 rpm for 20 h.

4. Harvest the bacterial culture by centrifugation at $7800 \times g$ for 30 min.

5. Remove and discard the supernatant and resuspend the pellet in 25 ml lysis buffer. Freeze overnight at −20 °C.

6. Thaw the frozen bacterial slurry at room temperature. Sonicate slurry for 2×30 s using a large sonication probe at amplitude 10.0 µm on ice.

7. Remove cell debris by centrifugation at 30,000 relative centrifugal force for 30 min at 4 °C. Remove and filter the supernatant using a 0.22 µm filter.

8. Equilibrate a 5 ml HiTrap FF column with binding buffer using an FPLC machine at 5 ml/min.

9. Load the cleared supernatant onto the column at 1.5 ml/min.

10. Wash the column with wash buffer until the UV trace reaches equilibrium, and then elute at 2 ml/min using a 45 min gradient into elution buffer, collecting 4 ml peak fractions. The protein should elute at 25–35 % elution buffer.

11. Run fractions on an 8 % polyacrylamide gel to visualize proteins:

    (a) Mix a 10 μl sample of each fraction with 2 μl 6× SDS loading dye.

    (b) Boil samples 5 min at 100 °C.

    (c) Load onto a pre-prepared 8 % polyacrylamide gel, include a pre-stained molecular weight marker sample, and run at 120 V in 1× SDS running buffer until the dye front has left the gel.

    (d) Stain for ~1 h in Coomassie blue staining buffer.

    (e) Rinse gel and incubate in destaining buffer until protein bands appear.

12. Pool appropriate fractions containing His_6-NS5 and dialyze overnight in 50 kDa MWCO dialysis tubing against 5 L dialysis buffer at 4 °C.

13. Remove protein from dialysis tubing and concentrate using a 75 kDa MWCO centrifugal filter device until volume is 4–5 ml.

14. Meanwhile equilibrate a Sephacryl H-200 26/60 size-exclusion chromatography column with SEC buffer at 1 ml/min for 1.5 column volumes.

15. Load protein onto column using a 5 ml loop.

16. Elute at 1 ml/min in SEC buffer, collecting 1.5 ml peak fractions.

17. Visualize fractions on an 8 % polyacrylamide gel as previously (**step 12**).

18. Pool fractions containing the purified His_6-NS5 and concentrate using a 75 kDa MWCO centrifugal filter device until volume is ~500 μl. Aliquot protein and store at –80 °C.

19. Determine the protein concentration using the Bio-Rad Bradford according to manufacturer's specifications:

    (a) Set up solutions in Eppendorf tubes containing 2 to 14 μg of protein standard BSA made up to 0.8 ml in distilled water (dH_2O).

    (b) Also dilute 0.5 μl, 1 μl, and 2 μl NS5 protein to 0.8 ml in dH_2O.

    (c) Add 0.2 ml of the protein-binding reagent to each Eppendorf tube and incubate for 10 min at room temperature.

(d) Measure the absorbance at 595 nm (blanked using reagent with no protein, prepared from 0.8 ml dH$_2$O and 0.2 ml of the protein-binding reagent).

(e) Calculate the protein concentration by extrapolating from the BSA standard curve.

3.15.2 Preparation of IMPα- and IMPβ-GST Fusion Proteins

1. Inoculate 5 ml LB containing 100 µg/ml ampicillin with a single colony of IMPα—or IMPβ—GST-transformed BL21(DE3), and grow overnight at 37 °C in a bacterial shaker incubator set at 220 rpm.

2. Transfer the culture to 2 L of LB containing 100 µg/µl ampicillin and incubate in a bacterial shaker incubator at 37 °C, 220 rpm until an OD$_{600}$ of 0.6 is reached.

3. Induce protein expression with the addition of 1 mM IPTG (final concentration) and incubate in a bacterial shaker incubator at 37 °C, 250 rpm for 8 h.

4. Harvest the bacterial culture by centrifugation at $7800 \times g$ for 30 min.

5. Remove and discard the supernatant and resuspend the pellet in 25 ml GST buffer A (*see* Subheading 2.14, **step 34**).

6. Add lysozyme (1 mg/ml) to lyse cells and incubate on ice for 30 min.

7. Remove cellular debris by centrifugation at $23,400 \times g$ for 30 min.

8. Wash 2 ml of Glutathione Sepharose matrix slurry in 25 ml GST buffer A. Pellet beads at $1,000 \times g$ and gently remove and discard supernatant.

9. Add protein extract to pre-washed slurry and incubate for 2 h at 4 °C on a rotator wheel.

10. Place the slurry/extract into an empty protein column and allow the supernatant to drip through (*see* **Note 32**).

11. Wash the beads three times with 50 ml GST buffer A to remove nonspecifically bound proteins.

12. Elute the protein from the column using 2.5 ml GST buffer B.

13. Exchange the buffer to PBS using a PD-10 column.

(a) Remove top cap from the PD-10 column and pour off the column storage buffer. Cut the sealed end of the column at the notch.

(b) Fill column with PBS and allow buffer to enter the gel bed completely. Repeat until a total of 25 ml of PBS has run through the column. Discard flow-through.

(c) Add the 2.5 ml protein sample directly to the column and allow the sample to enter the gel bed completely. Discard flow-through.

(d) Place a clean tube under the column.

(e) To elute protein, add 3.5 ml PBS to the top of the column and allow it to enter the gel bed completely. Collect eluates.

14. Concentrate proteins using a 75 kDa MWCO centrifugal filter device until volume is ~500 μl. Aliquot protein and store at −80 °C.

15. Determine the protein concentration using the Bio-Rad Bradford as described (*see* Subheading 3.15.1, **step 19**).

3.16 Biotinylation of GST-IMPα

1. Dilute 1 mg of Sulfo-NHS-Biotin in 150 μl H_2O.

2. Use 1 μl of diluted biotin per 4,145 pmol GST-IMPα. Combine protein and diluted biotin, and add dH_2O to a final volume of 400 μl.

3. Mix and incubate on ice for 2 h.

4. Remove excess unbound biotin using a NAP-5 column.

(a) Remove top and bottom caps from the NAP-5 column and allow liquid to drain out.

(b) Add 10 ml PBS to columns and allow column to drain completely.

(c) Add the 400 μl biotinylated protein sample directly to the column and allow the sample to enter the gel bed completely.

(d) Add 100 μl PBS to the column and allow it to enter the gel bed completely.

(e) Place a clean microcentrifuge tube under the column.

(f) To elute protein, add 1 ml PBS to the top of the column and allow it to enter the gel bed completely. Collect elution.

5. Concentrate protein using a 75 kDa MWCO centrifugal filter device until volume is ~500 μl. Aliquot protein and store at −80 °C.

6. Determine the protein concentration using the Bio-Rad Bradford as described in Subheading 3.15.1.

3.17 AlphaScreen Protein–Protein Binding Assay

The AlphaScreen binding assay described here is to determine the binding affinity of His_6-NS5 to IMPα/β. It requires that one binding partner contains a His_6 tag and that the other binding partner is biotinylated. Here we utilize His_6-NS5 and biotinylated-GST-IMPα heterodimerized with GST-IMPβ. The assay is performed in triplicate in a 384-well white opaque microtiter plate.

1. Pre-dimerize biotinylated-IMPα with GST-IMPβ. Combine the following in a microcentrifuge tube:

- 13.6 μM biotinylated-GST-IMPα.
- 13.6 μM GST-IMPβ.

- 1 mM DTT.
- 1× IB.
- Incubate at room temperature for 15 min.

2. Add 2 µl of 375 nM His$_6$-NS5 protein to each well (30 nM final concentration).

3. Prepare 12 microcentrifuge tubes for serial dilution of the IMPα/β heterodimer. Add 198.9 µl PBS to the first tube and add 100 µl of PBS to the remaining tubes.

4. Add 1.1 µl IMP α/β heterodimer to the first tube and mix gently.

5. Create 1:2 serial dilutions as follows: add 100 µl from the first tube to the second tube and mix. Repeat for the remaining tubes, leaving the final tube as a PBS-only control.

6. Add 20 µl of each IMP α/β dilution to separate triplicate His$_6$-NS5 containing wells of the 384 well microtiter plate (final IMP α/β concentration range of 60 nM to 0 nM).

7. Incubate for 30 min at room temperature. The remaining additions in this protocol must be conducted under subdued lighting conditions.

8. To each well, add 1 µl of a 1:10 dilution of streptavidin donor beads (in PBS) and 1 µl of PBS/2.5 % BSA, and incubate for 90 min at room temperature in the dark.

9. Add 1 µl of a 1:10 dilution of nickel chelate acceptor beads (in PBS) to give a final sample volume of 25 µl per well. Incubate for 2 h at room temperature in the dark.

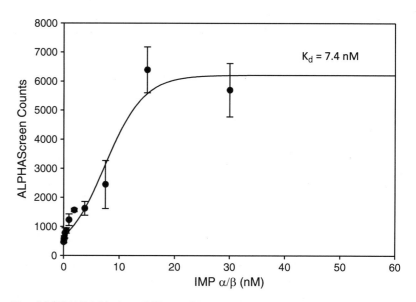

Fig. 4 DENV NS5 binds to IMPα/β with high affinity. AlphaScreen binding curve for DENV NS5 binding to IMPα/β performed as per Subheading 3.16, using 30 nM His$_6$-NS5 and the indicated concentrations of IMPα/β. Results represent the mean ± SEM for triplicate measurements

10. Read the plate on a PerkinElmer EnSpire plate reader fitted with Alpha technology.

11. Average triplicate results and fit a three-parameter sigmoidal titration curve using the SigmaPlot graphing program (*see* **Note 33**).

Using the AlphaScreen protein–protein binding assay, the binding affinity of His$_6$-NS5 for IMPα/β is examined. The apparent dissociation constant (K_d) of the interaction is determined to be 7.4 nM (Fig. 4), indicating a very high-affinity interaction. This is consistent with the imaging data that demonstrates NS5 localizes in the nucleus of host cells dependent on IMPα/β.

4   Notes

1. Release of cells can be visualized directly by examining the growth side of the flask. Trypsinization is complete if cells are no longer stuck to the surface, but flow from one side of the flask to the other when the flask is tilted.

2. For maintenance, cells are subcultured as a 1–8 split every Monday for the cells to be ready on the following Thursday, followed by a 1–12 split every Thursday to achieve confluency on the following Monday.

3. It is important to set up enough wells to do the entire plaque assay in duplicate.

4. For C6/36 cells, this is equivalent to 1×10^6 cells/well.

5. Virus stocks can be obtained with the proper permits from another laboratory.

6. An MOI of 1 indicates 1 infectious virus particle per cell.

7. A T-175 cm^2 flask at 80 % confluency contains approximately 6×10^7 cells.

8. Cytopathic effects of DENV infection include cell rounding, syncytium formation, and cell death.

9. Alternatively plasmid DNA can be used to generate the standard curve. In this case, omit **steps 1–5** and recalculate the number of copies/ng required to create the standard curve, taking into account the size of the plasmid.

10. A standard curve should be included in every qRT-PCR reaction. Each standard curve should include at least 4 known RNA standards (in vitro transcribed RNA prepared in Subheading 3.5, **step 2**) and a "no template control" (NTC; nuclease-free H$_2$O only). Every sample (unknown and standards) should be assayed in triplicate.

11. If doing a large number of plaque assays, it is best to do this step in small volumes to prevent the mixture from setting before it is used.

12. Release of cells can be visualized directly by examining the growth side of the flask. Trypsinization is complete if cells are no longer stuck to the surface, but flow from one side of the flask to the other when the flask is tilted.

13. For maintenance, cells are subcultured as a 1–10 split every Monday for the cells to be ready on the following Thursday, followed by a 1–20 split every Thursday to achieve confluency on the following Monday.

14. Gently push down coverslip with a sterile pipette tip after adding the media to remove any air bubbles trapped beneath the glass.

15. For Vero cells, this is equivalent to 1.5×10^5 cells/well, which enables 80 % confluency the next day.

16. An MOI of 1 indicates 1 infectious virus particle per cell.

17. One milliliter of fixative solution is used so that all the virus on the side of the wells is inactivated prior to removal of the plate from the tissue culture hood.

18. Images should be captured of ~50 cells per sample.

19. The ImageJ software is available in the public domain and can be downloaded from the NIH website (http://rsbweb.nih.gov/ij/).

20. Release of cells can be visualized directly by examining the growth side of the flask. Trypsinization is complete if cells are no longer stuck to the surface, but flow from one side of the flask to the other when the flask is tilted.

21. For maintenance, cells are subcultured as a 1–10 split every Monday for the cells to be ready on the following Thursday, followed by a 1–20 split every Thursday to achieve confluency on the following Monday.

22. Gently push down coverslip with a sterile pipette tip after adding the media to remove any air bubbles trapped beneath the glass.

23. For Cos-7 cells, this is equivalent to 3.5×10^5 cells/well for 80 % confluency the next day.

24. This can be performed in either a water bath set to 37 °C or in a 37 °C, 5 % CO_2 humidified incubator.

25. To maintain timings, it is best to stagger addition of the drug so that each sample can be imaged exactly 1 h after addition of the drug. A good starting point is to delay addition of the drug by 30 min per sample. Ensure that the cell chamber is washed thoroughly between samples to remove any residual drug.

26. A pair of fine-nosed tweezers helps with this procedure.

27. Close chamber finger tight only. Too much pressure will result in a breakage of the coverslip. Too little pressure will cause the assembly to leak.

28. A setup that includes a 37 °C, 5 % CO_2 chamber attached to the microscope stage is ideal as it maintains cells at the correct temperature throughout the experiment.

29. It is best to capture images of ~50 transfected cells per sample. The transfection efficiency of each sample will determine how many images are required to reach this target.

30. Expression clones encoding His_6-NS5, GST-mImpα, and GST-mImpβ are available on request.

31. For optimal bacterial growth conditions, use a 2 L Erlenmeyer flask that has been baffled.

32. Never allow the column to run dry. Add more buffer or stop the flow by capping the ends.

33. Values in the "hooking" zone, where artificial quenching of the signal has occurred, should be excluded from the plot prior to curve fitting.

Acknowledgments

This work was supported by the Australian National Health and Medical Research Council (Fellowship and Project grants ID#333013/384105 and #143710, respectively, to D.A.Jans) and the ARC (Discovery Project/Fellowship #110104437, to K.M.Wagstaff). We thank Cassandra David for assistance with tissue culture.

References

1. Pryor MJ et al (2007) Nuclear localization of dengue virus nonstructural protein 5 through its importin alpha/beta-recognized nuclear localization sequences is integral to viral infection. Traffic 8:795–807

2. Bartenschlager R, Miller S (2008) Molecular aspects of dengue virus replication. Future Microbiol 3:155–165

3. Murray CL, Jones CT, Rice CM (2008) Architects of assembly: roles of flaviviridae non-structural proteins in virion morphogenesis. Nat Rev Microbiol 6:699–708

4. Bartholomeusz A, Thompson P (1999) Flaviviridae polymerase and RNA replication. J Viral Hepatitis 6:261–270

5. Rawlinson SM, Pryor MJ, Wright PJ, Jans DA (2006) Dengue virus RNA polymerase NS5: a potential therapeutic target? Curr Drug Targets 7:1623–1638

6. Rawlinson SM, Pryor MJ, Wright PJ, Jans DA (2009) CRM1-mediated nuclear export of dengue virus RNA polymerase NS5 modulates interleukin-8 induction and virus production. J Biol Chem 284:15589–15597

7. Jans DA, Xiao CY, Lam MH (2000) Nuclear targeting signal recognition: a key control point in nuclear transport? Bioessays 22:532–544

8. Pemberton LF, Paschal BM (2005) Mechanisms of receptor-mediated nuclear import and nuclear export. Traffic 6:187–198

9. Nigg EA (1997) Nucleocytoplasmic transport: signals, mechanisms and regulation. Nature 386:779–787

10. Poon IK, Jans DA (2005) Regulation of nuclear transport: central role in development and transformation? Traffic 6:173–186

11. Wagstaff KM, Rawlinson SM, Hearps AC, Jans DA (2011) An AlphaScreen(R)-based assay for high-throughput screening for specific inhibitors of nuclear import. J Biomol Screen 16:192–200

12. Wagstaff KM, Sivakumaran H, Heaton SM, Harrich D, Jans DA (2012) Ivermectin is a specific inhibitor of importin alpha/beta-mediated nuclear import able to inhibit replication of HIV-1 and dengue virus. Biochem J 443:851–856

13. Wagstaff KM, Jans DA (2006) Intramolecular masking of nuclear localization signals: analysis of importin binding using a novel AlphaScreen-based method. Anal Biochem 348:49–56

14. Alen MM, Kaptein SJ, De Burghgraeve T, Balzarini J, Neyts J, Schols D (2009) Antiviral activity of carbohydrate-binding agents and the role of DC-SIGN in dengue virus infection. Virology 387:67–75

15. Bae HG, Nitsche A, Teichmann A, Biel SS, Niedrig M (2003) Detection of yellow fever virus: a comparison of quantitative real-time PCR and plaque assay. J Virol Methods 110:185–191

Part IV

In Vitro Enzyme Assays for Antiviral Screening

Small Molecule Inhibitor Discovery for Dengue Virus Protease Using High-Throughput Screening

Mark Manzano, Janak Padia, and Radhakrishnan Padmanabhan

Abstract

Dengue virus (DENV), a member of mosquito-borne flavivirus genus in the Flaviviridae family, is an important human pathogen of global significance. DENV infections are the most common arbovirus infections in the world, causing more than ~300 million cases annually. Although majority of infections result in simple self-limiting disease known as dengue fever which resolve in 7–10 days, ~500,000 cases lead to more severe complications known as dengue hemorrhagic fever/dengue shock syndrome, more frequently observed in secondary infections due to an antibody-dependent enhancement mechanism, resulting in ~25,000 deaths. Currently, there are no vaccines or antiviral drug available for the treatment of DENV infections. Several viral and host proteins have been identified as potential targets for drug development. Some of the viral targets have enzyme activities that play essential roles in viral RNA replication for which in vitro high-throughput screening (HTS) assays have been developed. In this chapter, we describe an in vitro assay for the viral serine protease that has been successfully adapted to HTS format and has been used to screen several thousand compounds to identify inhibitors of the viral protease.

Key words Trypsin-like serine protease, Fluorescence-based screening, Bovine pancreatic trypsin inhibitor, Z' factor

1 Introduction

The four serotypes of dengue virus (DENV) are members of mosquito-borne flavivirus genus in the *Flaviviridae* family. Mosquito-borne flaviviruses are broadly classified into two subfamilies. Flaviviruses that cause hemorrhagic fever are in one subgroup which includes the prototypical member yellow fever virus and the four serotypes of DENV (DENV1–4). The second subgroup also known as the Japanese encephalitis virus sero-complex consists of flaviviruses that cause encephalitis such as the West Nile virus, Japanese encephalitis virus, Murray Valley encephalitis virus, and St. Louis encephalitis virus [1]. Members of both subgroups have a common genomic organization and replication strategies, and yet the diseases caused by them are distinct. These differences are

Radhakrishnan Padmanabhan and Subhash G. Vasudevan (eds.), *Dengue: Methods and Protocols*, Methods in Molecular Biology, vol. 1138, DOI 10.1007/978-1-4939-0348-1_20, © Springer Science+Business Media, LLC 2014

at least in part attributable to virus–host interactions that play a role in distinct pathogenesis [2].

The flavivirus genome is a positive sense ~11 kb RNA that codes for a single open reading frame that is translated by the host translation machinery to yield a polyprotein [3, 4]. The polyprotein is processed by both viral and host proteases in the endoplasmic reticulum (ER) into 3 structural proteins, capsid, C; precursor membrane, prM; and envelope, E, and 7 nonstructural (NS) proteins, NS1, NS2A, NS2B, NS3, NS4A, NS4B, and NS5 [1]. The ER-resident host proteases are involved in the cleavages of C–prM, prM–E, E–NS1, and NS1–NS2A junctions (for reviews, see refs. [1, 5] and the references therein). On the other hand, the viral protease, NS2B/NS3 complex, is responsible for autoproteolysis of the NS2B–NS3 junction as well as the cleavages of NS2A–NS2B, NS3–NS4A, and NS4B–NS5 junctions. The cleavage of C–prM junction by the ER-resident signal peptidase results in the release of C anchored to the ER via its hydrophobic C-terminal amino acid residues; the product of cleavage of C–prM junction by the ER-resident signal peptidase is also catalyzed by the viral protease to produce mature C protein.

The viral protease belongs to the trypsin-like family of serine proteases consisting of NS2B cofactor-dependent NS3 serine protease complex that cleaves the polyprotein at sites consisting of two basic amino acids followed by a short-chain amino acid [6–13]. The cleavage efficiency of the viral protease is modulated by polyprotein conformation in the ER [14]. The 130 amino acids long cofactor NS2B is an integral membrane protein in the ER [15]. The hydrophobicity plot of NS2B shows three hydrophobic regions flanking a highly conserved hydrophilic domain of ~45 amino acid residues [15]. The NS2B–NS3 precursor is processed by autoproteolysis (*cis*-cleavage). NS2B interacts with NS3 protein at its N-terminal ~180 amino acid residues to form a stable complex [12, 13]. Experiments involving coupled in vitro transcription by T7 RNA polymerase and translation mediated by the rabbit reticulocyte lysate system for expression of NS2B–NS3 protease provide evidence that cotranslational membrane insertion of the NS2B–NS3 precursor is essential for optimal cleavage of the NS2B–NS3 junction site [15]. The deletion of the three hydrophobic regions and linking of the conserved hydrophilic region (NS2BH) directly to the protease domain also undergoes autoproteolysis but in the absence of membranes [15]. This observation is the basis for the expression of the NS2BH–NS3 protease domain (NS2BH–NS3pro) in *E. coli* that has proven useful for the development of in vitro protease assays using purified recombinant viral proteases and fluorogenic peptide substrates for kinetic and biochemical analysis of several flavivirus proteases [16–20]. The *E. coli*-expressed recombinant proteases are purified by affinity chromatography if either NS2BH at the N-terminus or NS3pro at its C-terminus bears an affinity tag (e.g., His-tag). The in vitro protease has been adapted for high-throughput screening (HTS) to identify inhibitors of the flaviviral proteases [21–23].

These HTS campaigns for flavivirus targets described in literature to date are based on traditional methods that test compounds at a single concentration of each compound in the library in duplicate (or triplicate) 384-well plates. An alternative method has been described known as "quantitative HTS (qHTS)" using the pyruvate kinase and glucocerebrosidase as the target enzymes [24, 25], performed at the NIH Chemical Genomics Center (NCGC), now part of the National Center for Advancing Translational Sciences (NCATS). In the qHTS, a dose–response curve is obtained for each of the >60,000 compounds in 1,536-well format at 7 or more concentrations in a single experiment [24]. qHTS allows identification of both activators and inhibitors of the target enzyme with various potencies and efficacies of each compound from which structure activity relationships could be established from the primary screen. In this chapter, we describe the traditional HTS assay for identification of DENV2 protease inhibitors.

2   Materials

1. Luria broth (LB).
2. Ampicillin.
3. LB agar plates with ampicillin (100 µg/mL).
4. Isopropyl β-D-1-thiogalactopyranoside (IPTG).
5. Imidazole.
6. Lysozyme.
7. TALON Metal Affinity Resin (Clontech).
8. Buffer A (50 mM HEPES, pH 7, and 500 mM NaCl).
9. Dialysis buffer; prepare fresh (50 mM Tris–Cl pH 7.5, 20 % glycerol, 300 mM NaCl).
10. Reaction buffer (200 mM Tris–Cl, pH 9.5, 30 % glycerol, 0.1 % CHAPS).
11. Triton-X100.
12. CHAPS   (3-[(3-Cholamidopropyl)dimethylammonio]-1-propanesulfonate).
13. Bz-Nle-Lys-Arg-Arg-AMC (MW:833; custom synthesized).
14. Aprotinin (also known as bovine pancreatic trypsin inhibitor).
15. Chemically competent *E. coli* TOP10F' cells (Life Technologies; Cat. C3030-03) or other suitable *E. coli* cells transformed by the protease expression plasmid.

16. DENV2pro expression vector pQE30-NS2BH(QR)-NS3pro [26] (MW ~26 kDa).

17. 17×100 mm polypropylene culture tube.

18. 0.8×4 cm poly-prep chromatography column (Bio-Rad Cat. 731-1550) or equivalent.

19. SnakeSkin dialysis tubing, 10 K MWCO (Thermo Scientific Cat. 88245) or equivalent.

20. Bradford reagent or equivalent for protein estimation.

21. Cuvettes.

22. 96-well plate, clear.

23. 96-well plate, black or white.

24. 384-well plates, black or white.

25. Matrix WellMate small-bore tubing assembly (Thermo Scientific Cat. 201-30002).

26. UV–Vis spectrophotometer.

27. Microplate dispenser.

28. Fluorescence microplate reader (for 96-well and 384-well plates).

29. Refrigerated shaking incubator.

30. Sonicator.

31. Tube rotator.

32. 4 L Erlenmeyer flasks.

3   Methods

3.1   Expression of DENV2pro

Day 1

1. Thaw a tube of chemically competent *E. coli* TOP10F' cells on ice (*see* **Note 1**). While thawing, prechill a 17×100 mm culture tube also on ice.

2. Transfer 50 µL of cells into the culture tube.

3. Add 50–100 ng of pQE30-NS2BH(QR)-NS3pro DNA.

4. Incubate for 20–30 min.

5. Heat shock at 42 °C for 30 s.

6. Immediately place on ice for 2 min.

7. Plate directly to LB + ampicillin plate.

8. Incubate overnight at 30 °C to prevent satellite colonies from growing.

Day 2

9. At the start of the day, pick a colony and culture in 2 mL LB + ampicillin at 37 °C.

10. At the end of the day, split the culture into two 50 mL tubes each containing 25 mL LB + ampicillin.

11. Incubate overnight at 37 °C.

Day 3

12. Add 20 mL of the seed cultures to at least 3×4 L Erlenmeyer flasks, each containing 1.2 L of LB + ampicillin with 0.1 % glucose (*see* **Note 2**).

13. Incubate at 37 °C until the OD_{600} reaches 0.5–0.6.

14. Spin down the cultures and replace the medium with 1.2 L of LB with ampicillin containing 1 mM IPTG.

15. Incubate ~16 h at 20 °C in the refrigerated incubator at 220 rpm.

Day 4

16. Pellet cells. You can either snap freeze the pellet in liquid nitrogen and store them at –80 °C or proceed with protein purification.

3.2 Purification of DENV2pro

1. Thaw bacterial pellets on ice.

2. Resuspend by pipetting or vortexing in the appropriate amount of buffer A containing 0.5 % Triton-X100 and lysozyme (30 mL of buffer for every 1.2 L of bacterial culture).

3. Incubate on ice for 30 min.

4. Sonicate on ice for a total of 20 min (excluding cooling time), with 15 s pulses and 45 s cooling.

5. Clarify lysate by centrifuging at $18,000 \times g$ for 45 min.

6. While centrifuging, wash the TALON resin slurry (1 mL for every 1.2 L of cell culture) by resuspending three times with 15 mL buffer A.

7. Add the clarified supernatant from Subheading 3.2, **step 5,** to the washed resin.

8. Incubate for 1 h in the cold room on a rotator.

9. Pellet resin and wash with 15 mL buffer A.

10. To facilitate quicker washes and elution in the chromatography columns, split the resuspended resin in two poly-prep columns.

11. Allow resin to pack by gravity flow.

12. Wash with 4 mL buffer A with 10 mM imidazole while collecting 1 mL fractions on ice. Monitor protein amounts in the eluate by adding 5 μL to 95 μL of 1× Bradford reagent in a clear 96-well plate.

13. Elute with 3 mL buffer A + 150 mM imidazole. Collect 0.5 mL fractions on ice.

14. Confirm quality of protein expression by running 5 μL aliquots on a denaturing SDS-PAGE. The His-tagged protein should run at ~26 kDa. Majority of the protein should elute in the second and third fractions of the eluates. The protein may still elute in up to fraction 5.

15. Pool the desired fractions and place in the appropriate length of SnakeSkin (or equivalent) dialysis tubing.

16. Dialyze in 1 L dialysis buffer for 1 h in the cold room with stirring.

17. Replace buffer and repeat dialysis for another hour in the cold room.

18. Replace the buffer again and continue dialysis overnight.

19. Quantify protein by Bradford or BCA assay using a standard curve of bovine serum albumin of known concentrations.

3.3 Activity of Recombinant DENV2pro (See Note 3)

1. Warm up the microplate reader and set temperature to 37 °C.

2. Based on the concentration of the protein preparation calculated from the BSA standard curve in Subheading 3.2, **step 19**, prepare a 10 μM dilution of the enzyme in dialysis buffer. You would need 0.5 μL of this dilution per reaction well.

3. Protein and substrate mastermixes in reaction buffers are prepared as follows:

    Three wells are used for DENV2pro and substrate, while another three are used for DENV2pro, substrate, and aprotinin (bovine pancreatic trypsin inhibitor). For each mastermix, prepare the following solutions calculated per well:

    3.1. Protease mastermix: contents/well (1×) consist of 0.5 μL of 10 μM DENV2pro + 44.5 μL reaction buffer (*see* **Note 4**); prepare for six wells with some excess (6 + 0.5×) by mixing 3.25 μL of 10 μM DENV2pro enzyme + 289.25 μL reaction buffer.

    3.2. Substrate mastermix: contents/well (1×) consist of 0.5 μL of 2 mM substrate Bz-Nle-Lys-Arg-Arg-AMC (in DMSO) and 44.5 μL of reaction buffer (for one well); prepare for six wells (6 + 0.5×) by mixing 3.25 μL of 2 mM substrate + 289.25 μL reaction buffer.

    3.3. Aprotinin mastermix: contents/well (1×) consist of 0.5 μL of 0.4 mM aprotinin in water and 9.5 μL reaction buffer (for one well); prepare for three wells (3 + 0.5×) by mixing 1.75 μL of 0.4 mM aprotinin in water + 33.25 μL reaction buffer.

    3.4. DMSO negative control mastermix: contents/well (1×) consist of 0.5 μL of DMSO and 9.5 μL of reaction buffer (for one well); prepare for three wells (3 + 0.5)× by mixing 1.75 μL of DMSO and 33.25 μL reaction buffer.

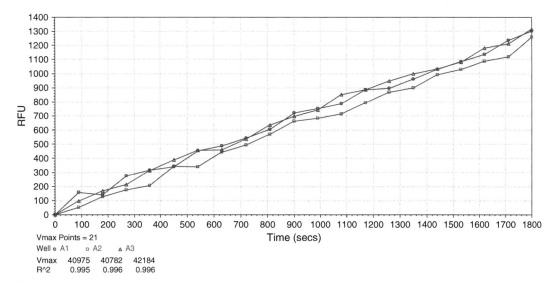

Fig. 1 An example of a time-course experiment of the in vitro protease assay performed in a 96-well plate format. Three wells were incubated with 50 nM purified DENV2pro and 10 µM Bz-Nle-Lys-Arg-Arg-AMC. Fluorescence intensities were read every 90 s for 30 min using the parameters $\lambda_{\text{excitation}}$ 380 nm and $\lambda_{\text{emission}}$ 460 nm

4. Dispense 45 µL of protein mastermix (Subheading 3.3, **step 3.1**) to each of 6 wells.

5. Dispense 10 µL of aprotinin mastermix (Subheading 3.3, **step 3.3**) to each of 3 wells. This is the positive control for an inhibitor.

6. Dispense 10 µL of DMSO mastermix (Subheading 3.3, **step 3.4**) to each of the remaining 3 wells. This is the negative control.

7. Add 45 µL of substrate mastermix (Subheading 3.3, **step 3.2**) to each well.

8. Read fluorescence intensity 37 °C for 30 min every 90 s at $\lambda_{\text{excitation}}$ 380 nm and $\lambda_{\text{emission}}$ 460 nm.

You should observe a linear increase in relative fluorescence intensity (RFU) over time for the negative control wells (Fig. 1). Aprotinin-treated wells should not exhibit any increase in fluorescence. If the RFU readings plateau before 30 min, it means that the substrate is limiting.

In this case, either verify the protein concentration using an alternate method or perform additional optimization assays using different protein dilutions.

3.4 Signal Optimization for Protease Assay in 384-Well Plate (See Note 5)

Methods under these sections are performed in an HTS screening facility equipped with fluidics systems to handle 384-well plates. Make sure that you have adequate amounts of reagents for optimization of the assay conditions and perform primary screens. Reaction volume for 384-well format is 30 µL/well.

1. Prepare the following stock solutions: 3 μM DENV2pro and 0.6 mM Bz-Nle-Lys-Arg-Arg-AMC.

2. Mix the following mastermixes:

    2.1. Protease mastermix: contents/well (1×) consist of 0.5 μL of 3 μM DENV2pro + 24.5 μL reaction buffer; prepare for 4 wells (4 + 0.5×) by mixing 2.25 μL of 3 μM DENV2pro enzyme + 110.25 μL reaction buffer.

    2.2. Substrate mastermix: contents/well (1×) consist of 0.5 μL of 0.6 mM substrate and 4.5 μL of reaction buffer; prepare for 4 wells (4 × 0.5×) by mixing 2.25 μL of the 0.6 mM substrate and 20 μL of reaction buffer.

3. Manually dispense 25 μL of DENV2pro mastermix to the four corner wells of the 384-well plate (wells A1, A24, P1, P24).

4. Add 5 μL of the substrate mastermix. Pipette up and down to mix.

5. Incubate at 37 °C for 30 min.

6. Run the optimization protocol in the Perkin Elmer EnVision plate reader for fluorescence intensity ($\lambda_{excitation}$ 380 nm, $\lambda_{emission}$ 460 nm). Save this protocol as "DENV2pro_RFU."

3.5 Z′ Factor Value Determination (See Note 6)

1. Attach the Matrix WellMate tubing assembly cartridge to the WellMate microplate dispenser.

2. Wash with 50 mL MilliQ H_2O. Empty out liquid back to container.

3. Wash with 50 mL 70 % ethanol. Empty out liquid back to container.

4. Rinse with 50 mL MilliQ H_2O. Empty out liquid back to container.

5. Choose "384" option and set "12.2 mm" as plate height.

6. Prepare protein and substrate mastermixes to fill one 384-well plate. Make sure to add ~10 mL to account for the "dead volume" attributed to the microplate dispenser.

    6.1. Protease mastermix: contents/well (1×) consist of 0.5 μL of 3 μM DENV2pro + 24.5 μL reaction buffer; prepare for 384 wells with a 10 mL excess for the dead volume (384× + 10 mL). The 384 reactions will have 192 μL of 3 μM DENV2pro in 9.408 mL of reaction buffer. The excess 10 mL will have 200 μL of 3 μM DENV2pro in 9.8 mL of reaction buffer.

    6.2. Substrate mastermix: contents/well (1×) consist of 0.5 μL of 0.6 mM substrate Bz-Nle-Lys-Arg-Arg-AMC (in DMSO) and 4.5 μL of reaction buffer (for one well); prepare for 384 wells with a 10 mL excess for the dead volume (384× + 10 mL). The 384 reactions will have

192 μL of 0.6 mM substrate in 1.728 mL of reaction buffer. The excess 10 mL will have 200 μL of 0.6 mM substrate in 1.8 mL of reaction buffer.

7. Dispense 25 μL of the DENV2pro mastermix to all the wells of the plate using the plate dispenser.

8. Manually add 1 μL of DMSO or 60 μM aprotinin to half the plate using a multichannel pipette.

9. Dispense 5 μL of the substrate mastermix to all the wells of the plate using the plate dispenser.

10. Incubate at 37 °C for 30 min.

11. Run the "DENV2pro_RFU" protocol in the Perkin Elmer EnVision plate reader for fluorescence intensity ($\lambda_{excitation}$ 380 nm, $\lambda_{emission}$ 460 nm).

12. Calculate the Z' factor using the means and standard deviations of the positive (aprotinin) and negative (untreated) controls:

$$Z' = \left| 1 - \frac{3\left(STDEV_{positive} + STDEV_{negative}\right)}{\left|AVE_{positive} - AVE_{negative}\right|} \right|$$

13. An excellent in vitro assay generally has a Z' of >0.7. A Z' factor of 0.4–0.6 is acceptable (*see* **Note 6**).

3.6 Primary HTS

In traditional facilities, HTS is usually performed in 384-well plates containing a final reaction volume of ~30 μL. The robotic pin tool transfer systems are set up to deliver 5 nL of each compound in a single concentration that could vary depending on the molecular weight of the compound (*see* **Note 7**).

1. Primary screens are done with not more than 60×384-well assay plates in a given time to avoid edge effects from evaporation. It is critical to have technical plate repeats (at least duplicate plates).

2. Thaw the compound library plates that you wish to test in a chamber with desiccator at room temperature (*see* **Note 8**). This usually takes ~30 min.

3. During this time, prepare mastermixes. Generally, columns 23 and 24 in the library plates of screening facilities are left blank. It is intended for these columns to serve as negative and positive controls for HTS assays. For our purposes, column 23 will be used for untreated samples, while column 24 will be used for aprotinin-treated samples (*see* **Note 9**).

4. Follow Subheading 3.5, **steps 1–5,** to prepare microplate dispenser.

5. Label the sides of the 384-well plates with the library plate number and plate replicate number.

6. Assemble the stacker chimney with the plate stacker base unit.

7. Place the plates into the chimney.

8. Dispense 25 µL of the DENV2pro mastermix in all wells of all plates. Make sure to account for the 10 mL dead volume required for the plate dispenser.

9. While the machine is dispensing, place the library plates in the Epson Compound Transfer Robot.

10. After all the DENV2pro mastermix has been dispensed, using a multichannel pipette, add 1 µL of 60 µM aprotinin in all of column 24 of the assay plates.

11. Place the stack of assay plates onto the Epson Compound Transfer Robot.

12. Perform the automated pin transfer in duplicate or triplicate plates.

13. During pin transfer, wash the microplate dispenser tubing as in Subheading 3.5, **steps 1–5**.

14. Working in batches of 10–15 assay plates, carefully remove the plates that have been processed by the transfer robot, and transfer back to the microplate dispenser.

15. Add 5 µL of the substrate mastermix to all the wells.

16. Incubate at 37 °C for 15 min.

17. Transfer the stack of plates to the Perkin Elmer EnVision plate reader, and run the "DENV2pro_RFU" protocol.

3.7 Data Analysis

1. For each assay plate, calculate the mean values of columns 23 (negative control) and 24 (positive control).

2. The mean of the aprotinin-treated sample is the background fluorescence. Subtract the mean of column 24 from all the raw RFU readings.

3. Express all the readings as a percentage of the mean of column 23.

4. Disregard readings that are >200 %. These are compounds that are autofluorescent at this wavelength or activators of protease.

5. Wells that have <50 % readings for all of the replicate assay plates will be considered as primary hits. Consider grouping the primary hits according to their degree of inhibition. This will be useful to decide what compounds to use for additional studies.

3.8 Verification and Characterization of Hits from Primary HTS

1. Identify active compounds from HTS screening with >50 % inhibitory activity (Fig. 2).

2. First chemical assessment of hits: select a set of "cherry-pick" compounds based on potency. Consideration should also be given to their "lead-like" and "drug-like" properties, Lipinski's

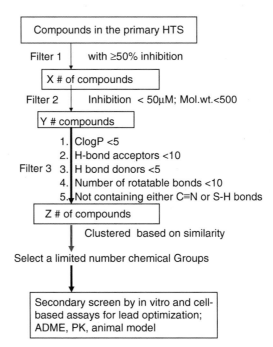

Fig. 2 Criteria for selection of compounds used for validation and further characterization

rule of 5 compliance (molecular weight <500, partition coefficient log P in −0.4 to +5.6 range, molar refractivity from 40 to 130, polar surface area no greater than 140 Å², hydrogen bond donor <5, hydrogen bond acceptors <10) [27, 28] and non-specific reactivity (based on previous reports), and potential reactive functionality that are known for toxic effects.

3. Acquire solid samples of the selected active compounds and their close analogs, and analyze them for their purity.

4. Perform additional biochemical tests to further validate and characterize the potency of these primary hits at 2 concentrations.

5. Perform dose response activity for DENV2 protease and determine IC50.

6. Perform hierarchical cluster analysis based on chemical structure and biological activity.

7. Second chemical assessment of confirmed hits: select 3–5 chemical classes based on structure activity relationship (SAR) range, synthetic tractability, intellectual property landscape, and relationship to other known pharmacophores and privileged structures.

8. Perform "Hit to Lead" optimization to explore optimization potential and ADME profile.

9. Select one chemical series for lead optimization.

4   Notes

1. Competent cells that were prepared in the lab will suffice. From our experience, freshly transformed cells give better protein expression than glycerol stocks.

2. pQE expression vector contains T5 promoter followed by two copies of lac operator sequences, ribosome binding site, translation initiation codon, 6× His-tag, multiple cloning sites, and stop codon. In spite of the overexpression of lac repressor due to lacI^q mutation in the host E. coli cells, the system is still leaky. While the bacteria are growing, as the plasmid copy number increases, the T5 promoter is turned on albeit at low level as the lac repressor is titrated out. The presence of glucose represses lac promoter by maintaining a low level of cAMP and cAMP activator protein (also known as cAMP receptor protein) resulting in a tight regulation of the promoter until the inducer of lac promoter, IPTG, is added.

3. Before proceeding to optimize the in vitro protease assay in the 384-well format, it is important to determine whether the purified DENV2pro is active. This quality control check is performed in a 96-well format.

4. The reaction buffer contains 0.1 % CHAPS. From our experience and our published results [29], the addition of 0.1 % CHAPS (or Brij-35 as described in ref. 21) in protein assays prevents false positives that are due to undesirable aggregation-based inhibition.

5. Signal optimization is done to calibrate the plate reader to the reaction plates that are intended to be used. This is to ensure that the maximum signal is detected in each well.

6. The Z' factor is a statistical value that measures the signal to noise ratio. In this case, we will compare the signal between samples that are untreated and treated with a known inhibitor of the DENV2pro, aprotinin. For in vitro assays, a Z' factor >0.7 is desired, while anywhere from 0.4 to 0.6 is acceptable for as long as duplicate plates are run in parallel.

7. This is a limitation of traditional HTS compared to qHTS. Including a variety of concentrations of each compound, the potential false-positive and false-negative hits are significantly minimized. However, qHTS is usually performed in 1,536-well format, and only some HTS facilities are equipped.

8. After 3 freeze–thaw cycles of the library plates, discard them. Freeze–thaw steps usually affect the stability of small molecule compounds.

9. Z' factor values are generally calculated from the readings of these wells in each assay plate.

Acknowledgments

The research was supported by NIH grants AI082068 and AI70791 to R. P., U54 AI057159 to NSRB, and the Cosmos Club Foundation Young Scholars Award to M.M.

References

1. Lindenbach D, Thiel HJ, Rice C (2007) Flaviviridae: the viruses and their replication. In: Knipe DM, Howley PM (eds) Field's virology, vol 1, 5th edn. Lippincott-Raven Publishers, Philadelphia, pp 1101–1152

2. Fernandez-Garcia MD, Mazzon M, Jacobs M, Amara A (2009) Pathogenesis of flavivirus infections: using and abusing the host cell. Cell Host Microbe 5(4):318–328. doi:S1931-3128(09)00102-4, [pii] 10.1016/j.chom.2009.04.001

3. Cleaves GR (1985) Identification of dengue type 2 virus-specific high molecular weight proteins in virus-infected BHK cells. J Gen Virol 66(Pt 12):2767–2771

4. Crawford GR, Wright PJ (1987) Characterization of novel viral polyproteins detected in cells infected by the flavivirus Kunjin and radiolabelled in the presence of the leucine analogue hydroxyleucine. J Gen Virol 68(Pt 2):365–376

5. Padmanabhan R, Strongin AY (2010) Translation and processing of the dengue virus polyprotein. In: Hanley KA, Weaver SC (eds) Frontiers in dengue virus research. Caister Academic Press, Norfolk, pp 14–33

6. Bazan JF, Fletterick RJ (1989) Detection of a trypsin-like serine protease domain in flaviviruses and pestiviruses. Virology 171(2):637–639

7. Chambers TJ, Weir RC, Grakoui A, McCourt DW, Bazan JF, Fletterick RJ, Rice CM (1990) Evidence that the N-terminal domain of nonstructural protein NS3 from yellow fever virus is a serine protease responsible for site-specific cleavages in the viral polyprotein. Proc Natl Acad Sci U S A 87(22):8898–8902

8. Wengler G, Czaya G, Farber PM, Hegemann JH (1991) In vitro synthesis of West Nile virus proteins indicates that the amino-terminal segment of the NS3 protein contains the active centre of the protease which cleaves the viral polyprotein after multiple basic amino acids. J Gen Virol 72(Pt 4):851–858

9. Chambers TJ, Grakoui A, Rice CM (1991) Processing of the yellow fever virus nonstructural polyprotein: a catalytically active NS3 proteinase domain and NS2B are required for cleavages at dibasic sites. J Virol 65(11):6042–6050

10. Zhang L, Mohan PM, Padmanabhan R (1992) Processing and localization of Dengue virus type 2 polyprotein precursor NS3-NS4A-NS4B-NS5. J Virol 66(12):7549–7554

11. Falgout B, Pethel M, Zhang YM, Lai CJ (1991) Both nonstructural proteins NS2B and NS3 are required for the proteolytic processing of dengue virus nonstructural proteins. J Virol 65(5):2467–2475

12. Falgout B, Miller RH, Lai C-J (1993) Deletion analysis of dengue virus type 4 nonstructural protein NS2B: Identification of a domain required for NS2B-NS3 protease activity. J Virol 67:2034–2042

13. Chambers TJ, Nestorowicz A, Amberg SM, Rice CM (1993) Mutagenesis of the yellow fever virus NS2B protein: effects on proteolytic processing, NS2B-NS3 complex formation, and viral replication. J Virol 67(11):6797–6807

14. Zhang L, Padmanabhan R (1993) Role of protein conformation in the processing of dengue virus type 2 nonstructural polyprotein precursor. Gene 129(2):197–205

15. Clum S, Ebner KE, Padmanabhan R (1997) Cotranslational membrane insertion of the serine proteinase precursor NS2B-NS3(Pro) of dengue virus type 2 is required for efficient in vitro processing and is mediated through the hydrophobic regions of NS2B. J Biol Chem 272(49):30715–30723

16. Yusof R, Clum S, Wetzel M, Murthy HM, Padmanabhan R (2000) Purified NS2B/NS3 serine protease of dengue virus type 2 exhibits cofactor NS2B dependence for cleavage of substrates with dibasic amino acids in vitro. J Biol Chem 275(14):9963–9969

17. Leung D, Schroder K, White H, Fang NX, Stoermer MJ, Abbenante G, Martin JL, Young PR, Fairlie DP (2001) Activity of recombinant dengue 2 virus NS3 protease in the presence of a truncated NS2B co-factor, small peptide

substrates, and inhibitors. J Biol Chem 276(49): 45762–45771

18. Chappell KJ, Nall TA, Stoermer MJ, Fang NX, Tyndall JD, Fairlie DP, Young PR (2005) Site-directed mutagenesis and kinetic studies of the West Nile Virus NS3 protease identify key enzyme-substrate interactions. J Biol Chem 280(4):2896–2903

19. Mueller NH, Yon C, Ganesh VK, Padmanabhan R (2007) Characterization of the West Nile virus protease substrate specificity and inhibitors. Int J Biochem Cell Biol 39(3):606–614

20. Li J, Lim SP, Beer D, Patel V, Wen D, Tumanut C, Tully DC, Williams JA, Jiricek J, Priestle JP, Harris JL, Vasudevan SG (2005) Functional profiling of recombinant NS3 proteases from all four serotypes of dengue virus using tetrapeptide and octapeptide substrate libraries. J Biol Chem 280(31):28766–28774

21. Johnston PA, Phillips J, Shun TY, Shinde S, Lazo JS, Huryn DM, Myers MC, Ratnikov BI, Smith JW, Su Y, Dahl R, Cosford ND, Shiryaev SA, Strongin AY (2007) HTS identifies novel and specific uncompetitive inhibitors of the two-component NS2B-NS3 proteinase of West Nile virus. Assay Drug Dev Technol 5(6):737–750

22. Mueller NH, Pattabiraman N, Ansarah-Sobrinho C, Viswanathan P, Pierson TC, Padmanabhan R (2008) Identification and biochemical characterization of small-molecule inhibitors of west nile virus serine protease by a high-throughput screen. Antimicrob Agents Chemother 52(9):3385–3393. doi:AAC.01508-07, [pii] 10.1128/AAC.01508-07

23. Yang CC, Hsieh YC, Lee SJ, Wu SH, Liao CL, Tsao CH, Chao YS, Chern JH, Wu CP, Yueh A (2011) Novel Dengue Virus-Specific NS2B/NS3 Protease Inhibitor, BP2109, Discovered by a High-Throughput Screening Assay. Antimicrob Agents Chemother 55(1):229–238. doi:AAC.00855-10, [pii] 10.1128/AAC.00855-10

24. Inglese J, Auld DS, Jadhav A, Johnson RL, Simeonov A, Yasgar A, Zheng W, Austin CP (2006) Quantitative high-throughput screening: a titration-based approach that efficiently identifies biological activities in large chemical libraries. Proc Natl Acad Sci U S A 103(31):11473–11478. doi:10.1073/pnas.0604348103

25. Zheng W, Padia J, Urban DJ, Jadhav A, Goker-Alpan O, Simeonov A, Goldin E, Auld D, LaMarca ME, Inglese J, Austin CP, Sidransky E (2007) Three classes of glucocerebrosidase inhibitors identified by quantitative high-throughput screening are chaperone leads for Gaucher disease. Proc Natl Acad Sci U S A 104(32):13192–13197. doi:10.1073/pnas.0705637104

26. Yon C, Teramoto T, Mueller N, Phelan J, Ganesh VK, Murthy KH, Padmanabhan R (2005) Modulation of the nucleoside triphosphatase/RNA helicase and 5′-RNA triphosphatase activities of dengue virus type 2 nonstructural protein 3 (NS3) by interaction with NS5, the RNA-dependent RNA polymerase. J Biol Chem 280:27412–27419

27. Lipinski CA (2000) Drug-like properties and the causes of poor solubility and poor permeability. J Pharmacol Toxicol Methods 44(1):235–249, S1056-8719(00)00107-6 [pii]

28. Lipinski CA, Lombardo F, Dominy BW, Feeney PJ (2001) Experimental and computational approaches to estimate solubility and permeability in drug discovery and development settings. Adv Drug Deliv Rev 46(1–3):3–26

29. Ezgimen MD, Mueller NH, Teramoto T, Padmanabhan R (2009) Effects of detergents on the West Nile virus protease activity. Bioorg Med Chem 17(9):3278–3282. doi:S0968-0896(09)00310-1, [pii] 10.1016/j.bmc.2009.03.050

Chapter 21

Construction of Dengue Virus Protease Expression Plasmid and In Vitro Protease Assay for Screening Antiviral Inhibitors

Huiguo Lai, Tadahisa Teramoto, and Radhakrishnan Padmanabhan

Abstract

Dengue virus serotypes 1–4 (DENV1–4) are mosquito-borne human pathogens of global significance causing ~390 million cases annually worldwide. The virus infections cause in general a self-limiting disease, known as dengue fever, but occasionally also more severe forms, especially during secondary infections, dengue hemorrhagic fever and dengue shock syndrome causing ~25,000 deaths annually. The DENV genome contains a single-strand positive sense RNA, approximately 11 kb in length. The 5′-end has a type I cap structure. The 3′-end has no poly(A) tail. The viral RNA has a single long open reading frame that is translated by the host translational machinery to yield a polyprotein precursor. Processing of the polyprotein precursor occurs co-translationally by cellular proteases and posttranslationally by the viral serine protease in the endoplasmic reticulum (ER) to yield three structural proteins (capsid (C), precursor membrane (prM), and envelope (E) and seven nonstructural (NS) proteins (NS1, NS2A, NS2B, NS3, NS4A, NS4B, and NS5). The active viral protease consists of both NS2B, an integral membrane protein in the ER, and the N-terminal part of NS3 (180 amino acid residues) that contains the trypsin-like serine protease domain having a catalytic triad of H51, D75, and S135. The C-terminal part of NS3, ~170–618 amino acid residues, encodes an NTPase/RNA helicase and 5′-RNA triphosphatase activities; the latter enzyme is required for the first step in 5′-capping. The cleavage sites of the polyprotein by the viral protease consist of two basic amino acid residues such as KR, RR, or QR, followed by short chain amino acid residues, G, S, or T. Since the cleavage of the polyprotein by the viral protease is absolutely required for assembly of the viral replicase, blockage of NS2B/NS3pro activity provides an effective means for designing dengue virus (DENV) small-molecule therapeutics. Here we describe the screening of small-molecule inhibitors against DENV2 protease.

Key words Dengue virus serotype 2, In vitro protease assay, Fluorogenic peptide substrate, Aprotinin, Trypsin inhibitor, NS2B cofactor, NS3 protease domain

1 Introduction

Dengue virus (DENV) is one of the most prevalent mosquito-borne flaviviruses affecting humans [1, 2]. According to a recent estimate, ~390 million cases of dengue infections occur annually [3]. Most of these infections either are asymptomatic or cause a simple

Radhakrishnan Padmanabhan and Subhash G. Vasudevan (eds.), *Dengue: Methods and Protocols*, Methods in Molecular Biology, vol. 1138, DOI 10.1007/978-1-4939-0348-1_21, © Springer Science+Business Media, LLC 2014

self-limiting disease, dengue fever. However, more severe forms of dengue disease can occur in 0.5 to 1 million cases, especially during secondary infection by a different serotype resulting in ~25,000 deaths annually [4]. Currently, there is no effective vaccine or chemotherapeutic agent for prevention or treatment of dengue infections [5, 6].

DENV NS3 is a multifunctional protein containing a trypsin-like serine protease domain within the N-terminal 185 amino acid (aa) residues (hereafter referred to as "NS3pro") with catalytic triad residues, H51, D75, and S135. The protease activity of NS3pro requires interaction with NS2B, an ER-resident integral membrane protein [7], consisting of three hydrophobic domains flanking a conserved hydrophilic domain of ~45 aa (NS2BH; in DENV2, the region between 49 and 92 aa). The hydrophilic domain of NS2B is sufficient for interaction with the NS3pro domain to form an active protease in vitro [7–9].

Since the viral protease is crucial for polyprotein processing and assembly of the viral replicase complex, it is an excellent target for development of anti-dengue viral therapeutics. However, to date only a small number of non-peptidyl inhibitor scaffolds for DENV proteases have been described and the available structure–activity relationship data are rather limited.

In this chapter, we describe expression and purification of DENV2 protease in *E. coli* and in vitro protease assays using DENV2 NS2BH-NS3pro. Using this protease assay, the potency of a class of compounds containing the 8-hydroxyquinoline scaffold is determined as potent inhibitors of the DENV2 and WNV proteases (general structure in Fig. 1a).

2   Materials

Use molecular biology grade chemicals and reagents and ultrapure water (prepared using Milli-Q or equivalent water purification system) for preparation of all solutions. Store all reagents at room temperature unless indicated otherwise. Follow regulatory guidelines established by the Institutional Biosafety Committee for recombinant DNA work and handling biohazard waste materials.

2.1   E. Coli Expression System

We use pQE30 *E. coli* expression vector (Qiagen). Other expression vectors are equally suitable:

1. PCR: Taq polymerase that adds an A at the 3′-end to facilitate cloning of the PCR fragment using a T-tailed vector (T/A cloning), oligodeoxynucleotide primers, and dNTP mixture.

2. Cloning: T/A ligation kit (2× ligation buffer, vector for T/A cloning, T4 DNA ligase) and regular ligation kit (10× ligation buffer and T4 DNA ligase).

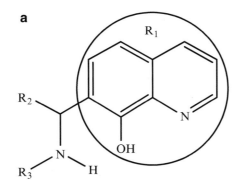

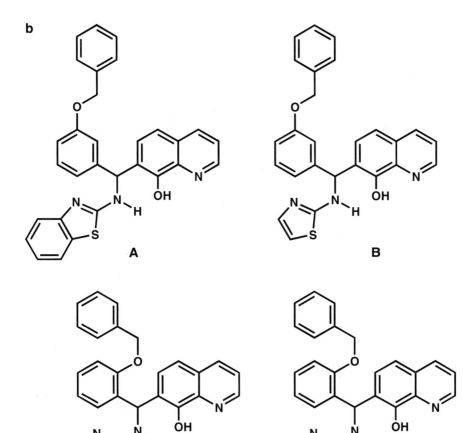

Fig. 1 Structures of 8-HQ derivatives. (**a**) The general structures of 8-HQ scaffold. (**b**) The structures of four compounds that are analyzed in this study are shown

3. Template for PCR: the plasmid clone encoding the full length cDNA of DENV2 New Guinea C strain in pRS424 vector is a gift from Dr. Barry Falgout (CBER, FDA, Bethesda, MD).

4. Restriction enzymes: appropriate restriction enzymes and the corresponding 10× buffers are from a suitable commercial source.

5. Bacterial transformation: *E. coli* Top10F′ competent cells (Invitrogen) are used for transformation. Super optimal catabolite (SOC) medium: add 20 g Bacto Tryptone, 5 g yeast extract, 0.58 g NaCl, and 0.186 g KCl in 1 L of H_2O. Mix well on a magnetic stir plate until all contents are dissolved. Autoclave the solution for 30 min and allow it to cool gradually to room temperature. Add 10 mL of sterile 2.0 M magnesium solution (1.0 M magnesium sulfate, 1.0 M magnesium chloride) and mix well. Add 10 mL of sterile 2.0 M glucose and mix well (final glucose concentration is 20 mM).

6. Luria-Bertani (LB) medium and LB agar for growth of *E. coli*: LB medium consists of 10.0 g tryptone, 5.0 g yeast extract, and 10.0 g NaCl in 1 L of H_2O. Dissolve the contents and autoclave. Ampicillin-containing LB is made by adding 100 μg of ampicillin/mL LB for growth of *E. coli* transformed by plasmids encoding ampicillin-resistant (Amp^r) gene. To make up LB-agar plates, add agar to a final amount of 1.5 %, autoclave the LB agar, and cool gradually to 40 °C before adding ampicillin to a final concentration of 100 μg of ampicillin/mL (stock ampicillin concentration is 100 mg/mL dissolved in H_2O) LB agar kept at 40 °C (*see* **Note 1**).

7. X-Gal and IPTG solutions are prepared at 20 mg/mL and 0.42 M, respectively. Add 100 μL X-Gal solution and 25 μL IPTG onto LB/Amp plate and spread evenly to make LB/Amp/X-Gal/IPTG plate.

8. Plasmid purification: a number of commercial kits are available to prepare plasmids in mini-, medium-, and large-scale preparations.

9. DNA fragment purification by agarose gel electrophoresis: agarose (molecular biology grade), TAE buffer (40 mM Tris base, pH 8.0, 20 mM acetic acid and 1 mM EDTA), and gel DNA recovery kit. To make a stock solution of 50×, add 242 g Tris base in water, 57.1 mL glacial acetic acid, and 100 mL of 500 mM EDTA (pH 8.0) solution. Adjust the final volume up to 1 L.

2.2 Expression and Purification of NS2BH-NS3pro

Talon metal affinity resin or equivalent such as Ni-NTA, protein storage buffer (100 mM Tris–HCl pH 8.0, 300 mM NaCl, 30 % glycerol), chromatography columns, 150 mM imidazole, and dialysis membranes.

2.3 Protease Assay Materials

1. A *tetra*-peptide substrate, benzoyl (Bz)-norleucine (Nle)-Lys-Arg-Arg-7-amino-4-methylcoumarin (AMC) [10] with a fluorogenic moiety at the C-terminus is purchased commercially or custom synthesized (*see* **Notes 2** and **3**).

2. Free AMC, the fluorescent product released from the peptide substrate, is purchased commercially. Make a stock solution in dimethylsulfoxide. Generate a standard curve in which X-axis represents different amounts of AMC and Y-axis, the relative fluorescence units (RFU). The amount released in a protease assay is estimated using this standard curve.

3. CHAPS, a zwitterionic detergent (3-[(3-cholamidopropyl) dimethylammonio]-1-propanesulfonate), is used in the protease assay to prevent nonspecific inhibition caused by aggregation of the compound [11, 12].

4. Spectrofluorometer: a spectrofluorometer with a plate reader which can accept either a black 96-well or 384-well plate (*see* **Note 4**).

5. SigmaPlot or GraphPad Prism software for nonlinear regression analysis and calculations of kinetic constants.

3 Methods

3.1 Construction of DENV2 NS2BH-(QR)- NS3pro

3.1.1 Design the Primers

NS2B hydrophilic region (NS2BH), 4276–4407 nt region in NS2BH, is selected for PCR, using forward primer **1** (5′-CGC GGATCCGCCGATTTGGAACTGGAGAGAGCCGCC-3′) and reverse primer **2** (5′-TTGGCGCGCTGTTCTTCCTCTTCGTTT TTTATCGAC-3′) (*see* **Notes 5–7**). The NS3 protease domain spans 4522–5076 nt, encoding 185 amino acids from the N-terminus, and is amplified using forward primer **3** (5′-TTGGCGCGCTGGA GTATTGTGGGATGTCCCTTCACC-3′) and reverse primer **4** (5′-CCCAAGCTTACTTTCGAAAAATGTCATCTTCGATC T-3′) (*see* **Note 8**).

3.1.2 The Components for Polymerase Chain Reactions (PCR) (50 μL)

PCR buffer (10×), 5 μL; DNA template (pRS424-FL-DENV2; 250 μg/mL), 2 μL; forward and reverse primers, **1** and **2** (100 μM), respectively, 2 μL each; dNTP mixture (25 μM each), 2 μL; Taq polymerase, 1 μL; deionized H_2O, 36 μL.

3.1.3 PCR Conditions

Initial denaturation for 5 min at 95 °C (1 cycle); subsequent cycles (~35) consist of denaturation for 30 s at 95 °C, annealing for 30 s at 55 °C, and extension for 60 s at 72 °C; the final step is extension for 10 min at 72 °C; keep at 4 °C until it is used at a subsequent step.

3.1.4 Analysis of PCR Products

1. Agarose gel (0.8 %) electrophoresis: add 0.4 g agarose in 50 mL 1× TAE buffer (a 50× stock is prepared containing 2.0 M Tris and 0.05 M EDTA in Milli-Q H_2O and adjusted to pH 8.3), heat in a microwave oven to dissolve and make a clear solution, gradually cool down, and add 2.0 μL ethidium

bromide (10 mg/mL) (*see* **Note 9**) before the solution is solidified. Pour the agarose solution into a gel electrophoresis tray. Let it stand for 30 min.

2. Add 10 μL DNA loading dye (6×) into the 50 μL PCR products.

3. Gently mix it and load into the wells.

4. Run the electrophoresis at 110 mA for 30 min.

5. View the gel under the UV light (*see* **Note 10**).

6. Cut the expected band (NS2BH PCR fragment) for extraction.

7. Extract PCR fragments using DNA recovery kit.

3.1.5 Restriction Enzyme Digestion and Ligation

1. Mix NS2BH PCR fragment (36 μL) with 10× NEBuffer 3 (New England Biolabs) (5.0 μL), 10× BSA (5.0 μL), and BamHI (2.0 μL, 20,000 U/mL); incubate at 37 °C for 2 h. Then add BssHII (2.0 μL) and raise the temperature to 50 °C for another 2 h.

2. Mix NS3pro PCR fragment (36 μL) with NEBuffer 2 (5.0 μL), 10× BSA (5.0 μL), and HindIII (2.0 μL); incubate at 37 °C for 2 h. Then add BssHII (2.0 μL) and raise the temperature to 50 °C for another 2 h.

3. Mix TA vector (36 μL) with 10× NEBuffer 2 (5.0 μL), 10× BSA (5.0 μL), BamHI (2.0 μL), and HindIII (2.0 μL); incubate at 37 °C for 2 h (*see* **Note 11**).

4. Purify the restriction digest by agarose gel electrophoresis (*see* **steps 1–4**, Subheading 3.1.4).

5. Ligation components: 10× T4 DNA ligase buffer, 1.5 μL; TA vector digested with BamHI/HindIII, 2.0 μL; "Insert 1" NS2BH cleaved by BamHI/BssHII, 5.0 μL; "Insert 2" NS3pro cleaved by BssHII/HindIII, 5.0 μL; and T4 DNA ligase, 1.5 μL; mix the components and incubate at 16 °C overnight.

3.1.6 Transformation of Competent E. Coli Cells

1. Add 5 μL ligation mixture (from **step 5**, Subheading 3.1.5) into 50 μL Top10F′ *E. coli* competent cells (or a suitable alternative *E. coli* cells); incubate on ice for 10 min, followed by heat shock at 42 °C for 45 s without shaking. Incubate on ice for 2 min and add 200 μL SOC medium into the tube. Incubate at 37 °C for 1 h with moderate shaking.

2. Add 50 μL of the transformed *E. coli* cells on to an X-Gal/IPTG/LB/Amp⁺ agar plate; evenly spread the cells on the plate and incubate the plate at 37 °C overnight (*see* **Note 12**).

3. Pick up single isolated white colonies and inoculate into 5 mL LB broth per colony taken in a 50 mL polypropylene centrifuge tube; add 5.0 μL 100 mg/mL ampicillin stock solution (final concentration is 100 μg/mL). Incubate the tubes at 37 °C overnight with moderate shaking for overnight.

3.1.7 Plasmid Extraction from the Bacterial Cultures

1. Add 600 μL of bacterial culture grown in LB medium to a 1.5 mL microcentrifuge tube.

2. Add 100 μL of 7× lysis buffer (blue) and mix by gently inverting the tube four to six times. Proceed to the next step within 2 min.

3. Add 350 μL of cold neutralization buffer (yellow, stored at 4 °C) and mix thoroughly (*see* **Note 13**).

4. Centrifuge at $11,000–15,000 \times g$ for 4 min.

5. Transfer the supernatant (~900 μL) into a column provided in a kit. While removing the supernatant, avoid disturbing the cell pellet.

6. Place the column into a collection tube provided and centrifuge for 30 s.

7. Discard the flow-through and place the column back into the same collection tube.

8. Add 200 μL wash buffer to the column. Centrifuge for 30 s.

9. Add 400 μL wash buffer to the column. Centrifuge for 30 s.

10. Discard the flow-through and place the column back into the same collection tube. Centrifuge for another 30 s.

11. Transfer the column to a clean 1.5 mL microcentrifuge tube, and then add 30 μL of water directly to the column matrix and let it stand for 1 min at room temperature.

12. Centrifuge at $14,000 \times g$ for 60 s to elute the plasmid DNA.

13. The sequence of NS2BH-NS3pro domain in the plasmid is confirmed by DNA sequencing.

3.1.8 Subcloning the NS2BH-NS3pro Fragment into pQE30 (or Suitable Alternative Vector) for Expression

1. Digest the plasmid (from **step 13**) and pQE30 plasmid using BamHI and HindIII (*see* Subheading 3.1.5 for details).

2. Purify the fragment and vector using gel electrophoresis (*see* Subheading 3.1.4 for details).

3. Ligate NS2BH-NS3pro and predigested pQE30 vector using T4 DNA ligase (*see* Subheading 3.1.5 for details).

4. Transform Top10F′ *E. coli* competent cells and spread onto LB/Amp⁺ (100 μg/mL) plate and incubate at 37 °C without shaking for overnight (*see* Subheading 3.1.6 for details).

5. Colonies are available overnight and the expected plasmid in pQE30 vector is used for our following protease assays.

3.2 Expression and Purification of NS2BH-NS3pro

1. Pick a single colony, and inoculate 5 mL LB broth; add 5.0 μL 100 mg/mL ampicillin stock solution (final concentration is 100 μg/mL) into a 50 mL polypropylene centrifuge tube. Incubate the tube overnight with moderate shaking.

2. Transfer the cell culture into a 1 L LB medium (with 100 μg/mL ampicillin) and continue to incubate at 37 °C with shaking until O.D. 600 reaches 0.6–1.0.

3. Add 1 mL, 1 M IPTG (final concentration 1 mM) into the cell culture (*see* **Note 14**).

4. Incubate the cell culture at 37 °C for 3 h.

5. Centrifuge the bacterial culture at $4,000 \times g$ for 10 min. Collect the cell pellet.

6. Resuspend the cell pellet with 30 mL protein storage buffer (100 mM Tris, pH 8.0, 300 mM NaCl, and 20 % glycerol).

7. Freeze the tube containing the cell suspension by dry ice/ethanol quickly and thaw it in water.

8. Repeat the freeze–thaw steps 3× and sonicate at 4 °C for 10 min (Misonix Sonicator 3000; program 30 W, 15 s on/45 s off).

9. Centrifuge at $14,000 \times g$ for 30 min.

10. Collect the supernatant into a 1.5 mL Talon resin or equivalent metal affinity beads.

11. Incubate the supernatant and resin at 4 °C for 1 h.

12. Collect the resin by centrifugation at $1,000 \times g$ for 1 min.

13. Wash the resin using protein storage buffer 5 times and transfer the resin to a suitable chromatography column.

14. Elute the protein using 150 mM imidazole.

15. Dialyze with Spectra/Por membrane (cut-off 1000 daltons, Spectrum Laboratories or suitable alternative source) through 500 mL storage buffer.

16. Aliquot the enzyme and store the aliquots at –80 °C.

3.3 Protease Assay to Determine the Potency of Small-Molecule Compounds

The detailed procedures are summarized below:

1. Weigh a specific amount of compounds (A, B, C, and D in Fig. 1b) and put them into four brown glass vials, respectively. Add dry DMSO to make a 10 mM final concentration (stock I, Table 1) (*see* **Note 15**).

2. Prepare stock II (50 μM) and III (20 μM) from stock I in buffer containing 200 mM Tris–HCl (pH 9.5), 6 mM NaCl, 30 % glycerol, and 0.1 % CHAPS (*see* **Note 16**):

Stock II: 5.0 μL (stock I) dilute to 1 mL buffer.

Stock III: 2.0 μL (stock I) dilute to 1 mL buffer.

3. Prepare 1 mL of DENV2 NS2BH/NS3pro protease (50 nM) in buffer (*see* **Note 17**).

4. Mix 50 μL inhibitor from stock II or stock III and 45 μL protease. Incubate at room temperature for 15 min (*see* **Note 18**) in a 96-well black flat bottom plate.

Table 1
Preparation of 10 mM inhibitors A–D stock solutions

Inhibitors	Mass (mg)	Chemical formula and molecular weight	Volume for 10 mM
A	10.1	$C_{30}H_{23}N_3O_2S$ 489	2.0654 mL
B	8.2	$C_{26}H_{21}N_3O_2S$ 439	1.8679 mL
C	8.9	$C_{26}H_{21}N_3O_2S$ 439	2.0273 mL
D	9.5	$C_{30}H_{23}N_3O_2S$ 489	1.9427 mL

5. Prepare a substrate stock (IV): 5.0 mM *tetra*-peptide substrate solution, benzyloxycarbonyl (Bz)-Nle-Gly-Arg-Arg-AMC, in DMSO.

6. Dilute 10 μL of 5.0 mM substrate stock (IV) to 500 μL buffer containing 200 mM Tris–HCl (pH 9.5), 6 mM NaCl, 30 % glycerol, and 0.1 % CHAPS to make substrate stock (V) with a final concentration of 100 μM.

7. Add 5.0 μL of 100 μM stock V to the mixture of inhibitor and protease in the plate in **step 4** above (*see* **Note 19**). The total volume is 100 μL.

8. Place the 96-well plate into the spectrofluorometer for measuring the increase in fluorescence when the substrate is hydrolyzed by the protease at 37 °C (*see* **Note 20**). Set the maximum excitation and emission wavelength for this substrate at 380 and 460 nm, respectively.

9. Convert the fluorescent intensity to percent inhibition (*see* **Note 21**). The results are summarized in Fig. 2.

3.4 IC$_{50}$ Determination

1. Make a serial dilution of inhibitors to prepare 20 nM, 100 nM, 200 nM, 1 μM, 4 μM, 8 μM, 12 μM, 16 μM, 20 μM, 40 μM, and 50 μM stock solutions.

2. Mix each prepared solution (50 μL) with 45 μL protease (50 nM) in a 96-well plate, and let it stand for 15 min at room temperature.

3. Add 5.0 μL of 100 μM stock of *tetra*-peptide substrate into the mixture. The final concentrations of the inhibitors are 0.010, 0.050, 0.1, 0.5, 1, 2, 4, 6, 8, 10, 20, and 25 μM.

4. Measure the fluorescence as described in **step 8**, Subheading 3.3.

5. Convert the fluorescent intensity to percent inhibition (*see* **Note 21**).

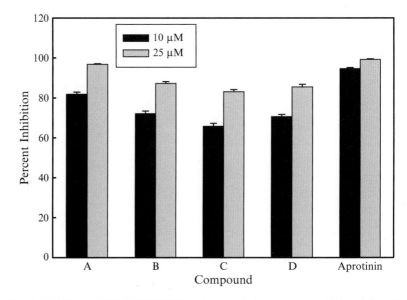

Fig. 2 Inhibition of DENV2 NS2B/NS3pro by selected compounds at 10 and 25 μM. The concentration of DENV2 proteases is 25 nM. The buffer contains 200 mM Tris–HCl, 6.0 mM NaCl, 30 % glycerol, and 0.1 % CHAPS, pH 9.5. The percent values are calculated from the relative fluorescence units obtained in the presence and absence of tested compound (*see* **Note 19**). BPTI (aprotinin) and DMSO are used as a positive and negative control, respectively. Letter below corresponds to compounds **A**, **B**, **C**, and **D**, where the molecular structure is shown in Fig. 1b. All assays are performed in triplicate and the average values are shown. Error bars indicate standard deviation of the means

Table 2
Percent inhibition and IC$_{50}$ values of compounds A, B, C, and D against DENV2 NS2BH/NS3 protease in the presence of 0.1 % CHAPS

Inhibitors	% Inhibition at 10 μM	% Inhibition at 25 μM	IC$_{50}$ (μM)	Hill slope
A	87.32 ± 1.25	96.39 ± 1.32	0.91 ± 0.02	1.05 ± 0.05
B	83.57 ± 1.21	92.92 ± 1.26	2.93 ± 0.07	1.27 ± 0.06
C	76.92 ± 1.52	92.62 ± 1.47	3.67 ± 0.08	1.18 ± 0.05
D	82.78 ± 1.06	91.79 ± 1.18	2.34 ± 0.14	1.02 ± 0.05

6. Plot the percentage protease activity against the Log of inhibitor concentrations to get the IC$_{50}$ values (*see* **Note 22**). The results are shown in Table 2.

3.5 Steady-State Kinetic Analysis

1. Prepare the assay buffer containing 200 mM Tris–HCl (pH 9.5), 6 mM NaCl, and 30 % glycerol containing 0.1 % CHAPS.

2. Make four different concentrations (0, 1.0, 3.0, and 5.0 μM) of inhibitor **A** in the above buffer (*see* **Note 23**).

3. Make a series of *tetra*-peptide, benzyloxycarbonyl (Bz)-Nle-Gly-Arg-Arg-AMC, concentrations of please replace with 0, 2, 4, 6, 8, 10, 15, 20, 25, 30, 40, and 50 μM.

4. Adjust the final volume to 100 μL by adding the buffer into the corresponding wells.

5. Measure the fluorescence as described in **step 8**, Subheading 3.3.

6. Monitor the fluorescent intensity of AMC released every 1.5 min for a total period of 30 min–2 h (*see* **Note 24**).

7. The results are analyzed using SigmaPlot 2001 v7.0 and the Michaelis–Menten equation to obtain the Michaelis–Menten constants (K_m) and maximal velocities, V_{max} (*see* **Note 25**). The experimental results are shown in Fig. 3.

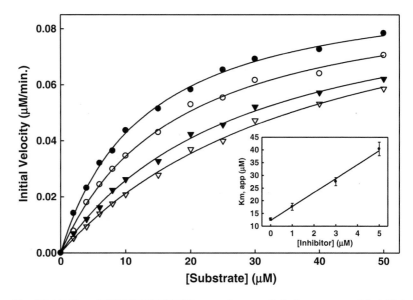

Fig. 3 Inhibition of DENV2 NS2B/NS3pro protease activity by compound **A**. Initial reaction rates of the substrate (Bz-Nle-Lys-Arg-Arg-AMC) cleavage catalyzed by DENV2 NS2B/NS3pro protease (25 nM) in 200 mM Tris–HCl (pH 9.5), 6.0 mM NaCl, 30 % glycerol, and 0.1 % CHAPS at 37 °C are determined by varying the substrate concentrations in the range of 0, 2, 4, 6, 8, 10, 15, 20, 25, 30, 40, and 50 μM at each concentration of inhibitor fixed at 0 (*solid circle*), 1.0 μM (*open circle*), 3.0 μM (*solid triangle*), and 5.0 μM (*open triangle*). The reactions are initiated by the addition of substrate and the fluorescence intensity at 460 nm is monitored with an excitation at 380 nm. Reactions are less than 5 % completion in all cases to maintain valid steady-state measurements. The *solid lines* are fitted lines using the Michaelis–Menten equation. Inset: secondary plot of $K_{m, app}$ against the concentration of selected compound **A** (*see* **Note 26**). Kinetic studies are carried out as described utilizing substrate concentrations of 0–50 μM Bz-Nle-Lys-Arg-Arg-AMC. Each experiment is performed in duplicate and repeated three times. Data are analyzed using SigmaPlot 2001 v7.0 software to determine values for apparent K_m and k_{cat}

Table 3
Kinetic parameters for the _tetra_-peptide substrate and compound A against DENV2 NS2BH-NS3pro at 37 °C

Inhibitor (μM)	K_m (μM)	k_{cat} (s^{-1})	k_{cat}/K_m (M$^{-1} \cdot$s^{-1})
0	12.78 ± 0.49	0.0974 ± 0.0014	$5,082 \pm 271$
1	17.61 ± 1.38	0.0952 ± 0.0032	$3,604 \pm 423$
3	27.54 ± 1.61	0.0972 ± 0.0029	$2,353 \pm 212$
5	40.39 ± 2.66	0.1071 ± 0.0041	$1,767 \pm 189$

3.6 Determination of K_i

K_i values are calculated from the steady-state kinetic data (_see_ **Note 25**) using a secondary plot of $K_{m,\,app}$ against the concentrations of selected compound **A** using SigmaPlot 2001 v7.0 software (_see_ **Note 26**). The results are summarized in Fig. 3 and Table 3.

4 Notes

1. It is important to ensure that the temperature is not warmer than 40 °C so that ampicillin is not inactivated. If the temperature is lower than 37 °C, the agar is likely to solidify before pouring into plates.

2. The substrate is also custom synthesized by companies specializing in peptide synthesis at a reasonable cost.

3. This substrate also works well in the West Nile virus protease assay.

4. Two types of plate readers are commercially available. In a filter-based instrument, high-intensity light is passed through a filter pair selected close to excitation/emission wavelengths of the fluorophore (e.g., AMC). The filter pair chosen may not be optimum for accurate measurements of fluorescence emission. In a monochromator-based instrument, the excitation and emission wavelengths can be precisely selected. Our laboratory uses a monochromator-based instrument.

5. The forward primer contains BamHI restriction site (underline) before NS2BH starts that gives two nonviral amino acids Gly(G) and Ser(S). The total numbers of nonviral sequences before these two are ten amino acids, MRGS<u>HHHHHH</u>, that are already present in the pQE protein expression vector. The six consecutive (6×) histidine residues (underlined) are included as a tag for the protein purification by metal affinity chromatography.

6. The reverse primer contains BssHII site (underlined) that is designed for connecting to NS3 protease fragment (_see_ below). The third nucleotides in the two codons at the carboxyl end of

NS2BH are altered from 5′-CA<u>A</u>CG<u>G</u> to 5′-CA<u>G</u>CG<u>C</u> to create BssHII site without affecting the original amino acid sequence (QR).

7. The linker region between the NS2BH and the NS3pro domain consists of the last two amino acid residues of NS2B, QR, at the NS2B-NS3 junction and is followed by A at the N-terminus of NS3. *Cis* cleavage at the C-terminus of QR by the viral protease results in a non-covalent complex of NS2BH/NS3pro, the active protease used in this assay.

8. The underlined sequences are BssHII and HindIII restriction sites at the N- and the C-terminal domains of NS3 protease, respectively.

9. Ethidium bromide is a carcinogen. Care should be exercised in handling the ethidium bromide solution. Wear new gloves before handling the solution.

10. For analytical gel electrophoresis, ethidium bromide is added during the preparation of agarose gel. For separation of DNA fragments and recovery from gels, it is preferable to add ethidium bromide after electrophoresis is completed so that incubation with ethidium bromide in dilute buffer or water could be for a short period of time without exposure to light to avoid damage to DNA bound to ethidium bromide.

11. The double digestion is not recommended in general. The digestion efficiency is variable in some cases. If the double digestion is not successful, a sequential digestion is required under optimum pH and ionic strength.

12. Three different amounts of transformed *E. coli* may be plated so that the plate containing well-isolated single colonies could be chosen for preparation of plasmid DNA.

13. The sample will turn yellow when the neutralization is complete and a yellowish precipitate will form. Invert the sample additional two to three times to ensure complete neutralization.

14. After 3 h induction by IPTG, the optimal level of protein expression usually is reached. The longer incubation time could result in decreased yield due to degradation.

15. The molarity calculator is available in GraphPad software (http://www.graphpad.com/quickcalcs/Molarityform.cfm).

16. Assays using two to three different concentrations of an inhibitor can give approximate IC_{50} values which could then be confirmed in subsequent experiments with a wide range of inhibitor concentrations to determine more accurate IC_{50} values for selected compounds.

17. The buffer contains 200 mM Tris–HCl, 6.0 mM NaCl, 30 % glycerol, and 0.1 % CHAPS (3-[(3-<u>c</u>holamidopropyl) dimethyl<u>a</u>mmonio]-1-<u>p</u>ropane <u>s</u>ulfonate), pH 9.5. Make at

least 1 L buffer for multiple experiments. CHAPS is a zwitterionic detergent used in the laboratory to solubilize biological macromolecules. CHAPS is also used to prevent any nonspecific, aggregation-based inhibition of the protease, especially at higher concentrations of compounds. Long term storage of the protease in CHAPS containing buffer is not recommended.

18. Preincubate the inhibitors with the protease for 15 min to ensure binding. The final concentrations of inhibitors are 25 μM and 10 μM when stock II and stock III are used, respectively.

19. In the mixture, the final concentrations are 25 nM and 5 μM for the protease and the *tetra*-peptide substrate, respectively.

20. Run the experiments in triplicate and get the average percentage inhibition values. Repeat at least once more to verify the results.

21. Inhibition (%) $\dfrac{\text{RFU}\,(-\,\text{inhibitor})-\,\text{RFU}\,(+\,\text{inhibitor})}{\text{RFU}\,(-\,\text{inhibitor})}\times 100$

(RFU: relative fluorescence units)

Fluorescence values obtained with the no-inhibitor control are taken as 100 %, and those in the presence of inhibitors are calculated as the percentage of inhibition of the control using Microsoft Excel and plotted using SigmaPlot 2001 v7.0 software. The background of AMC in the absence of protease is subtracted before the data analysis.

22. IC_{50} value represents the concentrations of inhibitor that is required for 50 % inhibition in vitro. The lower the number, the greater the potency. IC_{50} values are calculated using the SigmaPlot 2001 v7.0 software. Compounds with more than 50 % percent inhibition at 25 μM are chosen to determine the IC_{50} values.

23. Three different concentrations are required in order to get the straight line for a competitive inhibition mechanism. We choose four different concentrations (0, 1.0, 3.0, and 5.0 μM) for clarity of the straight line.

24. We need to know the initial rate of the substrate cleavage in the presence of protease and inhibitors. If only the amount of substrate cleavage versus the time keeps a linear relationship during a specific time, we can calculate the initial rate. Thirty minutes is usually enough in this protease assay. Under certain circumstances, especially when the substrate concentration is very low, running a longer time may be necessary in order to obtain the initial rate.

25. All reactions are carried out to less than 5 % substrate completion to make sure that the initial velocity is still in a linear range of protease activity.

26. Calculation of K_i

K_i values of inhibitors are calculated assuming Michaelis–Menten kinetics Eq. 1:

$$v = \frac{V_{max}[S]}{K_m + [S]} \tag{1}$$

For a competitive inhibition, the equation is written as Eq. 2:

$$v = \frac{V_{max}[S]}{K_m\left(1 + \frac{[I]}{K_i}\right) + [S]} \tag{2}$$

Equation 2 can be rewritten as Eq. 3 with a standard Michaelis–Menten equation except the expression of K_m:

$$v = \frac{V_{max}[S]}{K_m\left(1 + \frac{[I]}{K_i}\right) + [S]} = \frac{V_{max}[S]}{K_m' + [S]} \tag{3}$$

Then, K_m' can be obtained in Eq. 4 based on Eq. 3 using a secondary plot:

$$K_m' = K_m\left(1 + \frac{[I]}{K_i}\right) = \frac{K_m[I]}{K_i} + K_m \tag{4}$$

K_m' against [I], a linear relationship with the slope and intercept of the linear equation, can be obtained directly from the Eq. 4:

$$Slope = \frac{K_m}{K_i}, Intercept = K_m, and, K_i = \frac{K_m}{Slope} = \frac{Intercept}{Slope}$$

Acknowledgments

The research was supported by an NIH grant, AI082068 to R.P.

References

1. Lindenbach D, Thiel HJ, Rice C (2007) Flaviviridae: the viruses and their replication. In: Knipe DM, Howley PM (eds) Field's Virology, vol 1, 5th edn. Lippincott-Raven Publishers, Philadelphia, pp 1101–1152

2. Lambrechts L, Scott TW, Gubler DJ (2010) Consequences of the expanding global distribution of aedes albopictus for dengue virus transmission. PLoS Negl Trop Dis 4(5):e646. doi:10.1371/journal.pntd.0000646

3. Mitka M (2013) Dengue more prevalent than previously thought. JAMA 309(18):1882. doi:10.1001/jama.2013.4903

4. Guzman MG, Halstead SB, Artsob H, Buchy P, Farrar J, Gubler DJ, Hunsperger E, Kroeger A, Margolis HS, Martinez E, Nathan MB, Pelegrino JL, Simmons C, Yoksan S, Peeling RW (2010) Dengue: a continuing global threat. Nat Rev Microbiol 8(12 Suppl):S7–S16. doi:nrmicro2460, [pii] 10.1038/nrmicro2460

5. Sampath A, Padmanabhan R (2009) Molecular targets for flavivirus drug discovery. Antiviral Res 81(1):6–15. doi:S0166-3542(08)00386-0, [pii] 10.1016/j.antiviral.2008.08.004

6. Noble CG, Chen YL, Dong H, Gu F, Lim SP, Schul W, Wang QY, Shi PY (2010) Strategies for development of dengue virus inhibitors. Antiviral Res 85(3):450–462. doi:S0166-3542(10)00004-5, [pii] 10.1016/j.antiviral.2009.12.011

7. Clum S, Ebner KE, Padmanabhan R (1997) Cotranslational membrane insertion of the serine proteinase precursor NS2B-NS3(Pro) of dengue virus type 2 is required for efficient in vitro processing and is mediated through the hydrophobic regions of NS2B. J Biol Chem 272(49):30715–30723

8. Falgout B, Miller RH, Lai C-J (1993) Deletion analysis of dengue virus type 4 nonstructural protein NS2B: identification of a domain required for NS2B-NS3 protease activity. J Virol 67:2034–2042

9. Chambers TJ, Nestorowicz A, Amberg SM, Rice CM (1993) Mutagenesis of the yellow fever virus NS2B protein: effects on proteolytic processing, NS2B-NS3 complex formation, and viral replication. J Virol 67(11):6797–6807

10. Li J, Lim SP, Beer D, Patel V, Wen D, Tumanut C, Tully DC, Williams JA, Jiricek J, Priestle JP, Harris JL, Vasudevan SG (2005) Functional profiling of recombinant NS3 proteases from all four serotypes of dengue virus using tetrapeptide and octapeptide substrate libraries. J Biol Chem 280(31):28766–28774

11. Feng BY, Shoichet BK (2006) A detergent-based assay for the detection of promiscuous inhibitors. Nat Protoc 1(2):550–553. doi:10.1038/nprot.2006.77

12. Ezgimen MD, Mueller NH, Teramoto T, Padmanabhan R (2009) Effects of detergents on the west nile virus protease activity. Bioorg Med Chem 17(9):3278–3282. doi:S0968-0896(09)00310-1, [pii] 10.1016/j.bmc.2009.03.050

Chapter 22

Construction of Plasmid, Bacterial Expression, Purification, and Assay of Dengue Virus Type 2 NS5 Methyltransferase

Siwaporn Boonyasuppayakorn and Radhakrishnan Padmanabhan

Abstract

Dengue virus (DENV), a member of mosquito-borne flavivirus, causes self-limiting dengue fever as well as life-threatening dengue hemorrhagic fever and dengue shock syndrome. Its positive sense RNA genome has a cap at the 5′-end and no poly(A) tail at the 3′-end. The viral RNA encodes a single polyprotein, C-prM-E-NS1-NS2A-NS2B-NS3-NS4A-NS4B-NS5. The polyprotein is processed into 3 structural proteins (C, prM, and E) and 7 nonstructural (NS) proteins (NS1, NS2A, NS2B, NS3, NS4A, NS4B, NS5). NS3 and NS5 are multifunctional enzymes performing various tasks in viral life cycle. The N-terminal domain of NS5 has distinct GTP and S-adenosylmethionine (SAM) binding sites. The role of GTP binding site is implicated in guanylyltransferase (GTase) activity of NS5. The SAM binding site is involved in both N-7 and 2′-O-methyltransferase (MTase) activities involved in formation of type I cap. The C-terminal domain of NS5 catalyzes RNA-dependent RNA polymerase (RdRp) activity involved in RNA synthesis. We describe the construction of the MTase domain of NS5 in an *E. coli* expression vector, purification of the enzyme, and conditions for enzymatic assays of N7- and 2′O-methyltransferase activities that yield the final type I 5′-capped RNA (^{7Mc}GpppA$_{2'OMe}$-RNA).

Key words Methyltransferase assay, DENV2 NS5, Thin layer chromatography

1 Introduction

The mosquito-borne flaviviruses are classified as positive-strand RNA viruses as their genomes serve as mRNAs in the infected host and are translated by the host translational machinery. The 5′-end of flavivirus RNA has a conserved dinucleotide AG and a type I cap structure (^{7Mc}GpppA$_{2'OMe}$) in which the guanine base at 7-position and the 2′-OH of the ribose moiety of 5′-A are methylated [1] like most eukaryotic mRNAs (reviewed in [2]). Mimicking the host mRNA, the viral capped RNA can utilize the host's cap-dependent translational machinery. The 5′-cap also protects the genome from 5′-exonuclease-mediated degradation and evades innate immunity (RIG-I/MDA5 pathway) that could target the uncapped ssRNA [3]. The 5′-capping of flavivirus RNA consists of four

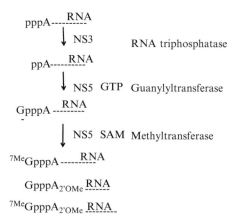

Fig. 1 Schematics of steps in 5′-capping

sequential steps: removal of γ-phosphate of triphosphorylated RNA to diphosphorylated RNA (pp-A-RNA), addition of guanine monophosphate from GTP substrate to the diphosphorylated RNA to form GpppA-RNA, and the two methylations catalyzed by NS5 MTase domain (N7 MTase and 2′-O-MTase) in the presence of SAM to form ^{7Me}GpppA$_{2'OMe}$-RNA, ^{7Me}GpppA-RNA, and GpppA$_{2'OMe}$-RNA [4–10] (Fig. 1).

The N-terminal domain (265 amino acids, Fig. 2a) of the nonstructural protein 5 (NS5) is sufficient for both N7 and 2′-O-methylations (reviewed in [11]). Methylations occurred at N7 followed by 2′OH in both WNV [8] and DENV2 MTases [12]. RNA substrate of at least 74 nucleotides in length is required for N7-MTase activity [7], whereas short RNA oligonucleotides (^{7Me}GpppAC$_{(5)}$ and GpppAC$_{(5)}$) are sufficient to serve as substrates for the 2′-O-MTase activity [13]. Mutational analysis [5,8] revealed distinct mechanisms involved in the two catalytic steps; D146 is critical for N7-methylation, whereas the KDKE motif (K61-D146-K181-E217 in DENV2 and K61-D146-K182-E218 in WNV) is important for 2′-O-methylation. Substrate reposition model has been proposed [14] for sequential N7- and 2′-O-methylation reactions.

In previous studies, the two enzymatic assays use buffers at different pH and ionic strengths, pH 7 for the N7-MTase and pH 10 for the 2′-O-MTase. However, more recently, a pH 7.5 buffer with slight differences in ionic strengths is described for both methylations [12]. In this study, we use the buffer (50 mM Tris–HCl, pH 7.5, 10 mM KCl, 2 mM MgCl$_2$, and 2 mM DTT) for both 2′O- and N7-methylations (replacing 20 mM NaCl with 10 mM KCl, 2 mM MgCl$_2$). To detect the methylated species, [α-^{32}P]-labeled GTP is used as the substrate for the vaccinia virus-encoded GTase-catalyzed addition of guanine cap to the diphosphorylated

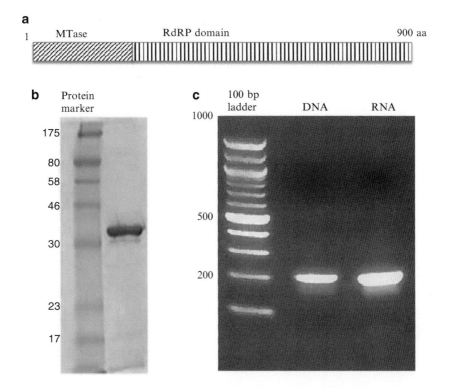

Fig. 2 Purified DENV2 NS5$_{1-265}$ protein. (**a**) The purity of *E. coli*-expressed DENV2 NS5$_{1-265}$ is verified by running a 12 % SDS-PAGE and Coomassie Blue staining. (**b**) Purity of PCR product and RNA template. The PCR product encoding the 5′-end DENV2 RNA sequence (200 nucleotides) is amplified as described in the text (**c**). It is used as a template for in vitro transcription and the integrity and purity are verified by 2 % native agarose gel electrophoresis and stained by ethidium bromide

RNA prior to the two methylation steps. Gp*ppA-RNA (* represents [α-^{32}P]-label) that is formed prior to the two methylations has an advantage of visualizing directly the unmethylated, monomethylated, and dimethylated species (Gp*ppA, Gp*ppA$_{2'OMe}$, ^{7Me}Gp*ppA, and ^{7Me}Gp*ppA$_{2'OMe}$) by autoradiography using PhosphorImager. Thin layer chromatography (TLC) is used to separate the unmethylated, monomethylated, and dimethylated species [15] using solvents such as 1.2 M LiCl or 0.4 M (NH$_4$)$_2$SO$_4$ [15]. In this study, we use a solvent of 0.45 M (NH$_4$)$_2$SO$_4$ for separation of GpppA, GpppA$_{2'OMe}$, ^{7Me}GpppA, and ^{7Me}GpppA$_{2'OMe}$ by TLC using PEI cellulose-coated plates. The cap species formed using the vaccinia virus N7- and/or 2′-O-methyltransferases are used as mobility markers for detection of corresponding species formed in DENV2 NS5 MTase reaction. We describe the construction of the expression plasmid encoding the DENV2 MTase domain (NS5$_{1-265}$). The MTase domain is expressed in *E. coli* and the protein is purified (Fig. 2b). The MTase activities of the purified proteins use both unmethylated and N7-methylated capped RNA (Fig. 3).

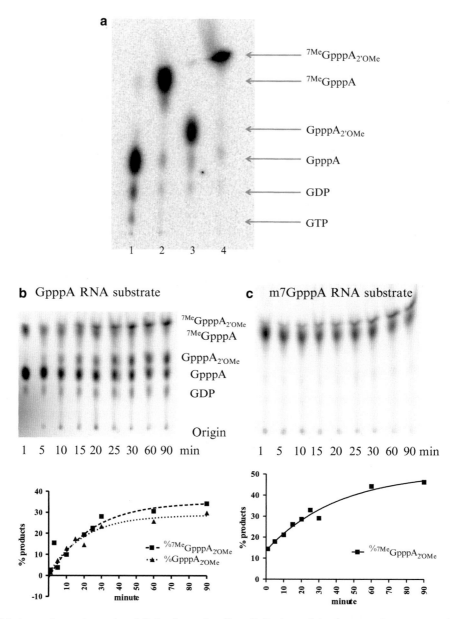

Fig. 3 Thin layer chromatography of Gp*ppA species. Two distinct vaccinia virus capping enzymes have been characterized: one catalyzes N7-methylation and the other 2′-O-methylation. These are used to produce markers for characterizing the unmethylated, monomethylated, and dimethylated species in the DENV2 NS5$_{1-265}$ enzyme-catalyzed reactions. (**a**) Markers produced from (1) no MTase enzyme; (2) vaccinia N7-MTase; (3) vaccinia 2′-O-MTase; (4) vaccinia N7-MTase and 2′-O-MTase. * represents 5′-[α-^{32}P]. (**b**) Time course of the formation of mono- and dimethylated Gp*ppA species produced from unmethylated Gp*ppA-RNA. The details of the experiments are described in the text. * represents 5′-[α-^{32}P]. (**c**) Time course of the formation of fully methylated Gp*ppA species produced from N7-methylated RNA substrate. The details of experiments are described in the text. * represents 5′-[α-^{32}P]

2 Materials

Prepare stock solutions of buffers and reagents using molecular biology grade chemicals and Milli-Q H_2O. Use RNase-free water in all assays involving RNA substrate. All reagents are stored at room temperature unless indicated otherwise. All recombinant DNA work is done following established protocols and regulations by the Institutional Biosafety Committee for handling biohazard materials such as recombinant DNA and plasmid-transformed *E. coli* waste disposal.

1. PCR kits and reagents: Phusion (High-Fidelity) DNA Polymerase (*see* **Note 1**), oligodeoxynucleotides primers (Table 1), deoxyribonucleoside triphosphates mixture (10 mM each).

2. Cloning: restriction enzymes (NdeI and HindIII-HF), 100 bp and 1 kb ladders as size markers, DH5α and BL21 (DE3) competent cells, T4 DNA ligase, kits to purify DNA and RNA, and Luria-Bertani (LB)-kanamycin agar plates (LB-K). Mix 6.25 g LB powder, 3.75 g agar, and 250 mL deionized H_2O. Autoclave, and after the LB-agar medium cools down to ~40 °C, add 250 μL of 25 mM filter-sterilized kanamycin to the medium. Pour into Petri dishes and let the plates sit at room temperature for 3 h. Store the plates at 4 °C until use (*see* **Note 2**).

3. Protein expression reagents: LB Broth-Miller formula (LB-Miller) is made by mixing 25 g of LB powder in 1 L of water and sterilized by autoclave. Although LB powder contains premixed components and sold as such, the components are 10 g of enzymatic digest of casein, 5 g yeast extract, and 10 g NaCl per L of H_2O. 50 % w/v glucose (100×) is also sterilized by autoclave for 15 min or by filtration through 0.45 μm filter fitted to a syringe if small volumes are required; kanamycin (25 mM; 1,000×) and IPTG (1 M; 1,000×) are sterilized by filtration as mentioned above (*see* **Note 3**).

Table 1
PCR primer sequences

Primer name	Primer sequence
Primer A	5′-TGGGAACTGGCAACATAGGAGAGACGC-3′
Primer B	5′-TAATAAAAGCTTGATGTTACGGGGTTCCGCTTCCG-3′
Primer C	5′-CAGTAATACGACTCACTATTAGTTGTTAGTCTACGTG-3′
Primer D	5′-AGTGAGAATCTCTTTGTCAGCTG-3′

4. Protein purification reagents: buffer A (50 mM Tris–HCl, pH 7, 300 mM NaCl, 10 % glycerol) is prepared by mixing autoclaved 2× buffered salt solution (100 mM Tris–HCl, pH 7, 600 mM NaCl) with 50 % glycerol (5×) and deionized H$_2$O. Lysis buffer (50 mL per 1 L bacterial culture) is freshly prepared from buffer A (50 mL) and 500 μL NP-40 (Nonidet P-40; octylphenoxypolyethoxyethanol) to a final concentration of 1 %, a tablet or 500 μl of protease inhibitor cocktails, and 2–3 mg lysozyme. Imidazole (1 M in buffer A) is freshly prepared just before use. To make 500 mM imidazole elution buffer, the stock is diluted in buffer A (1:1). Wash buffers (buffer A with 5, 10, 15, 20, 25 mM imidazole) are prepared by serial dilutions. Protease inhibitor cocktails (without EDTA), Talon resin, and a column are also needed.

5. Dialysis buffer: 50 mM Tris–HCl, pH 7.5, 50 mM NaCl, 2 mM MgCl$_2$, and 40 % glycerol is prepared by mixing autoclaved buffer, salt, and glycerol. DTT (to 1 mM final concentration) is added to the buffer just before the dialysis step. The dialysis membrane (10,000 Da cutoff) is prepared according to the manufacturer's protocol.

6. Quantification of RNA, DNA, and protein: spectrophotometry for DNA and RNA, Pierce BCA (bicinchoninic acid) protein assay kit, 0.7–2 % native agarose electrophoresis, 7 M urea-8 % TBE (89 mM Tris–HCl, pH 8.3, 89 mM Boric acid, 2 mM EDTA)-polyacrylamide gel electrophoresis, and 12 % SDS-PAGE confirming the band size, purity, and quantity of DNA, RNA, and protein, respectively.

7. RNA substrate preparation: MEGAshortscript from Ambion (Life Technologies, Grand Island, NY) or a suitable alternative source is used for T7 RNA polymerase-catalyzed in vitro transcription. Micro Bio-Spin 30 Columns from Bio-Rad (Hercules, CA) or an alternative gel filtration column is used to remove unincorporated nucleotides. Acid-Phenol/Chloroform (RNase-free) is purchased from commercial sources.

8. MTase assay: 10× MTase buffer (500 mM Tris–HCl, pH 7.5, 100 mM KCl, 20 mM MgCl$_2$, and 20 mM DTT) is prepared with RNase-free water and stored in aliquots (10 μL) at –20 °C. SAM is purchased and stored as aliquots (10 μL) at –20 °C. ScriptGuard RNase inhibitor (40 U/μL) is purchased from CELLSCRIPT (Madison, WI). Nuclease P1 and PEI cellulose plates are needed. Nuclease P1 is diluted with RNase-free water to give aliquots of 4 U/μL; store aliquots at –20 °C.

9. Vaccinia virus capping enzymes: ScriptCap m7G capping system and ScriptCap 2′-O-methyltransferase kit are purchased from CELLSCRIPT (Madison, WI).

10. PhosphorImager screen and the Storm 840 scanner.

3   Methods

3.1   Construction of pET28b DENV2 NS5₁₋₂₆₅ Expression Plasmid

3.1.1   Preparation of the Insert

1. Follow standard PCR protocols. For initial PCR, the reaction mixture (50 µL) consists of the following components in 0.2 mL microfuge tube: H_2O (33.5 µL), 5× Phusion PCR buffer (10 µL), 10 mM each of dNTP mixture (1 µL; the final concentration of each dNTP is 200 µM), primers A and B (Table 1) (100 µM; 1 µL each), 20 µg/µL of pQE32 DENV2 NS5 [16] as DNA template (1 µL), 3 % DMSO (1.5 µL), 0.04 U/µL Phusion DNA Polymerase (1 µL). Vortex well to mix, and then spin briefly in a microfuge. Keep the DMSO concentration below 1 %.

2. Perform PCR using the following conditions: initial denaturation at 98 °C for 30 s, followed by 35 cycles of denaturation at 98 °C for 10 s, annealing at 55 °C for 20 s, and extension at 72 °C for 30 s and final extension incubated at 72 °C for 10 min.

3. The PCR product is purified using standard 2 % agarose gel electrophoresis and recovered by gel extraction. Elute the PCR product from the matrix in minimum volume (25–30 µL).

4. Measure the concentration of DNA using a spectrophotometer. Store at −20 °C until use.

3.1.2   Cloning of the PCR Fragment into pET28b Vector

1. Clone the PCR fragment from **step 4**, Subheading 3.1.1, into pET28b vector between NdeI and HindIII sites using standard recombinant DNA techniques (*see* **Notes 4** and **5**). The His-tag is at the C-terminus of the DENV2 NS5₁₋₂₆₅. The vector has kanamycin resistance gene and selection of transformants of competent *E. coli* DH5α cells is done on LB-K plates.

2. Isolate and purify the plasmid DNA encoding DENV2 NS5₁₋₂₆₅ on a mid-scale.

3. Verify the coding sequence of DENV2 NS5₁₋₂₆₅ by DNA sequencing.

4. Transform BL21(-DE3) competent cells (25 µL) with the expression plasmid following the manufacturer's protocol.

5. Store the plasmid at −20 °C. Prepare a glycerol stock of the transformed *E. coli* cells in LB broth containing 15 % glycerol at −70 to −80 °C.

3.2   Expression and Purification of DENV2 NS5₁₋₂₆₅

3.2.1   Expression of DENV2 NS5₁₋₂₆₅

Inoculate the seed culture in 100 mL LB containing 0.5 % glucose and 25 µM kanamycin with the *E. coli* transformed overnight in a 37 °C shaker set at 200 rpm. Use 20–50 mL of the overnight culture to inoculate a 2 L LB medium containing 0.5 % glucose and 25 µM kanamycin. Incubate at 37 °C in a shaker set at 200 rpm until the OD_{600} reaches 0.5–0.6.

1. Centrifuge the *E. coli* cells at $5{,}000 \times g$ and 4 °C for 20 min. Discard the spent LB medium with 2 L fresh LB kept at 4 °C containing 1 mM IPTG and 25 μM kanamycin; incubate the culture in a shaker at 16 °C and 200 rpm speed for 20–24 h.

2. Centrifuge at $5{,}000 \times g$ and 4 °C for 20 min; wash once with 50 mM Tris–HCl, pH 7.5, 200 mM NaCl (*see* **Note 6**). The pellet is incubated for 30 min on ice with 30 mL of cold lysis buffer (50 mM Tris–HCl, pH 7, 300 mM NaCl, 10 % glycerol, 1 % NP-40, protease inhibitor cocktails, and lysozyme) (*see* Subheading 2, **item 4**). Sonicate for 15 min with 15 s ON, 45 s OFF per cycle; centrifuge at $18{,}000 \times g$ and 4 °C for 1 h.

3.2.2 Purification Using Metal Affinity Chromatography

1. Wash 1 mL of Talon resin 5×10 mL each with H_2O; wash 2×10 mL with a buffer containing 50 mM Tris–HCl, pH 7, 300 mM NaCl, 10 % glycerol, 1 % NP-40, and protease inhibitor cocktails (*see* Subheading 2, **item 4**). Centrifuge at 500–$700 \times g$ for 5 min each and discard the wash solutions.

2. Incubate the supernatant with Talon resin (or equivalent) at 4 °C overnight on a rocking machine. Transfer the protein-bound resin to a column kept at 4 °C; wash the column with 20 mL buffer A (*see* Subheading 2, **item 4**) and then with 10 mL each of wash buffers (buffer A containing 5, 10, 15 20, 25 mM imidazole). Sample the flowthrough from each wash to check for successful removal of nonspecifically bound proteins and subsequent analysis by SDS-PAGE (12 %) and Coomassie Blue staining (Fig. 2b).

3. Elute the protein from the column with 500 mM imidazole in buffer A. Dialyze the eluate against 500 mL of a buffer containing 50 mM Tris–HCl, pH 7.5, 50 mM NaCl, 1 mM $MgCl_2$, 40 % glycerol, and 1 mM DTT (*see* Subheading 2, **item 5**) at 4 °C for 3 h and 1.5 L of the same buffer overnight.

4. Quantify the protein (*see* Subheading 2, **item 6**) and store in aliquots (10 μL) at −70 °C.

3.3 Preparation of RNA Substrate

1. Amplify by PCR the DENV2 mini genome template from the plasmid (pSY2) [17]. The reaction mixture (50 μL) consists of the following components in 0.2 mL microfuge tube: RNase-free H_2O (36 μL), 10× AmpliTaq buffer (5 μL), 25 mM $MgCl_2$ (4 μL), 10 mM each of dNTP mixture (1 μL; the final concentration of each dNTP is 200 μM) (*see* **Note 6**), 1 μL each primers C and D (100 μM stock) (*see* Table 1 for sequences), pSY2 plasmid (10 μg/μL; 1 μL), AmpliTaq DNA polymerase (0.02 U/μL; 1 μL). Vortex well to mix, and then spin down briefly.

2. The conditions for PCR are initial denaturation at 95 °C for 5 min, followed by 35 cycles of denaturation at 95 °C for 30 s,

annealing at 57 °C for 30 s, and extension at 72 °C for 30 s and final extension incubated at 72 °C for 10 min to produce the insert.

3. The PCR product is purified using standard 2 % agarose gel electrophoresis (Fig. 2c) and recovered by gel extraction. Use a minimum volume to elute (25–30 μL) the PCR product from the matrix.

4. Quantify using a spectrophotometer. Store at –20 °C until use.

5. In vitro transcription of 200 nucleotides PCR-amplified DNA fragment. We use MEGAshortscript-T7 RNA polymerase kit. The reaction mixture (20 μL) contains RNase-free H_2O (2 μL), 10× T7 RNA polymerase buffer provided in the kit (2 μL), 75 mM each of NTP mixture (8 μL; the final concentration of each NTP is 7.5 mM) (*see* **Note** 7), 375 ng DNA template (5 μL), T7 RNA polymerase enzyme mix (2 μL), and 40 U/μL RNase Inhibitor (0.5 μL). Incubate at 37 °C for 6 h; digest DNA template with 2 U/μL DNase (1 μL) at 37 °C for 15 min.

6. The unincorporated nucleotides are removed by gel filtration (e.g., P-30 column from Bio-Rad); RNA is recovered using the RNA clean and concentrator kit. The 200-nt RNA sequence is from the 5′-end of the DENV2 genome.

7. Verify the integrity of the RNA by 2 % agarose electrophoresis (Fig. 2c). Quantify using spectrophotometry. Store the RNA substrate in aliquots (1 μg/tube) at –80 °C until use.

3.4 Methyltransferase Assay

Both N7- and 2′-O-MTase reactions are carried out using the buffer condition (50 mM Tris–HCl, pH 7.5, 10 mM KCl, 2 mM $MgCl_2$, and 2 mM DTT) reported in a previous study [12]

1. Preparation of 5′-Cap RNA substrate. The reaction (20 μL) is started with 1 μg RNA (4 μL), 10× ScriptCap m7G reaction buffer (500 mM Tris–HCl, pH 8.0, 60 mM KCl, and 12.5 mM $MgCl_2$) (2 μL), 10 mM GTP (2 μL), 10 μCi/μL [α-^{32}P]GTP (2 μL) (adjusted to yield 1 μCi per reaction calculated using the half life of [α-^{32}P] as 14 days), 40 U/μL RNase Inhibitor (0.5 μL), 10 U/μL ScriptCap m7G enzyme (1 μL). Incubate 1 h at 37 °C.

2. The unincorporated nucleotides are removed as described above (*see* **step 6**, Subheading 3.3). Elute RNA with 25–30 μL with RNase-free water.

3. Mix the components of the methyltransferase assay in 10 μL volume: RNase-free water (1.5 μL), 10× MTase buffer (500 mM Tris–HCl, pH 7.5, 100 mM KCl, 20 mM $MgCl_2$, and 20 mM DTT) (1 μL), 8 mM SAM (0.5 μL), 40 U/μL RNase Inhibitor (0.5 μL), 5′-cap labeled RNA (5 μL), and 5 μM DENV2 NS5$_{1-265}$ (1 μL). Incubate for 1 h at 37 °C, or as indicated.

4. Pass the samples through gel filtration columns (e.g., P-30 column from Bio-Rad) and recover the RNA using the kit (**step 6**, Subheading 3.3). Elute 6 µL from the column and add 1 U nuclease P1 in 0.75 M sodium acetate solution (1 µL) (*see* **Note 8**). Incubate for 1 h at 37 °C for a complete digestion.

5. Separation of unmethylated, monomethylated, and dimethylated species by TLC: heat inactivates the nuclease P1 reaction mixture (95 °C for 5 min) to prevent trapping of the MTase products at the origin of TLC plate. Draw a line using a pencil gently about an inch from the edge of a PEI cellulose plate without scratching the surface of the matrix. Apply the contents of each of the MTase assay mixtures including the controls slowly, 1–2 µL at a time, using an Eppendorf pipet fitted with a fine tip; use air dryer to fasten the drying time until the entire contents of each assay are applied. Develop the PEI plate by placing it in a tank containing H_2O until waterfront raises above the line on which the samples are applied; the plate is then placed in a second TLC chamber or the same chamber after replacing the water with 0.45 M ammonium sulfate until the solvent front reaches the top of the TLC plate (*see* **Note 9**).

6. Remove the plate from the chamber and dry using a hair dryer; [α-^{32}P]-labeled species from the MTase reaction are visualized and quantified using PhosphorImager (Storm 840 scanner or equivalent; *see* Subheading 2, **item 10**) overnight. Spot intensities are quantified using ImageJ (http://rsbweb.nih.gov/ij/) and GraphPad Prism 5.

7. In order to identify the unmethylated, monomethylated, and dimethylated Gp*ppA species, well-characterized vaccinia virus methyltransferase 5′-capping system is used to produce markers. Incubate the 5′-[α-^{32}P] GMP capped RNA with (1) no-enzyme control, (2) 2′-O ScriptCap enzyme, (3) m7 ScriptCap enzyme, and (4) both m7G and 2′ O ScriptCap enzymes (Fig. 3). These four reactions would yield the cap species: GpppA (lane 1), GpppA$_{2'OMe}$ (lane 2), ^{7Me}GpppA (lane 3), and ^{7Me}GpppA$_{2'OMe}$ (lane 4) (Fig. 3a). Commercially available guanosine analogs, GpppA and ^{7Me}GpppA, are also applied onto the same TLC plate and visualized under UV$_{265nm}$ light.

8. Time course of MTase activity using unmethylated Gp*ppA-RNA substrate. Prepare radiolabeled substrate RNA as described in **steps 1** and **2**, Subheading 3.4. Assemble the components of the methyltransferase assay in 10 µL volume: RNase-free water (1.5 µL), 10× MTase buffer (500 mM Tris–HCl, pH 7.5, 100 mM KCl, 20 mM $MgCl_2$, and 20 mM DTT) (1 µL), 40 U/µL RNase Inhibitor (0.5 µL), 5′-[α-^{32}P] GMP labeled RNA (5 µL), and 5 µM DENV2 NS5$_{1-265}$ (1 µL). Add 8 mM SAM (0.5 µL) to initiate the reaction, incubate at

37 °C, and stop at designated time points by heating to 95 °C for 5 min. After purification of RNA, add nuclease P1 and separate the capped species by TLC as described (**steps 4–6**, Subheading 3.4). Spots are quantified by ImageJ, analyzed, and plotted using GraphPad Prism 5 (Fig. 3b). Rates of generation of both $^{7Me}Gp^*ppA_{2'OMe}$ and $Gp^*ppA_{2'OMe}$ are equal over time.

9. Time course of MTase using $^{7Me}Gp^*ppA$-RNA substrate

   (a) Prepare radiolabeled $^{7Me}Gp^*ppA$-RNA substrate using the vaccinia virus m7G ScriptCap enzyme under conditions described in **steps 1** and **2**, Subheading 3.4, with addition of 8 mM SAM (0.5 μL) to the reaction mixture.

   (b) Assemble the components of the methyltransferase assay in 10 μL volume: RNase-free water (1.5 μL), 10× MTase buffer (500 mM Tris–HCl, pH 7.5, 100 mM KCl, 20 mM $MgCl_2$, and 20 mM DTT) (1 μL), 40 U/μL RNase Inhibitor (0.5 μL), 5′-N7-methylated cap labeled RNA (5 μL), and 5 μM DENV2 NS5$_{1–265}$ (1 μL).

   (c) Add 8 mM SAM (0.5 μL) to initiate the reaction.

   (d) Incubate at 37 °C and stop the reaction at designated time points by heating to 95 °C for 5 min. After all samples are taken, follow the **steps 4–6**, Subheading 3.4, for purification of RNA, nuclease P1, and analysis of the capped species by TLC. Spots are quantified by ImageJ, analyzed, and plotted by GraphPad Prism 5 (Fig. 3c).

4 Notes

1. Phusion HF is used in the construction of the NS5$_{1–265}$ expression plasmid because it is a high-fidelity polymerase that is required for amplification of the protein-coding region free of mutations.

2. The LB-kanamycin plates can be kept in 4 °C for 6 months.

3. Kanamycin, IPTG, and dithiothreitol (DTT) stock solutions are made in high concentrations (1,000×), stored in 1 mL aliquots at –20 °C. The buffers containing ATP and/or dithiothreitol (DTT) are stored in 10–20 μL aliquots at –20 °C. LB broth and glucose stocks are kept at room temperature and have to be carefully checked for microbial growth. Buffers containing glycerol (buffer A, dialysis buffer) are prepared from autoclaved 2× stock solutions.

4. Overnight incubation with NdeI is required to ensure completion of digestion.

5. The vector is blunt ended after NdeI digestion to generate CATA sequence at the termini. When this is ligated to the

insert ending with TG sequence, NdeI site is restored providing the translation initiator codon.

6. The washing step with 50 mM Tris-HCl, pH7.5 and 200 mM NaCl before bacterial cell lysis can prevent contaminants from the LB medium.

7. dNTP and NTP mixtures are stored in aliquots at –20 °C for short-term storage (<6 months). Store NTP stocks at –80 °C.

8. Nuclease P1 activity is enhanced under acidic pH. Before use, the nuclease P1 stock (4 U/μL) in H_2O is diluted to 1 U/μL in 0.75 M sodium acetate (pH 5.5). Use 1 μL to the RNA sample (6 μL).

9. The higher concentration of ammonium sulfate that we use for TLC seems to yield better separation of the capped species.

Acknowledgments

The research was supported by NIH grants, AI-087856 and AI082068.

References

1. Wengler G, Wengler G, Gross HJ (1978) Studies on virus-specific nucleic acids synthesized in vertebrate and mosquito cells infected with flaviviruses. Virology 89(2):423–437

2. Decroly E, Ferron F, Lescar J, Canard B (2012) Conventional and unconventional mechanisms for capping viral mRNA. Nat Rev Microbiol 10(1):51–65. doi:10.1038/nrmicro2675

3. Xagorari A, Chlichlia K (2008) Toll-like receptors and viruses: induction of innate antiviral immune responses. Open Microbiol J 2:49–59. doi:10.2174/1874285800802010049

4. Egloff MP, Benarroch D, Selisko B, Romette JL, Canard B (2002) An RNA cap (nucleoside-2′-O-)-methyltransferase in the flavivirus RNA polymerase NS5: crystal structure and functional characterization. EMBO J 21(11):2757–2768

5. Ray D, Shah A, Tilgner M, Guo Y, Zhao Y, Dong H, Deas TS, Zhou Y, Li H, Shi PY (2006) West Nile virus 5′-cap structure is formed by sequential guanine N-7 and ribose 2′-O methylations by nonstructural protein 5. J Virol 80(17):8362–8370

6. Egloff MP, Decroly E, Malet H, Selisko B, Benarroch D, Ferron F, Canard B (2007) Structural and functional analysis of methylation and 5′-RNA sequence requirements of short capped RNAs by the methyltransferase domain of dengue virus NS5. J Mol Biol 372(3):723–736

7. Dong H, Ray D, Ren S, Zhang B, Puig-Basagoiti F, Takagi Y, Ho CK, Li H, Shi PY (2007) Distinct RNA elements confer specificity to flavivirus RNA cap methylation events. J Virol 81(9):4412–4421. doi:10.1128/JVI.02455-06

8. Zhou Y, Ray D, Zhao Y, Dong H, Ren S, Li Z, Guo Y, Bernard KA, Shi PY, Li H (2007) Structure and function of flavivirus NS5 methyltransferase. J Virol 81(8):3891–3903

9. Liu L, Dong H, Chen H, Zhang J, Ling H, Li Z, Shi PY, Li H (2010) Flavivirus RNA cap methyltransferase: structure, function, and inhibition. Front Biol 5(4):286–303. doi:10.1007/s11515-010-0660-y

10. Dong H, Chang DC, Hua MH, Lim SP, Chionh YH, Hia F, Lee YH, Kukkaro P, Lok SM, Dedon PC, Shi PY (2012) 2′-O Methylation of internal adenosine by flavivirus NS5 methyltransferase. PLoS Pathog 8(4):e1002642. doi:10.1371/journal.ppat.1002642

11. Bollati M, Alvarez K, Assenberg R, Baronti C, Canard B, Cook S, Coutard B, Decroly E, de Lamballerie X, Gould EA, Grard G, Grimes JM, Hilgenfeld R, Jansson AM, Malet H, Mancini EJ, Mastrangelo E, Mattevi A, Milani M, Moureau G, Neyts J, Owens RJ, Ren J, Selisko B, Speroni S, Steuber H, Stuart DI, Unge T, Bolognesi M (2010) Structure and

functionality in flavivirus NS-proteins: perspectives for drug design. Antiviral Res 87(2):125–148. doi:10.1016/j.antiviral.2009.11.009

12. Chung KY, Dong H, Chao AT, Shi PY, Lescar J, Lim SP (2010) Higher catalytic efficiency of N-7-methylation is responsible for processive N-7 and 2′-O methyltransferase activity in dengue virus. Virology 402(1):52–60. doi:10.1016/j.virol.2010.03.011

13. Selisko B, Peyrane FF, Canard B, Alvarez K, Decroly E (2010) Biochemical characterization of the (nucleoside-2′O)-methyltransferase activity of dengue virus protein NS5 using purified capped RNA oligonucleotides (7Me) GpppAC(n) and GpppAC(n). J Gen Virol 91(Pt 1):112–121. doi:10.1099/vir.0.015511-0

14. Dong H, Ren S, Zhang B, Zhou Y, Puig-Basagoiti F, Li H, Shi PY (2008) West Nile virus methyltransferase catalyzes two methylations of the viral RNA cap through a substrate-repositioning mechanism. J Virol 82(9):4295–4307. doi:10.1128/JVI.02202-07

15. Rahmeh AA, Li J, Kranzusch PJ, Whelan SP (2009) Ribose 2′-O methylation of the vesicular stomatitis virus mRNA cap precedes and facilitates subsequent guanine-N-7 methylation by the large polymerase protein. J Virol 83(21):11043–11050. doi:10.1128/JVI.01426-09

16. Ackermann M, Padmanabhan R (2001) De novo synthesis of RNA by the dengue virus RNA-dependent RNA polymerase exhibits temperature dependence at the initiation but not elongation phase. J Biol Chem 276(43):39926–39937

17. You S, Padmanabhan R (1999) A novel in vitro replication system for Dengue virus. Initiation of RNA synthesis at the 3′-end of exogenous viral RNA templates requires 5′- and 3′-terminal complementary sequence motifs of the viral RNA. J Biol Chem 274(47):33714–33722

Part V

Dengue Animal Model for Vaccines and Antivirals

Chapter 23

Animal Models in Dengue

Emily Plummer and Sujan Shresta

Abstract

Validation of a mouse model of dengue virus (DENV) infection relies on verification of viremia and productive replication in mouse tissues following infection. Here, we describe a quantitative assay for determining viral RNA levels in mouse serum and tissues. For the purpose of confirming DENV replication, we outline a fluorescence immunohistochemistry (FIHC) protocol for staining a nonstructural protein of DENV.

Key words Dengue virus, Virus replication, qRT-PCR, 18S RNA qRT-PCR, Fluorescence immunohistochemistry

1 Introduction

Our limited understanding of dengue virus (DENV) pathogenesis results from a lack of an adequate animal model. Successful control of DENV replication in mice depends on an intact IFN response, so wild-type mice do not allow replication of human clinical DENV strains. We have developed an alternative model in mice deficient in both type I and II interferon (IFN) receptors (AG129 mice) or type I IFN receptor alone (A129 or AB6 mice).

Validation of a model of DENV infection in mice requires confirmation of viremia and successful replication of DENV in various tissues. Confirmation of viremia requires analysis of viral titers in the serum. Real-time RT-PCR assay for the detection of DENV RNA levels is both sensitive and quantitative [1, 2]. Also important is verification of virus replication in the tissues, which requires a control for RNA degradation. 18S rRNA levels remain relatively stable during viral infection and are thus used as a control [3]. Although qRT-PCR can be used to show increasing viral RNA levels in tissues, another method is critical to exclude the possibility that virus is present in the tissue due to means other than replication. One way of confirming replication is to visualize the presence

Radhakrishnan Padmanabhan and Subhash G. Vasudevan (eds.), *Dengue: Methods and Protocols*, Methods in Molecular Biology, vol. 1138, DOI 10.1007/978-1-4939-0348-1_23, © Springer Science+Business Media, LLC 2014

of nonstructural viral proteins, which are only produced when the virus is replicating. We use a previously described fluorescence immunohistochemistry (FIHC) on frozen sections [4–7] to stain the nonstructural protein 3 (NS3 protein) of DENV. Although flow cytometry of NS3 positive cells would be faster and more quantitative, the antibodies that are available have not been successfully validated in that assay.

This study was carried out in strict accordance with the recommendations in the Guide for the Care and Use of Laboratory Animals of the National Institutes of Health, the US Public Health Service Policy on Humane Care and Use of Laboratory Animals, and the Association for Assessment and Accreditation of Laboratory Animal Care International (AAALAC). All experimental procedures were pre-approved and performed according to the guidelines set by the La Jolla Institute for Allergy and Immunology Animal Care and Use Committee.

2 Materials

2.1 RNA Isolation from Serum

1. QIAamp Viral RNA Mini Kit from Qiagen, Cat # 52906.
2. RNase-/DNase-free PBS.
3. Ethanol.
4. 1.1 mL Z-Gel serum collection tube from Sarstedt, Catalog # 41.1500.005.

2.2 RNA Isolation from Tissue

1. RNAlater from Qiagen, Cat # 76106.
2. RLT Buffer from Qiagen RNeasy Mini Kit 250, Cat # 74106.
3. Stainless steel beads 5 mm from Qiagen, Cat # 69989.
4. 2-mercaptoethanol (BMe).
5. Qiagen TissueLyser or other tissue homogenization device.
6. Ethanol.

2.3 qRT-PCR

1. DNase-/RNase-free water.
2. One-Step qRT-PCR Mix, from Quanta or equivalent.
3. One-Step qRT-PCR Reverse Transcriptase Enzyme.
4. Dengue primers.
   (a) DENV2 (PLO46) 3′UTR forward: CAT ATT GAC GCT GGG AAA GA.
   (b) DENV2 (PLO46) 3′UTR reverse: AGA ACC TGT TGA TTC AAC.
5. Dengue probe.
   (a) DENV2 (PLO46) 3′UTR: CTG TCT CCT CAG CAT CAT TCC AGG CA.

6. Dengue standard-frozen aliquots of 1E9 GE DENV2, thawed and diluted serially to 1E7, 1E5, and 1E3. For more information, please *see* **Note 1**.

7. Mouse 18S RNA primers.

    (a) CGGCTACCACATCCAAGGAA.

    (b) GCTGGAATTACCGCGGCT.

8. Mouse 18S RNA probe.

    (a) TGCTGGCACCAGACTTGCCCTC.

9. 8 strip PCR tubes: DNase-/RNase-/pyrogen-free tubes.

10. Multiplate 96-well PCR plate clear from Bio-Rad, Cat # MLP-9601.

11. Bio-Rad qRT-PCR thermocycler.

2.4 FIHC

1. Cryomolds from Sakura Tissue-Tek, Cat # 4557 (large), # 4566 (small).

2. Optimum cutting temperature (OCT) compound from Sakura Tissue-Tek, Cat # 4583.

3. Sodium phosphate dibasic heptahydrate from Sigma, Cat # S9390.

4. L-Lysine from Sigma, Cat # L5501.

5. pH meter.

6. 0.1 M hydrochloric acid concentrate from Sigma, Cat # 38280.

7. Sodium (meta) periodate from Sigma, Cat # S1878.

8. Slides from Fisher Scientific, Cat # 12-550-15.

9. Slide box from Fisher Scientific, Cat # 03-446.

10. Acetone from Fisher Scientific, Cat # 67-64-1.

11. Formaldehyde from Acros, Cat # 11969-0010.

12. Super PAP pen from Invitrogen, Cat # 00-8899.

13. Avidin/Biotin Blocking Kit from Vector, Cat # SP-2001.

14. Albumin from bovine serum from Sigma, Cat # A3294.

15. Goat serum donor herd from Sigma, Cat # G6767.

16. Rabbit anti-NS3 polyclonal antibody.

17. Dy649 goat anti-rabbit IgG.

18. Mounting media.

19. Microscope.

3 Methods

3.1 RNA Isolation from Serum

1. Sacrifice mice by an approved method and remove serum from mouse and place in serum collection tube. Alternatively, bleed live mice using an approved survival method (*see* **Notes 2** and **3**).

2. Spin serum in collection tube at full speed for 15 min at 4 °C.

3. Remove 30 µL of serum and freeze or use immediately.

4. Prepare Buffer AVL with carrier RNA, as in Qiagen protocol. Store aliquots of 560 µL at –20 °C. Thaw at –80 °C for 5 min.

5. To 30 µL serum, add 110 µL PBS. Mix and add to 560 µL warmed aliquot of AVL and vortex.

6. Incubate at RT for 10 min.

7. Spin to re-pool and add 560 µL 100 % ethanol. Vortex and spin to re-pool.

8. Transfer 630 µL of sample to QIAamp Mini Spin column in a 2 mL collection tube. Centrifuge at 10,000 rpm for 1 min. Discard flow-through.

9. Repeat **step 8** until the complete volume of sample has been passed through Mini Spin column.

10. Add 500 µL Buffer AW1 and centrifuge 8,000 rpm for 1 min. Discard flow-through.

11. Add 500 µL Buffer AW2 and centrifuge 14,000 rpm for 3 min. Transfer to a clean collection tube and centrifuge 13,200 rpm for 1 min.

12. Place column in a fresh 1.5 mL microfuge tube for recovery, and add 60 µL room temperature Buffer AVE and incubate for 1 min RT. Centrifuge for 1 min at 8,000 rpm.

13. Store at –80 °C until processed further.

3.2 Tissue Homogenization

1. Sacrifice mice by an approved method, harvest tissue from mouse, and remove the fat. Chop tissues into pieces smaller than 5×5 mm. For example, a healthy mouse spleen should be cut into at least five to six pieces. The process described below applies to many tissues, including spleen, lymph node, and kidney (*see* **Note 5**).

2. Place cut tissue immediately in RNAlater of volume specified in Table 1 and store at 4 °C until processed, within 7 days (*see* **Note 4**).

3. For each tissue sample, add 1 mL of 1 % BMe in RLT buffer to a 2 mL seal cap tube with one stainless steel bead and place on ice.

4. Remove tissue pieces from RNAlater and place in tube with bead on ice, with minimal transfer of RNAlater liquid.

5. Beat in homogenizer for 3–5 min at 3.0/s frequency.

6. Transfer samples to ice immediately.

7. Spin full speed for 5 min in microfuge at 4 °C.

8. Transfer 50 µL to a new tube with 300 µL of 1 % BMe in RLT for RNA extraction.

Table 1
Mouse tissues in RNAlater

Tissue	RNAlater (mL)	Tissue	RNAlater
Liver	7	Lymph nodes	300 µL
Spleen	1	Lg intestine	3 mL
Kidney	3	Sm intestine	5 mL
Lungs	3	Bone marrow	750 µL
Stomach	1.2	Brain	3 mL

3.3 RNA Isolation from Tissue

1. To 350 µL of homogenized sample, add 350 µL of 70 % ethanol and mix.

2. Transfer 700 µL of sample to RNeasy spin column. Centrifuge 15 s at 10,000 rpm. Discard flow-through.

3. Add 700 µL Buffer RW1 to column. Spin 15 s at 10,000 rpm. Discard flow-through.

4. Add 500 µL Buffer RPE to column. Spin 15 s at 10,000 rpm. Discard flow-through.

5. Add 500 µL Buffer RPE to RNeasy column. Spin 2 min at 10,000 rpm.

6. Move column to new collection tube and spin 1 min at 13,200 rpm to eliminate Buffer RPE.

7. Place column in new 1.5 mL microfuge tube. Add 30 µL RNase-free water to column membrane. Spin 1 min at 10,000 rpm.

8. Store at −80 °C until processed further.

3.4 Dengue Virus qRT-PCR Assay

Carry out all procedures on ice, unless otherwise specified. Keep all areas clear of RNase.

1. Thaw samples on ice. Vortex each sample and load 10 µL of each isolated RNA sample into PCR tubes.

2. Thaw and vortex dengue standard (1E12 GE/mL). Add 10 µL to PCR tube and 10 µL into 990 µL RNase-/DNase-free water (first serial dilution, 1:100, 1E7 GE/mL). Repeat for 1E5 and 1E3 samples.

3. To final empty well, add 10 µL RNase-/DNase-free water as negative control.

4. Make master mix for number of samples +3 extra, using recipe in Table 2.

5. Vortex master mix and spin to re-pool.

Table 2
Master mix for DENV2 qRT-PCR

Component	Amount per sample (µL)
ddH$_2$O	10.5
2× qRT-PCR mix	25
Dengue primer mix	3
Dengue probe	1
qRT-PCR enzyme	0.5

Table 3
DENV2 thermocycler program

Cycle	Repeats	Step	Rate	Dwell time	Set point
1	1	1	Max	10:00	48
2	1	1	Max	5:00	95
3	50	1	Max	0:15	95
		2	Max	0:30	57
		3	Max	0:30	72

6. To each experimental and standard RNA sample, add 40 µL master mix.

7. Pipet up and down 10× to mix, and transfer 22 µL of each sample to PCR plate, and repeat with same tips to new wells for duplicate.

8. Seal plate using Microseal B Film. Place in thermocycler.

9. Run thermocycler program in Table 3, using FAM-490 as the fluorochrome.

10. Set threshold in the exponential zone. Record cycle where the threshold is crossed for each experimental and standard sample (*see* **Note 6**).

11. Make a standard curve by graphing log of the standard DENV samples on the *x*-axis and the cycle for the standards on the *y*-axis. Determine slope, intercept, and R-squared.

12. Average the duplicate experimental samples and calculate DENV copies in the sample using the formula: $10^{((\text{average}-\text{intercept})/\text{slope})}$.

13. Determine copies per mL by correcting for the amount of sample that was analyzed.

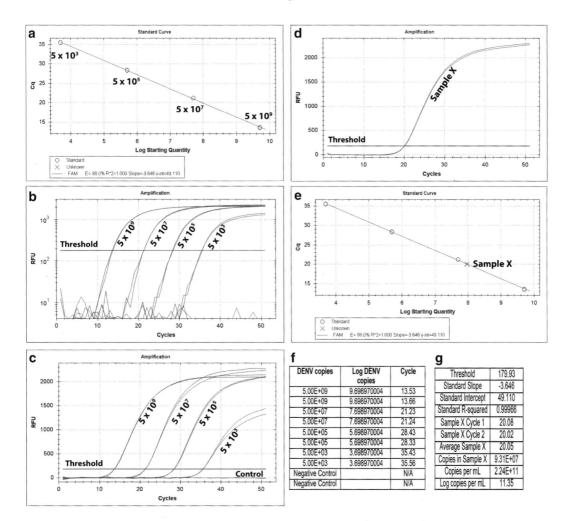

Fig. 1 DENV2 qRT-PCR Analysis. (**a**) Example standard curve with log of the standard DENV samples on the *x*-axis and the cycle for the standards on the *y*-axis. Slope is −3.646, intercept is 49.110, and R-squared is 0.99966. (**b**) Log view of standards with threshold set in linear zone. Threshold set by Bio-Rad CFX Manager 3.0 is 179.93 relative fluorescence units (RFU). (**c**) Linear view of standards and water control with threshold set above control in the exponential zone. (**d**) Linear view of two replicates of Sample X, which cross the threshold at 20.02 and 20.08 cycles. (**e**) Sample X on the standard curve. (**f**) Threshold cycles for DENV2 standards. (**g**) Values for analysis of Sample X

For serum:

14. Determine copies per mL by correcting for the amount used initially.

15. To determine limit of detection: we monitor for 50 cycles, so our limit of detection is set at 50.1 cycles. Use the same formulas to determine the lowest number of copies per mL you can detect. Examples of standard curve and determining copies per mL in sample can be found in Fig. 1 (*see* **Note 7**).

    For tissues, continue on to the 18S protocol to correct for amount of tissue used and possible degradation of RNA in the sample.

**3.5 18S RNA
qRT-PCR Assay**

1. Because of the high levels of 18S RNA in tissue, the sample must be diluted 1:10,000. To do this, first place 5 μL of sample *into* 495 μL RNase-/DNase-free water and vortex thoroughly. Further dilute this by placing 10 μL in 990 μL RNase-/DNase-free water and vortex thoroughly.

2. Thaw and vortex 18S RNA standard (1E5 GE/mL). Add 10 μL to PCR tube and 10 μL into 90 μL RNase-/DNase-free water (first serial dilution, 1:10, 1E4 GE/mL). Repeat for 1E3 and 1E2 samples.

3. To final empty well, add 10 μL RNase-/DNase-free water as negative control.

4. Make master mix for number of samples +3 extra, using recipe in Table 4.

5. Vortex master mix and spin to re-pool.

6. To each experimental and standard RNA sample, add 40 μL master mix.

7. Pipet up and down 10× to mix, and transfer 22 μL of each sample to PCR plate, and repeat with same tips to new wells for duplicate.

8. Seal plate using Microseal B Film. Place in thermocycler.

9. Run thermocycler program in Table 5, using FAM-490 as the fluorochrome.

Table 4
Master mix for 18S RNA qRT-PCR

Component	Amount (μL)
ddH$_2$O	12.5
2× qRT-PCR mix	12.5
18S primer mix	3
18S probe	3
qRT-PCR enzyme	0.5

Table 5
18S thermocycler program

Cycle	Repeats	Step	Rate	Dwell time	Set point
1	1	1	Max	10:00	48
2	1	1	Max	5:00	95
3	40	1	Max	0:15	95
		2	Max	1:00	60

16. As before, set threshold and record cycle where threshold is crossed.

17. Make a standard curve by graphing log of the standard DENV samples on the *x*-axis and the cycle for the standards on the *y*-axis. Determine slope, intercept, and R-squared.

18. Average the duplicate 18S experimental samples and calculate 18S copies in the sample using the formula: $10^{((\text{average-intercept})/\text{slope})}$.

19. Determine DENV per 18S by dividing DENV copies by 18S copies.

3.6 FIHC Protocol

1. Remove organ and remove any excess fat.

2. Wash tissue in PBS.

3. Quickly place in an appropriately sized labeled cryomold partially filled with OCT. If the tissue is too large, such as liver, cut in pieces. Avoid unnecessary loss of tissue morphology (*see* **Note 8**).

4. Add additional OCT until cryomold is full and tissue is completely immersed.

5. Place the cryomold on dry ice with a barrier of aluminum foil (*see* **Note 9**).

6. When OCT *solidifies* completely (about 20 min), store blocks at –80 °C.

7. Equilibrate tissue blocks to cryostat temperature before sectioning.

8. Cut 6 μm sections and place on clean slides.

9. Store slides at –80 °C until staining (*see* **Note 10**).

10. Prepare fresh paraformaldehyde/lysine/periodate (PLP) buffer.

    (a) Add 1.34 g $Na_2HPO_4 \cdot 7H_2O$ and 0.45 g L-lysine to ~45 mL ddH_2O in a 50 mL conical.

    (b) Calibrate pH meter and adjust pH of solution to 7.40 w/~3.0 mL of 10 % HCl; add ddH_2O to 50 mL.

    (c) Add 73 mg of sodium (meta)periodate, mix, and keep on ice.

11. Quickly after removing slides from freezer, fix in acetone for 10 min at room temperature.

12. While slides are fixing in acetone, dilute 3 % paraformaldehyde to 1 % in PLP buffer (one part paraformaldehyde to two parts PLP buffer)—make ~0.5 mL per slide.

13. Draw a circle around tissue with PAP pen.

14. Fix with PLP for 8 min on ice.

15. Wash with PBS for 3 min at room temperature; repeat two more times.

16. If using biotinylated antibodies:

    (a) Incubate with Avidin D solution (1%BSA in PBS+ 4 drops Avidin D/mL) 10 min, R/T.

    (b) Wash briefly with PBS.

    (c) Incubate with Biotin solution (1%BSA in PBS+4 drops Biotin/mL) 10 min, R/T.

    (d) Wash with PBS for 3 min at room temperature; repeat 2 more times.

17. Incubate in blocking buffer (1%BSA in PBS with 5 % normal goat serum) 30 min at room temperature.

18. Remove blocking buffer briefly.

19. Add primary antibody: rabbit anti-NS3 (1:2,000) in PBS.

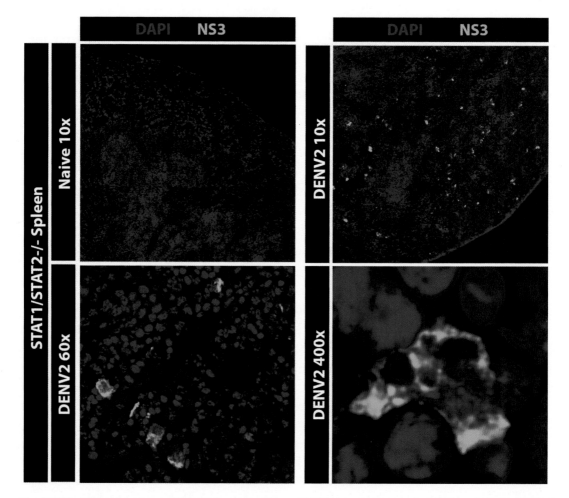

Fig. 2 NS3 staining in naïve and infected mouse spleen. 129/Sv mice lacking both STAT1 and STAT2 (STAT1/STAT2−/−) were injected intravenously with PBS or DENV2 strain S221 and spleens harvested 72 h postinfection. Naïve spleen (*top left*) has no NS3 staining (*green*). Spleens from infected mice have NS3 positive cells. Nuclei are stained with DAPI (*blue*)

20. Incubate overnight at room temperature.

21. Wash with PBS for 3 min at room temperature; repeat two more times.

22. Incubate with blocking buffer (1 % BSA in PBS with 5 % normal goat serum) 10 min at room temperature.

23. Remove blocking buffer briefly.

24. Add secondary antibody: Dy649 goat anti-rabbit IgG (1:500).

25. Wash with PBS for 3 min at room temperature; repeat two more times.

26. Add mounting media.

27. Place coverslip over tissue and gently press out all excess mount.

28. Lie flat 24 h to dry.

29. Image on fluorescence or confocal microscope.

30. Analyze images using ImageJ. Example staining of NS3 in spleen can be seen in Fig. 2 (*see* **Note 11** and **12** for important information regarding FIHC controls).

4   Notes

1. Serial dilution of known DENV RNA copies is important for formation of a standard curve in qRT-PCR. DENV RNA standards are made by in vitro synthesis of RNA transcripts from DNA templates. After purifying the RNA, the number of in vitro transcripts is determined using spectrophotometry. Standards are frozen and stored at known concentrations and then serially diluted for use on each DENV qRT-PCR plate.

2. An important consideration to include when studying host-pathogen interaction is the prior health of the mice. For this reason, we use specific pathogen-free mice, and the vivarium is checked regularly for a variety of pathogens. Any infection can change the immune system, which can subsequently alter the course of a viral infection. This could result in inconsistent or invalid results.

3. Previous to terminal blood collection, mice are sacrificed by isoflurane overdose, and blood is collected immediately after breathing stops, while the heart is still beating. During cardiac puncture for blood harvest, it is best to puncture the chest cavity only once with a 26 gauge, ½ inch needle on a 1 mL syringe and pull back blood in one smooth motion. An adult mouse should yield around 0.8 mL of blood depending on the strain. Repeated entry and delayed or slow harvest can result in clotting in the needle and broken red blood cells (serum will be red). This can affect accuracy of RNA analysis.

Other methods of blood collection, such as opening the chest for direct puncture of the heart, are useful as well. Survival blood collection protocols, including eye bleeding and cheek bleeding, are methods for obtaining small amounts of blood (i.e., a single collection of 10 % blood volume or daily collection of 1 % blood volume) from temporarily anesthetized mice. However, in the context of a viral infection with high viremia, removal of blood, and therefore virions, can alter the progress of the infection and should therefore be limited. If necessary, take as little blood as possible for analysis.

4. For titering of virus in tissue, RNA degradation is a constant concern. Once removed, tissues should be completely immersed in RNAlater immediately. Although tissues can be stored at 4 °C in RNAlater, they should be homogenized as soon as possible. A week of storage can result in loss of half of a log of virus. Analyzing 18S RNA in the tissue is an important control for degradation, as well as correcting for amount of tissue. However, 18S RNA is very stable compared to viral RNA and cannot be depended on to rule out degradation. It is important to be very consistent in treatment and storage of tissue prior to homogenization.

5. For homogenization of tissues, the following variations apply. For liver, homogenize once as above, then transfer 200 µL homogenate into a new seal cap tube with bead and 1 mL 1 % BMe in RLT. Homogenize and spin as above. Transfer 350 µL of second homogenate to an empty 1.5 mL tube for RNA extraction. For small intestine, remove from mouse, remove chyme, and fillet longitudinally before placing in RNAlater. Remove from RNAlater and chop finely. Follow liver homogenization protocol. For large intestine, remove from RNAlater. In Petri dish, fillet intestine and carefully remove fecal matter. Blot dry and chop finely. Follow standard homogenization protocol. For bone marrow, place entire femur bone in RNAlater. Remove whole femur from RNAlater and clean bone of tissue. Chop bones finely. Follow standard homogenization protocol.

6. For qRT-PCR analysis, we set a threshold line on the curve of fluorescence versus cycle number. This line is always set above the background (based on the blank, water-only sample). Where the curves for each sample meet the threshold line is called the threshold cycle. For all 50 cycles that we are monitoring, the line for water never crosses that threshold. We set the threshold in the exponential phase of amplification (which is linear in log view), where there should be a doubling of product every cycle since the reagents are fresh and available.

7. Our background (water sample) never passes the threshold for the 50 cycles we are monitoring. Therefore, we set the threshold

cycle for the limit of detection just beyond that, at 50.1 cycles. We calculate the log copies per mL at that threshold (50.1), exactly the same way we calculate log copies per mL from the threshold cycle for positive samples. The resulting number is our limit of detection. Anything below that cannot be distinguished from our background.

8. Tissues harvested for FIHC should be treated to preserve tissue morphology. Tissues should be placed in a cryomold that allows sufficient space in all directions so that it can be completely surrounded by OCT, without folding or crowding that does not reflect the positioning in vivo. If tissue must be cut to fit in a cryomold, as is often the case for the liver, care must be taken while cutting to preserve the morphology.

9. We use frozen tissue to perform multicolor FIHC for staining of NS3 and various cell type-specific markers. Freezing tissues in OCT on dry ice is a successful alternative to flash freezing in liquid nitrogen because it requires less "hands on" time, allowing faster work, and provides reproducibly efficient sectioning.

10. FIHC tissue slides should be stored in a slide box to prevent frost accumulation. For long-term storage, the box should be placed in a plastic bag to protect from drying and degradation.

11. For FIHC staining, the potential for nonspecific staining cannot be ignored. Appropriate controls must be used to rule this out. A key control is a virus incapable of replication. For example, the virus can be inactivated by treatment with ultraviolet (UV) rays. Plaque assay should be used to confirm that the UV-treated virus is not capable of replication. Another important control is to exclude primary antibody only to confirm the specificity of the secondary antibody binding. If in this control, there are large spots of random fluorescence, the secondary antibody can be spun prior to dilution. Another important control is to complete the entire staining process on naïve tissue. This will indicate whether the primary antibody is binding to anything in the tissue. If there is a high level of background in general, the blocking buffer can be extended or altered.

12. In all cases, varying dilutions of both primary (e.g., 1:50, 1:100, 1:500) and secondary (e.g., 1:500, 1:1,000, 1:2,000) antibody should be tested to reduce background and to reduce the amount of antibody needed.

Acknowledgements

We would like to thank the contribution of Tyler Prestwood, Malika Morar, and Monica May, past members of our laboratory who contributed to the development of the methods presented here.

References

1. Houng HH, Hritz D, Kanesa-thasan N (2000) Quantitative detection of dengue 2 virus using fluorogenic RT-PCR based on 3′-noncoding sequence. J Virol Methods 86(1):1–11

2. Prestwood TR, Prigozhin DM, Sharar KL, Zellweger RM, Shresta S (2008) A mouse-passaged dengue virus strain with reduced affinity for heparan sulfate causes severe disease in mice by establishing increased systemic viral loads. J Virol 82(17):8411–8421. doi:10.1128/JVI.00611-08

3. Kuchipudi SV, Tellabati M, Nelli RK, White GA, Perez BB, Sebastian S, Slomka MJ, Brookes SM, Brown IH, Dunham SP, Chang KC (2012) 18S rRNA is a reliable normalisation gene for real time PCR based on influenza virus infected cells. Virol J 9:230. doi:10.1186/1743-422X-9-230

4. Benedict CA, De Trez C, Schneider K, Ha S, Patterson G, Ware CF (2006) Specific remodeling of splenic architecture by cytomegalovirus.

PLoS Pathog 2(3):e16. doi:10.1371/journal.ppat.0020016

5. Swirski FK, Nahrendorf M, Etzrodt M, Wildgruber M, Cortez-Retamozo V, Panizzi P, Figueiredo JL, Kohler RH, Chudnovskiy A, Waterman P, Aikawa E, Mempel TR, Libby P, Weissleder R, Pittet MJ (2009) Identification of splenic reservoir monocytes and their deployment to inflammatory sites. Science 325(5940):612–616. doi:10.1126/science.1175202

6. Prestwood TR, May MM, Plummer EM, Morar MM, Yauch LE, Shresta S (2012) Trafficking and replication patterns reveal splenic macrophages as major targets of dengue virus in mice. J Virol 86(22):12138–12147. doi:10.1128/JVI.00375-12

7. Zellweger RM, Prestwood TR, Shresta S (2010) Enhanced infection of liver sinusoidal endothelial cells in a mouse model of antibody-induced severe dengue disease. Cell Host Microbe 7(2):128–139. doi:10.1016/j.chom.2010.01.004

Chapter 24

Evaluation of Dengue Antiviral Candidates In Vivo in Mouse Model

Satoru Watanabe and Subhash G. Vasudevan

Abstract

In vivo evaluation of antiviral compounds can serve as criteria in the drug discovery process for selection of compounds that are suitable to enter late preclinical studies and further development. Dengue virus serotypes 1–4 can infect and replicate in the interferon type I and type II receptor deficient mice (AG129). Here we describe the use of a mouse-adapted dengue 2 virus strain (S221) that has been used to develop a robust lethal model of infection. Treatment with small molecule inhibitors of DENV replication at the time of infection or delayed treatment up to 48 h post infection can result in measurable protection that reflects the efficacy of the tested compound.

Key words Dengue mouse model, Dengue antiviral testing, Virus quantification by plaque assay, Antibody-dependent enhanced infection, 4G2 antibody

1 Introduction

Dengue is a global public health threat caused by infection with any of the four related viral serotypes (DENV1–4). Clinical manifestations range from self-limiting febrile illness, known as dengue fever (DF), to the life-threatening severe diseases, such as dengue hemorrhagic fever (DHF) or dengue shock syndrome (DSS) [1, 2]. Most cases of DHF/DSS are associated with secondary heterotypic infections, probably due to the phenomenon known as antibody-dependent enhanced (ADE) infection [2, 3]. More than two billion people are at risk of infection worldwide, and there are more than 400 million human infections and several hundred thousand cases of DHF/DSS per year [2, 4]. At present, however, there are no approved preventive vaccines or antiviral drugs against DENV infection. In vitro cell-based assays [5–7] are routinely used to determine concentration of drug that results in 50 % reduction in infection (EC50). The testing of compounds in suitable in vivo models provides an opportunity to optimize the potency of the compound in terms of pharmacokinetics, pharmacodynamics, and toxicity [6, 8, 9].

Radhakrishnan Padmanabhan and Subhash G. Vasudevan (eds.), *Dengue: Methods and Protocols*, Methods in Molecular Biology, vol. 1138, DOI 10.1007/978-1-4939-0348-1_24, © Springer Science+Business Media, LLC 2014

1.1 Animal Models for DENV Infection

The lack of a suitable animal model has hampered the evaluation of novel antiviral candidates for DENV infection. The AG129 (Sv/129 mice deficient in type I and II IFN receptors) viremia model was used to show that animals infected with a clinical (non-lethal) isolate of DENV2 resulted in reduction in viral replication and suppression of the inflammatory response after treatment with antiviral drug [10]. The infected animals showed a transient increase in serum viral levels that peaked 3 days post infection (dpi) and also showed an increase in virus-expressed antigen NS1, thus confirming that the unadapted virus was able to replicate in the AG129 mouse model. Similar to human infection, the viral load and NS1 antigen levels decreased over time. Furthermore, infected mice had significantly enlarged spleens as well as higher levels of pro-inflammatory cytokines. Directly acting antivirals which target the viral RNA-dependent RNA polymerase (RdRp) activity of NS5 protein, such as 7-deaza 2'-C-methyl adenosine, and host alpha-glucosidase inhibitors, such as N'-nonyl-deoxynojirimycin or 6-O-butanoyl castanospermine (celgosivir) reduced viral load in a dose-dependent manner in this murine model. Most importantly, levels of pro-inflammatory cytokines and the extent of splenomegaly were also reduced with the drug treatment thus demonstrating the value of antiviral treatment, albeit in a mouse infection model. More recently, AG129 mouse was used as a lethal dengue mouse model that recapitulated several of the major pathologies of human infection including high level of viremia, elevated levels of cytokines, vascular leakage, intestinal bleeding, thrombocytopenia, and death [11, 12], which further supports the use of this mouse model to select the suitable candidates and treatment strategy before the clinical study. In this chapter, we describe the method for the in vivo drug evaluation using the example of celgosivir treatment in mice [6, 9].

2 Materials

2.1 Generation of DENV Stocks

1. Cells of the *Aedes albopictus* C6/36 line (ATCC catalogue number: CRL-1660).

2. RPMI1640 (10 % FBS): RPMI1640 medium (GIBCO) supplemented with 10 % heat-inactivated fetal bovine serum (FBS; Gibco), 2 mM L-glutamine, 100 U/mL penicillin and streptomycin (PenStrep; GIBCO), 25 mM HEPES.

3. RPMI1640 (2 % FBS): RPMI1640 medium supplemented with 2 % heat-inactivated FBS, 2 mM L-glutamine, 100 U/mL penicillin and streptomycin, 25 mM HEPES.

4. Incubator: No CO_2 atmosphere at 28 °C.

5. 1× PBS: 137 mM NaCl, 6.25 mM Na_2HPO_4, 2.5 mM Na_2PO_4; sterilized by autoclave.

6. 0.25 % Trypsin–EDTA.

7. Dengue virus stock: DENV-2 mouse-adapted S221.

8. Sterile 2 ml microcentrifuge tubes (screw cap and safe for storage at −80 °C; Sarstedt or similar).

9. −80 °C freezer.

2.2 Determination of Viral Titer by Plaque Assay

1. Cells of the baby hamster kidney cell line, BHK-21 (ATCC catalogue number: CCL-10).

2. RPMI1640 (10 % FBS): RPMI1640 medium supplemented with 10 % heat-inactivated FBS; Gibco, 100 U/mL penicillin and streptomycin, 25 mM HEPES.

3. RPMI1640 (no serum): RPMI1640 medium supplemented with 2 mM L-glutamine.

4. Humidified incubator with 5 % CO_2 atmosphere at 37 °C.

5. 1× PBS.

6. 0.25 % Trypsin–EDTA.

7. Dengue virus samples to titer (collected from infected C6/36 cell supernatant).

8. 0.8 % Methyl-cellulose medium with 2 % FBS.

9. 3.7 % Formaldehyde.

10. 1 % Crystal violet.

2.3 Preparation of a-DENV E Protein Antibodies (4G2) for the Induction of Antibody-Dependent Enhanced (ADE) Infection

1. Hybridoma cells; 4G2 (ATCC catalogue number: HB-112).

2. RPMI1640 (10 % FBS): RPMI1640 medium supplemented with 10 % heat-inactivated FBS; Gibco, 100 U/mL penicillin and streptomycin, 25 mM HEPES.

3. PFHM-II: Protein-Free hybridoma Medium (GIBCO) supplemented with 2 mM L-glutamine, 100 U/mL penicillin and streptomycin.

4. Humidified incubator with 5 % CO_2 atmosphere at 37 °C.

5. 1× phosphate-buffered saline (PBS), filtered, pH 7.2.

6. 0.1 M glycine (pH 2.7).

7. 1 M Tris–HCl (pH 7.2).

8. 5 ml Protein G column (GE Healthcare).

9. AKTApurifier™ UPC 10 (GE Healthcare).

2.4 DENV Infection and Drug Treatment in AG129 Mice

1. Sv/129 mice deficient in type I and II IFN receptors (AG129 mice).

2. Purified 4G2 Abs.

3. S221 virus stock.

4. Test compounds.

5. Dimethyl sulfoxide, DMSO.

6. 29G insulin syringe (BD).

7. 27G needle (BD).

2.5 Determination of Serum Viral Load by Plaque Assay

1. BHK-21 cells.

2. RPMI1640 (10 % FBS).

3. RPMI1640 (no serum).

4. Humidified incubator with 5 % CO_2 atmosphere at 37 °C.

5. 1× PBS.

6. 0.25 % Trypsin–EDTA.

7. Mouse serum samples.

8. 0.8 % Methyl-cellulose medium with 2 % FBS.

9. 3.7 % Formaldehyde.

10. 1 % Crystal violet.

3 Methods

3.1 Generation of DENV Stocks

3.1.1 Preparation of C6/36 Insect Cells for Generation of Viral Stocks

1. Maintain C6/36 cells in RPMI1640 (10 % FBS) media in sterile tissue culture flasks (*see* **Note 1**).

2. Dislodge healthy cells in 0.25 % Trypsin–EDTA for less than 5 min. Add culture medium to resuspend cells and transfer into a centrifugal tube.

3. Centrifuge cells at $900 \times g$ for 5 min at room temperature.

4. Discard medium and resuspend cells in fresh growth medium.

5. Seed cells into 175 cm² sterile tissue culture flasks in 25 ml of culture media and incubate at 28 °C under non-CO_2 atmosphere condition.

3.1.2 Generation of Viral Stocks

1. Once cells have reached 90 % confluency in 175 cm² tissue culture flasks, remove the existing media and infect virus with 0.1 MOI in 5 ml RPMI1640 (2 % FBS).

2. Incubate for 1 h at 28 °C under non-CO_2 atmosphere condition.

3. Remove virus and add 25 ml RPMI1640 (2 % FBS).

4. Incubate for 4–7 days at 28 °C under non-CO_2 atmosphere condition (*see* **Note 2**).

5. Scrape cells and transfer into centrifugal tubes, then centrifuge cells at $1,800 \times g$ for 20 min at 4 °C.

6. Collect supernatant by syringe and transfer into fresh tubes by filtration with 0.45 mm filter.

7. Aliquot virus into 2 ml cryotubes and store in liquid nitrogen until use.

3.2 Determination of Viral Titer by Plaque Assay

3.2.1 Preparation of BHK-21 Cells for Plaque Assay

1. Maintain BHK-21 cells in RPMI1640 (10 % FBS) media in sterile tissue culture flasks.

2. Dislodge healthy cells in 0.25 % Trypsin–EDTA for less than 5 min. Add culture medium to resuspend cells and transfer into a centrifuge tube.

3. Centrifuge cells at $900 \times g$ for 5 min at room temperature.

4. Discard medium and resuspend cells in fresh growth medium.

5. Count cell number and seed cells at 2×10^5 cells/well/500 ml in 24-well plate (*see* **Note 3**).

6. Incubate cells overnight at 37 °C in 5 % CO_2 incubator to allow cells to adhere plate.

3.2.2 Plaque Assay

1. Serially dilute virus tenfold in RPMI1640 (no serum).

2. Remove culture supernatant of BHK-21 cells and add 200 ml of diluted virus into each well (*see* **Note 4**).

3. Incubate plate for exactly 1 h at 37 °C in 5 % CO_2 incubator.

4. Remove virus and add 500 ml of 0.8 % Methyl-cellulose medium supplemented with 2 % FBS (*see* **Note 5**).

5. Incubate plate for 4–5 days at 37 °C in 5 % CO_2 incubator (*see* **Note 6**).

6. Fix cells with 3.7 % formaldehyde for 20 min.

7. Rinse plate with copious volume of water in a container. Shake plate robustly to remove Methyl-cellulose medium completely.

8. Add 500 ml of 1 % crystal violet into each well and stain for 1 min.

9. Rinse plate with copious volume of water in a container and shake plate to remove excess water.

10. Dry plate and count number of plaques to determine virus titer.

3.3 Preparation of a-DENV E Protein Antibodies (4G2) for the Induction of ADE Infection

3.3.1 Collection of 4G2 Hybridoma Cell Culture Supernatant

1. Culture 4G2 hybridoma cells in 50–75 ml of PFHM-II (Protein-Free hybridoma Medium) in sterile 175 cm^2 tissue culture flasks (*see* **Note 7**).

2. Once the color of culture media turns to orange or yellow, collect cell suspension into a centrifuge tube.

3. Centrifuge cells at $900 \times g$ for 5 min at room temperature.

4. Collect supernatant (*see* **Note 8**).

5. If more supernatant is required, continue culture of cells and repeat **steps 1–4**.

3.3.2 Purification of Antibodies (See Note 9)

1. Filter combined supernatant through a 0.45 mm membrane.
2. Load the 4G2 supernatant onto a 5 ml Protein G column pre-equilibrated in pH 7.2 PBS.
3. Wash the column with PBS using 5× the column volume (i.e., 25 ml).
4. Prepare a 96-well block containing 60 ml 1 M Tris–HCl (*see* **Note 10**).
5. Elute using 100 % 0.1 M glycine and 1 ml fractions are collected into the wells of the block.
6. Check the quality of purified antibody by running a SDS PAGE.
7. Collect fractions of similar quality into a dialysis membrane and dialysed against PBS overnight.
8. Quantitate the concentration of the purified antibody using nanodrop.

3.4 DENV Infection and Drug Treatment in AG129 Mice

3.4.1 Administration of 4G2 to Induce ADE Infection

1. Prepare male AG129 mice aged 6–10 weeks (6–10 mice per group) (*see* **Note 11**).
2. Adjust purified 4G2 Abs at the concentration of 1–20 mg/200 ml in PBS (*see* **Note 12**).
3. Administer of 200 ml of Abs into AG129 mice intraperitoneally by 29G insulin syringe 1 day prior to virus infection.

3.4.2 Virus Infection and Drug Treatment

1. Dilute S221 virus stock with PBS into 2×10^5 pfu/200 ml.
2. Dissolve solid test compound in DMSO to 100 mg/ml stock (*see* **Note 13**).
3. Dilute test compounds with PBS (*see* **Note 14**).
4. Inoculate with 200 ml of virus intraperitoneally into mice by 29G insulin syringe.
5. Administer test compound intraperitoneally or orally into mice according to appropriate dosage regimen (*see* **Note 15**).
6. Observe mouse disease status and survival rate until day 10 post infection.
7. Collect blood samples at periodical time points to determine viremia level.

3.5 Determination of Serum Viral Load by Plaque Assay

3.5.1 Preparation of Mouse Serum Samples

1. Obtain blood samples from the facial vein of mice by using 27G needles.
2. Store blood samples overnight at 4 °C before separation of serum.
3. Centrifuge blood samples at 12,000 rpm for 5 min.
4. Collect serum and store at −80 °C until use.

3.5.2 Plaque Assay

1. Seed BHK-21 cells at 2×10^5 cells/well/500 ml in 24-well plate.

2. Incubate cells overnight at 37 °C in 5 % CO_2 incubator to allow cells to adhere plate.

3. Serially dilute serum samples (up to 10,000-fold) in RPMI1640 (no serum).

4. Remove culture supernatant of BHK-21 cells and add 200 ml of diluted serum samples into each well.

5. Incubate plate for exactly 1 h at 37 °C in 5 % CO_2 incubator.

6. Remove diluted serum and add 500 ml of 0.8 % Methyl-cellulose medium supplemented with 2 % FBS.

7. Incubate plate for 4–5 days at 37 °C in 5 % CO_2 incubator.

8. Fix cells and stain plate as indicated in Subheading 3.2.2.

9. Count plaque number and calculate viremia as *pfu/ml* (*see* **Note 16**).

4   Notes

1. Growth media supplemented with 20 % FBS improves cell growth if C6/36 cells do not proliferate desirably.

2. Keep incubation until the cytopatic effect (syncytium) can be clearly seen by microscopy. Some virus strains do not induce syncytium. In this case, check virus titer in the supernatant of different time point to determine the optimal day to harvest.

3. Cells become confluent on the following day by seeding 2×10^5 cells into each well. Two days incubation after seeding of 1×10^5 cells is also viable.

4. Virus should be added immediately after removing culture supernatant to avoid cells dried out.

5. Prepare 0.8 % methyl-cellulose medium as follows;
   Add 8 g of methyl-cellulose powder into 500 ml water followed by autoclave to dissolve powder. Prepare 500 ml of 2× RPMI1640 media by dissolving RPMI1640 powder in water followed by supplement with 2 % heat-inactivated FBS, 2 mM L-glutamine, 100 U/mL penicillin and streptomycin and 25 mM HEPES. After filtration of 2× RPMI1640 media by 0.2 mm membrane, mix well with 500 ml of prepared methyl-cellulose and store at 4 °C.

6. The rate of plaque formation is affected by virus replication rate of the strain. Check the plaque size by visual observation before fixation.

7. Recommend the use of RPMI1640 (10 %) during the initial period of a couple of days after thawing cells. Once cells grow well, replace the media gradually with PFHM-II media.

8. The supernatant can be stored at 4 °C without filtration until the enough amount of supernatant can be obtained.

9. 4G2 antibody is purified using the AKTApurifer (GE). Please refer to the manufacturer's instruction guides regarding sample loading specifications.

10. The standard ratio of Tris–HCl to glycine (100:6) for neutralization is subjected to change depending on the concentration of buffers prepared. Volume of Tris–HCl to be added required for neutralization (pH 7) can be adjusted by pH paper testing.

11. AG129 mice were purchased from B&K Universal (UK) and maintained under specific pathogen-free (SPF) conditions in the animal facility. Maintenance of 20-breeding pairs produce more than 100 mice per month. Male mice are suitable for the drug evaluation since female mice are more sensitive to some kind of drugs.

12. Administration of 4G2 Abs ranged from 1 mg to 20 mg induce 100 % lethal infection by subsequent intraperitoneal inoculation with 2×10^5 pfu of S221 or intravenous inoculation with 2×10^4 pfu of S221.

13. Test compound should be dissolved in compatible solvent, generally DMSO, to high concentration stock.

14. If test compound is not soluble in PBS, appropriate solvent should be selected.

15. For celgosivir treatment, a twice-a-day regimen (BID) of 50 mg/kg achieved 100 % protective efficacy, while a single daily dose (QD) of 100 mg/kg for 5 days failed to protect mice from lethal infection (Fig. 1a).

16. Successful treatment will show clear reduction of viremia level compared with vehicle control (Fig. 1b).

Acknowledgments

We thank Sujan Shresta for providing DENV 2S221 strain. This work is in part supported by a DUKE-NUS Signature Research Program (funded by the Ministry of Health, Singapore), and the National Medical Research Council, Singapore (http://www.nmrc.gov.sg), under grant NMRC/1315/201.

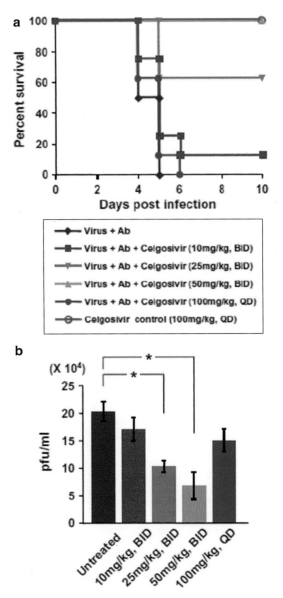

Fig. 1 Dose and schedule effects of celgosivir against lethal DENV infection in mice. AG129 mice were inoculated i.p. with 2×10^5 pfu of S221 in the presence of 20 mg of DENV anti-E Ab (4G2). Mice were treated with vehicle or different concentrations and doses of celgosivir at the time of infection and daily for 5 days. Mouse survival rates were monitored until day 10 post infection (**a**) and survival significance was evaluated using log-rank test. Viremia on day 3 post infection was measured by plaque assay (**b**). Significant differences between data groups were determined by two-tailed Student t-test analysis and P value less than 0.05 was considered significant (*$P < 0.05$). A number of mice per group are 7–8

References

1. Gubler DJ (2006) Dengue/dengue haemorrhagic fever: history and current status. Novartis Found Symp 277:3–165

2. Halstead SB (2007) Dengue. Lancet 370(9599):1644–1652

3. Fink J, Gu F, Vasudevan SG (2006) Role of T cells, cytokines and antibody in dengue fever and dengue haemorrhagic fever. Rev Med Virol 16:263–275

4. Bhatt S, Gething PW, Brady OJ, Messina JP, Farlow AW, Moyes CL, Drake JM, Brownstein JS, Hoen AG, Sankoh O, Myers MF, George DB, Jaenisch T, Wint GR, Simmons CP, Scott TW, Farrar JJ, Hay SI (2013) The global distribution and burden of dengue. Nature 496:504–507

5. Wang QY, Patel SJ, Vangrevelinghe E, Xu HY, Rao R, Jaber D, Schul W, Gu F, Heudi O, Ma NL, Poh MK, Phong WY, Keller TH, Jacoby E, Vasudevan SG (2009) A small-molecule dengue virus entry inhibitor. Antimicrob Agents Chemother 53:1823–1831

6. Rathore AP, Paradkar PN, Watanabe S, Tan KH, Sung C, Connolly JE, Low J, Ooi EE, Vasudevan SG (2011) Celgosivir treatment misfolds dengue virus NS1 protein, induces cellular pro-survival genes and protects against lethal challenge mouse model. Antiviral Res 92:453–460

7. Luo D, Wei N, Doan DN, Paradkar PN, Chong Y, Davidson AD, Kotaka M, Lescar J, Vasudevan SG (2010) Flexibility between the protease and helicase domains of the dengue virus NS3 protein conferred by the linker region and its functional implications. J Biol Chem 285:18817–18827

8. Yin Z, Chen YL, Schul W, Wang QY, Gu F, Duraiswamy J, Kondreddi RR, Niyomrattanakit P, Lakshminarayana SB, Goh A, Xu HY, Liu W, Liu B, Lim JY, Ng CY, Qing M, Lim CC, Yip A, Wang G, Chan WL, Tan HP, Lin K, Zhang B, Zou G, Bernard KA, Garrett C, Beltz K, Dong M, Weaver M, He H, Pichota A, Dartois V, Keller TH, Shi PY (2009) An adenosine nucleoside inhibitor of dengue virus. Proc Natl Acad Sci U S A 106: 20435–20439

9. Watanabe S, Rathore AP, Sung C, Lu F, Khoo YM, Connolly JE, Low J, Ooi EE, Lee HS, Vasudevan SG (2012) Dose- and schedule-dependent protective efficacy of celgosivir in a lethal mouse model for dengue virus infection informs dosing regimen for a proof of concept clinical trial. Antiviral Res 96:32–35

10. Schul W, Liu W, Xu HY, Flamand M, Vasudevan SG (2007) A dengue fever iremia model in mice shows reduction in viral replication and suppression of the inflammatory response after treatment with antiviral drugs. J Infect Dis 195:665–674

11. Balsitis SJ, Williams KL, Lachica R, Flores D, Kyle JL, Mehlhop E, Johnson S, Diamond MS, Beatty PR, Harris E (2010) Lethal antibody enhancement of dengue disease in mice is prevented by Fc modification. PLoS Pathog 6:e1000790

12. Zellweger RM, Prestwood TR, Shresta S (2010) Enhanced infection of liver sinusoidal endothelial cells in a mouse model of antibody-induced severe dengue disease. Cell Host Microbe 7:128–139

ERRATUM TO

Analysis of Affinity of Dengue Virus Protein Interaction Using Biacore

Yin Hoe Yau and Susana Geifman Shochat

Radhakrishnan Padmanabhan and Subhash G. Vasudevan (eds.), *Dengue: Methods and Protocols*, Methods in Molecular Biology, vol. 1138, DOI 10.1007/978-1-4939-0348-1_17, © Springer Science+Business Media, LLC 2014

DOI 10.1007/978-1-4939-0348-1_25

The inline equation in page 278 is incorrect. The correct information is given below:

5. Estimate the rate of dissociation (k_d) from the result by calculating the linear decline in response after injection end. The rate of dissociation can be converted to half-life ($t_{1/2}$) through the equation $t_{1/2} = \dfrac{ln2}{k_d}$, the half life will determine if regeneration is necessary for assay set up (*see* **Note 7**).

The online version of the original chapter can be found at
http://dx.doi.org/10.1007/978-1-4939-0348-1_17

INDEX

Radhakrishnan Padmanabhan and Subhash G. Vasudevan (eds.), *Dengue: Methods and Protocols*, Methods in Molecular Biology,
vol. 1138, DOI 10.1007/978-1-4939-0348-1, © Springer Science+Business Media, LLC 2014